Dahlem Workshop Reports
Physical, Chemical, and Earth Sciences Research Report 5
Patterns of Change in Earth Evolution

The goal of this Dahlem Workshop is:
to identify temporal variations in the nature
and intensity of important Earth processes,
and to assess the effects of these variations
on the evolution of the Earth and its biota

Physical, Chemical, and Earth Sciences Research Reports
Editor: Silke Bernhard

Held and published on behalf of the
Stifterverband für die Deutsche Wissenschaft

Sponsored by:
Senat der Stadt Berlin
Stifterverband für die Deutsche Wissenschaft

Patterns of Change in Earth Evolution

H. D. Holland and A. F. Trendall,
Editors

Report of the Dahlem Workshop on
Patterns of Change in Earth Evolution
Berlin 1983, May 1 – 6

Rapporteurs:
W. H. Berger · K. S. Deffeyes · K. Padian
F. M. Richter · A. B. Thompson

Program Advisory Committee:
H. D. Holland and A. F. Trendall, Chairpersons
K. S. Deffeyes · H. Füchtbauer · D. P. McKenzie
D. M. Raup · E. R. Oxburgh · F. M. Richter
A. Seilacher · A. B. Thompson · G. J. Wasserburg

Springer-Verlag
Berlin Heidelberg New York Tokyo 1984

Copy Editors: M. A. Cervantes-Waldmann, K. Geue
Photographs: E. P. Thonke

With 4 photographs, 48 figures, and 12 tables

ISBN-13: 978-3-642-69319-9 e-ISBN-13: 978-3-642-69317-5
DOI: 10.1007/978-3-642-69317-5

CIP-Kurztitelaufnahme der Deutschen Bibliothek

Patterns of change in earth evolution :
report of the Dahlem Workshop on Patterns of Change in Earth Evolution,
Berlin 1983, May 1–6 /
H. D. Holland, ed. Rapporteurs: W. H. Berger
[Held and publ. on behalf of the Stifterverb. für d. Dt. Wiss.
Sponsored by: Senat d. Stadt Berlin ; Stifterverb. für d. Dt. Wiss.].
– Berlin ; Heidelberg ; New York ; Tokyo : Springer, 1984. –
(Physical, chemical, and earth sciences research reports ; 5)
(Dahlem Workshop reports)
NE: Holland, Heinrich D. [Hrsg.]; Berger, Wolfgang H. [Mitverf.];
Workshop on Patterns of Change in Earth Evolution ⟨1983, Berlin, West⟩;
1. GT

Softcover reprint of the hardcover 1st edition 1984

2131/3020-5 4 3 2 1 0

Table of Contents

The Dahlem Konferenzen

Founders

Recognizing the need for more effective communication between scientists, especially in the natural sciences, the Stifterverband für die Deutsche Wissenschaft*, in cooperation with the Deutsche Forschungsgemeinschaft**, founded Dahlem Konferenzen in 1974. The project is financed by the founders and the Senate of the City of Berlin.

Name

Dahlem Konferenzen was named after the district of Berlin called "Dahlem", which has a long-standing tradition and reputation in the arts and sciences.

Aim

The task of Dahlem Konferenzen is to promote international, interdisciplinary exchange of scientific information and ideas, to stimulate international cooperation in research, and to develop and test new models conducive to more effective communication between scientists.

Dahlem Workshop Model

Dahlem Konferenzen organizes four workshops per year, each with a limited number of participants. Since no type of scientific meeting proved effective enough, Dahlem Konferenzen had to create its own concept. This concept has been tested and varied over the years, and has evolved into its present form which is known as the *Dahlem Workshop Model*. This model provides the framework for the utmost possible interdisciplinary communication and cooperation between scientists in a given time period.

**The Donors Association for the Promotion of Sciences and Humanities*

***German Science Foundation*

The main work of the Dahlem Workshops is done in four interdisciplinary discussion groups. Lectures are not given. Instead, selected participants write background papers providing a review of the field rather than a report on individual work. These are circulated to all participants before the meeting to provide a basis for discussion. During the workshop, the members of the four groups prepare reports reflecting their discussions and providing suggestions for future research needs.

Topics
The topics are chosen from the fields of the Life Sciences and the Physical, Chemical, and Earth Sciences. They are of contemporary international interest, interdisciplinary in nature, and problem-oriented. Once a year, topic suggestions are submitted to a scientific board for approval.

Participants
For each workshop participants are selected exclusively by special Program Advisory Committees. Selection is based on international scientific reputation alone, although a balance between European and American scientists is attempted. Exception is made for younger German scientists.

Publication
The results of the workshops are the Dahlem Workshop Reports, reviewed by selected participants and carefully edited by the editor of each volume. The reports are multidisciplinary surveys by the most internationally distinguished scientists and are based on discussions of new data, experiments, advanced new concepts, techniques, and models. Each report also reviews areas of priority interest and indicates directions for future research on a given topic.

The Dahlem Workshop Reports are published in two series:
1) Life Sciences Research Reports (LS), and
2) Physical, Chemical, and Earth Sciences Research Reports (PC).

Director
Silke Bernhard, M.D.

Address
Dahlem Konferenzen
Wallotstrasse 19
1000 Berlin 33
F.R. Germany

Patterns of Change in Earth Evolution, eds. H.D. Holland and A.F. Trendall, pp. 1-4. Dahlem Konferenzen 1984. Berlin, Heidelberg, New York, Tokyo: Springer-Verlag.

Introduction

H.D. Holland* and A.F. Trendall**
*Dept. of Geological Sciences, Harvard University
Cambridge, MA 02138, USA
**Geological Survey of Western Australia
Perth, W.A. 6000, Australia

Recent discoveries have forced a reassessment of the essential nature of Earth history. The development of plate tectonics has led geologists to rethink the entire structural and chemical evolution of the Earth; the discovery of iridium anomalies in sediments at the Cretaceous-Tertiary boundary has had a profound impact on theories of extinction and on our views of the role of sudden events in Earth history. Major catastrophes have now come to be considered normal events in Earth history. These developments led one of us (A.F.T.) to propose that specialists from a wide range of subdisciplines in the Earth Sciences be brought together at a Dahlem Workshop to address the question whether Earth history has been essentially smooth or importantly spasmodic. The topic won approval, and a Dahlem Workshop entitled "Earth History: How Smooth, How Spasmodic?" was held in Berlin in May of 1983. At the conclusion of the workshop a number of alternative titles were suggested for the workshop volume. "Patterns of Change in Earth Evolution" was proposed by Oxburgh; this entry narrowly outpolled "Trends, Cycles, and Sudden Events in Earth History" and a number of other suggested titles that contained the words catastrophe, spasmodic, episodic, and uniformitarian.

The Program Advisory Committee spent what seemed an interminable day in 1982 choosing the best way to constitute the four working groups of the workshop. A division of the workshop topic based largely on a variety of time scales was finally agreed upon. Group I was asked to deal with the possible physical influences of sudden events on biological radiations and extinctions; Group II with short-term changes that affected the atmosphere, oceans, and sediments in the Phanerozoic; Group III with events on a 10^7 to 10^9 year time scale controlled by tectonism or volcanism, and Group IV with the long-term evolution of the crust and mantle.

Inevitably, Group 1 gravitated toward the effects of bolide impacts on the Earth's biota. This is a subject shared by Raup, Shoemaker, Toon, and Hsü (all this volume). Virtually all of the pertinent new data that have come to light during the past several years have supported the proposition that a bolide impact was responsible either directly or indirectly for the massive extinctions at the Cretaceous-Tertiary boundary. This contradicts one of the most cherished tenets of uniformitarian doctrine. Deffeyes suggested during the course of the workshop that the Uniformitarian Creed should read:

"As it was in the beginning,
Is now, and ever shall be
World without end. Amen, amen,"

which is only a slight caricature of the philosophy proposed by Lyell ((1), p. 318):

> "The course directly opposed to this method of philosophising consists in an earnest and patient enquiry, how far geological appearances are reconcilable with the effect of changes now in progress, or which may be in progress in regions inaccessible to us, but of which the reality is attested by volcanos and subterranean movements. It also endeavors to estimate the aggregate result of ordinary operations multiplied by time, and cherishes a sanguine hope that the resources to be derived from observation and experiment, or from the study of Nature such as she now is, are very far from being exhausted. For this reason all theories are rejected which involve the assumption of sudden and violent catastrophes and revolutions of the whole Earth, and its inhabitants - theories which are restrained by no reference to existing analogies, and in which a desire is manifested to cut, rather than patiently to untie, the Gordian knot."

As Shoemaker makes clear in his paper (this volume), Lyell's "sanguine hope" was misplaced. There are processes and events in Earth history that are not part of the experience of the human race, let alone part of the experience of the last few generations. The group of happily unexperienced events includes large bolide impacts with the Earth. The evidence for the occurrence of such impacts at intervals of some tens of millions of years is quite convincing, and Lyell stands admonished by Hamlet: "There are more things in heaven and earth, Horatio, than are dreamt of in your philosophy."

The role of bolide impacts on the history of life during other portions of the Phanerozoic Eon is less clear (see Raup and Fischer, both this volume), and catastrophic changes unrelated to extraterrestrial processes may have been important (see Holser, this volume). Changes in the later Precambrian biota are still difficult to interpret, in part because the preservation of soft-bodied animals from this period of Earth history is so unusual (see Seilacher, this volume).

During the past billion years or so, bolide impacts have exerted a significant effect on the Earth's surface and its inhabitants, but not on its interior. The 3800 Ma rocks at Isua in West Greenland are the oldest terrestrial rocks that are currently available for inspection (see Dymek, this volume). They contain abundant evidence for the operation of chemical and physical processes that are similar to those of the present day. This situation could not have prevailed during the entire 700 Ma preceding the formation of the Isua rocks. The history of the very early Earth must have been dominated by the effects of the infall of planetary material and by the degassing of the Earth (see Holland, this volume). The Lyellian view of Earth history cannot possibly do justice to such an Earth. The transition of the Earth from an infall-dominated state to its present condition has been accompanied by a decrease in the rate of heat generation within the Earth and almost certainly by a decrease in heat flow from the Earth. Geophysical models of the consequences of this change for the evolution of tectonism and volcanism are hard to construct (see Richter, this volume), and the geologic record is so obscure and poorly studied that there is still a good deal of room for differences of opinion regarding changes in the style and intensity of tectonism during the past 3800 Ma. These differences are well illustrated by the divergent opinions expressed in several of the papers in this book. It is still not clear, for instance, whether the average geothermal gradient during the Archean was significantly steeper than today (see Dymek, Thompson, and Weber, all this volume). The geology of the preserved Archean regions is surely different from that of the Phanerozoic parts of the globe, but it is not yet clear which of these differences are due to initial differences in the nature of the terrains, and which are due to selective preservation (see Bickle, this volume). Nevertheless, a consensus on some of the central questions of Earth history is emerging. Archean geology does seem to have differed in important respects from Proterozoic and Phanerozoic geology. The inferred changes in the nature of tectonic and plutonic processes can be interpreted in terms of a gradual cooling of the Earth (see Richter, this volume), and the changes in the patterns of sedimentation can probably be linked to the growth of stable cratons (see Knoll, this volume), although the timing of these changes is still somewhat debatable (see Trendall, this volume). Whether episodes of mountain building have been synchronous on a worldwide basis is a question that has divided geologists since it was warmly debated by Elie de Beaumont and Charles Lyell during the last century. The question is still unanswered (see Moorbath, this volume), but it is now capable of resolution and will presumably be settled during the next few decades.

The workshop turned out to be firmly in the Dahlem mold. The discussions were animated, dealt with many of the central problems of Earth history, circled around presently intractable questions, and sent participants away with new ideas and plans for tackling problems that are now solvable (see Berger et al., Deffeyes et al., Padian et al., and Richter, Thompson et al., all this volume). The workshop seems to have scored two "firsts."

According to several of the participants, it was the first conference during which the causes of the Cretaceous-Tertiary extinction were discussed in a non-adversary manner. According to the workshop staff, the members of this Dahlem Workshop were the first to finish off all of the desserts at all of the meals. The scientific discussions at lunch were obviously so intense that waist-line concerns were completely submerged in pudding and torte. Much of the success of the conference was due to the untiring efforts of the Dahlem staff: Gloria Custance, Kelly Geue, Myra Lax, and Renate Rosing. Silke Bernhard guided and presided over the proceedings with great skill and charm. K. Geue saw the manuscript of this book through to publication. The Editors are particularly grateful to her for her efficiency in extracting manuscripts from so many parts of the world and in converting them into a well organized whole.

Patterns of Change in Earth Evolution, eds. H.D. Holland and A.F. Trendall, pp. 5-14.
Dahlem Konferenzen 1984. Berlin, Heidelberg, New York, Tokyo: Springer-Verlag.

Evolutionary Radiations and Extinctions

D.M. Raup
Dept. of Geophysical Sciences, University of Chicago
Chicago, IL 60637, USA

INTRODUCTION

Evolutionary turnover is pervasive in the fossil record. Species appear, persist for some interval of time (with or without change), and then become extinct. Mean durations in the Phanerozoic fossil record vary from 1-10 Ma (6); the mean residence time of species is therefore short, relative to geologic time. The frequency distribution of durations is highly skewed; short-lived taxa are far more common than long-lived taxa. Since many short duration species are not preserved, the actual mean duration is probably considerably shorter than the measured estimates. As Van Valen (13) and others have shown, the extinction of species is often rather similar to the decay of radioactive isotopes; that is, the proportion of species that become extinct in successive time intervals of equal length is essentially constant. Major exceptions to this pattern occur as a consequence of mass extinction events.

When we speak of extinction at the species level we must make the distinction between true extinction and pseudoextinction. Pseudoextinction occurs when a species is transformed by phyletic evolution through a sequence of genetically changing populations so that, at some point in time, the organism differs from an ancestral form at the species level. In this case there is no extinction of the lineage, and the phenomenon is called pseudoextinction. In this paper, I will be concerned only with true lineage extinction, where the lineage is the line through time representing a sequence of populations of a species or series of transformations of a species.

Related groups of species, such as genera and families, also have high turnover rates relative to the span of geologic time. The mean duration of genera rarely exceeds 20-30 Ma, and the range of families is only somewhat longer.

Measured rates of origination and extinction vary greatly from group to group and from one geologic situation to another. Some of the differences are undoubtedly artefacts of factors which bias our estimates of temporal duration. For example, if the taxonomy of a particular fossil group has been oversplit, its stratigraphic duration may be shortened. Species may be called genera, genera called families, and so on. The ability of paleobiologists to discriminate species and higher taxonomic groups depends to some extent on the complexity of the morphology. As a general rule, the simpler the morphology of an organism, the less likely it is that paleobiologists will be able to discriminate subgroups; this leads to an overestimation of geologic ranges (9). In this connection it has been noted that over "monographic" time the ranges of biologic groups tend to become shorter as more and more fine distinctions are made (8).

Despite the difficulties of measuring rates of evolutionary activity, there seems to be little question that rates do vary from group to group and from time to time. The extremes in these variations are the sudden bursts of speciation, known as radiations, and the sudden increases in the incidence of extinction, known as mass extinctions. This paper is concerned primarily with a review of the current state of knowledge regarding radiations and mass extinctions. A major question which is yet to be answered is this: Are marked changes in evolutionary activity predictable in terms of some general model, or are they simply accidents of a very complex physical and biological history?

PROCESSES

Speciation and species extinction are totally different processes. One is a branching, the other the termination of a branch. They are as different as birth and death. Surprisingly little is actually known about either process in spite of the large amount of research that has been done. We know that in eukaryotic organisms, speciation must involve the division of a gene pool into reproductively isolated units such that independent evolution can occur; however, the mechanisms of this process are poorly understood. In the classic view, geographic isolation is a prerequisite for speciation, but a number of apparently viable models for speciation without geographic isolation have been developed, and it is not certain that the geographic breakup of a species is essential for speciation.

Extinction must involve the death of all members of a species. This much is straightforward. The implication is, of course, that the species is in some way poorly adapted to its environment or is unable to cope with particular stresses. But questions abound. For example, are species interactions important in the extinction process? Is extinction primarily a result of competition for space or resources between species with the loser dying out, or are species interactions of relatively trivial importance? The conventional wisdom on this point is that interspecies competition is important in normal background extinction but less so in mass extinction where an external physical or chemical stress forces the event.

The processes of speciation and extinction may be linked through controls on total diversity. It may even be argued that some sort of linkage is essential. Consider the following equation for growth of a group of species under conditions of birth and death:

$$S_t = S_o e^{(p-q)t}, \tag{1}$$

where p and q are speciation and extinction rates, respectively. S_o is the number of species present at some time = 0 and S_t is the number of species at time = t. The process is one of exponential growth, decay, or stability, depending upon whether (p-q) is positive, negative, or zero. It may take a short time for the population of species either to explode to near infinite numbers or to decrease to zero. For example, if at some time in the late Cambrian there were 10,000 species of organisms and if the speciation rate (p) exceeded the extinction rate (q) by one event per 10 Ma, then the expected number of species existing after 500 Ma (late Cambrian to Recent) would be approximately 5 x 10^{25}. We are thus virtually driven to the conclusion that p and q must be linked, not as interacting processses but through the medium of limitations on resources or on space. The only alternative is to suggest that life on this planet has been extraordinarily lucky to have escaped Malthusian explosion on the one hand, or total extinction on the other.

The processes producing the birth and death of higher taxonomic groupings are also important. The origin of a genus or family presumably results from a speciation event. Such speciation events may be somewhat unusual in producing a marked change in the organism such that its descendants form a taxonomically distinctive group; but except for the magnitude of the speciation that marks the founding of a new biologic group, such an event is presumably no different from any other speciation. If the group is to persist beyond the duration of the founding species, branching events must outnumber extinction events so that diversity develops. Most analysis of radiations and extinctions in the fossil record perforce involve higher taxa. This is in part because the fossil record of species is generally too spotty to be amenable to rigorous analysis.

The extinction of a higher taxon requires, by definition, the extinction of all constituent species. This may occur gradually through an excess of q over p in Eq. 1, or it may occur suddenly from a burst of extinction at the species level. A troublesome problem with higher taxa is to know whether the extinctions of species are linked to each other. If a given stress produces the demise of one species, is it probable that all species in the group will be similarly subject to extinction because they have common morphological, physiological, or functional characteristics?

The growth equation is useful in analyzing evolution only if the extinction and speciation rates are stochastically constant through time. This is clearly a reasonable assumption in some biologic groups over some spans of geologic time, but the desired constancy is definitely not observed in some other situations (such as extinction events of short duration).

RADIATIONS

An evolutionary radiation is defined as a sudden increase in standing diversity in a biologic group. Diversity may be measured by numbers of species or by numbers of higher taxa within some larger group. It is generally presumed that radiations are due to bursts of cladogenetic activity, that is, sudden increases in speciation rate. But a radiation could also be caused by a decrease in extinction rate. This would allow longer survival of the existing species and therefore more opportunities for branching. One must therefore be careful to leave open the possibility that evolutionary radiations may be due as much to failure of the extinction process as to an acceleration of the speciation process.

Radiations are commonly termed "adaptive" radiations. In fact, one rarely sees the word radiation without this qualifying adjective. The use of the word adaptive is at best a tautology. It is assumed that a group that suddenly increases in diversity does so for reasons of adaptive success, but we cannot recognize nonadaptive or inadaptive radiations were such to occur. At present we know very little about the mechanisms or causes of sudden bursts of cladogenetic activity.

Some radiations in the geologic record have been labeled "exponential," and the doubling time of the number of species has been used as a metric (11, 14). It should be noted, however, that any evolutionary radiation, whether rapid or slow, is by its nature an exponential process (see Eq. 1). When the quantity (p-q) is greater than zero, a radiation will occur and growth of the species population will be exponential. Presumably radiations in the fossil record are labeled "exponential" when they appear to be particularly rapid.

Most analyses of evolutionary radiations concentrate on one or a very few biologic groups, such as the mammals of the early Cenozoic or the echinoderms of the early to middle Paleozoic. This implies that radiations in the several biologic groups are relatively independent of one another. Although this may be true to some extent, simultaneous radiations often take place immediately following mass extinction. Sepkoski's work (10) on the phenomenon of rebound from mass extinction develops this theme. The causes are presumably related to the availability of geographic or resource space following a major extinction. Many of the radiations of the geologic record are as spectacular as the mass extinctions, although they have not attracted as much attention.

EXTINCTIONS

This subject is presently dominated by work on the several mass extinctions in the Phanerozoic record. According to a recent analysis by Raup and Sepkoski (7), the five most pronounced mass extinctions in the marine realm came in the late Ordovician, the middle late Devonian, the late Permian, the late Triassic, and the late Cretaceous. In the terrestrial realm, some mass extinction occurred at the same time, but there were other extinctions of substantial magnitude, especially in the early Mesozoic. It is no accident that nearly all mass extinctions coincide with major stratigraphic boundaries. The reason for this is that the stratigraphic time scale was established originally by using points of major faunal turnover as markers. Thus, the early stratigraphers recognized the mass extinctions as fundamental in developing the classification of geologic time.

The extinction of a biologic group is most likely caused by an increase in q of Eq. 1. But it is also possible, although perhaps not probable, that an extinction event could be caused by a decrease in the branching rate (p). Anything which lowers the numerical value of (p-q) in Eq. 1 will lower diversity and can be seen as an extinction.

The amount by which q increases in a mass extinction is difficult to determine. Raup and Sepkoski (7) showed increases in the absolute rate of family extinctions (that is, extinctions per Ma) by factors of 2-5 for the five major mass extinctions. But such estimates are closely tied to the length of the interval of geologic time over which the rates are measured (3). In the Raup and Sepkoski study, the unit of time employed was the stratigraphic stage with an average duration of about 7 Ma. The number of family extinctions occurring during the total length of a stage was divided by the duration of the stage to arrive at an estimate of the rate of extinction. This assumed that extinctions occurred throughout the time interval in question. If the extinctions are actually concentrated in a very small part of the time interval, as might be the case during a sudden environmental crisis, the rate per milllion years calculated in this fashion is meaningless. The actual q during such a catastrophic event might become virtually infinite. In fact, if mass extinctions are typically short-lived events, then the measurement of q or even the use of Eq. 1 becomes meaningless. This raises a general and fundamental question regarding extinction. Is extinction a continuous process and therefore amenable to analysis in terms of Eq. 1, or is it an episodic or point process? The conventional wisdom in paleobiology for many years has been that extinction is a continuous process, that species are always at risk of extinction, but that this risk may increase or decrease through time or from biologic group to biologic group. The contrasting viewpoint is that of a discontinuous process wherein species are not at risk of extinction most of the time, but that the probability of extinction during certain, short intervals in their history is very high.

This raises a related question: Is there a qualitative difference between mass extinctions and what has been called background extinction? This must be considered an open question. We know that substantial extinctions of intermediate intensity are buried in the background. They become part of the noise, because the resolving power of stratigraphic dating is limited. If our general understanding of extinctions is to improve, an important objective should be to devise more rigorous methods of analyzing background extinction.

The intensity of extinction may be measured at different levels in the taxonomic hierarchy. Extinctions can be measured at the level of numbers of individuals, and this is the measure of choice for some of the major extinction events. The mass extinction in the middle late Devonian, for example, is expressed most clearly in the sudden disappearance of a number of especially common species. In other cases, counts of species and higher taxon extinctions provide useful metrics of the severity of extinction. Where the concern is to observe patterns of extinction intensity over tens or hundreds of millions of years, one is almost forced to use higher taxon counts because data on abundance of individual organisms or species are too prone to the vagaries of preservation and discovery.

By far the largest amount of synoptic work on extinction over long spans of geologic time is done at higher taxonomic levels, primarily the generic, familial, and ordinal levels. This is in part because sampling at the higher levels is far better, and in part due to the general interest of paleobiologists in the extinction of major types of organisms such as trilobites and dinosaurs. It should be noted that the record of extinctions at high taxonomic levels is a highly damped signal compared to that seen at the species or individual level. For example, at the end of the Permian, marine extinctions at the family level are estimated to have been approximately 52% (7). This inevitably translates into a much higher extinction rate for species, because to eliminate a family one must eliminate all its species. In the Permian, the extinction at the species level has been variously estimated to have been between 77 and 96% (5, 12).

Another important aspect of extinction is its selectivity. Are extinctions random in the sense that all species of all biologic groups are equally at risk, or is extinction concentrated in certain kinds of organisms, certain geographic areas, or certain habitats? In the first of these interpretations, species may be viewed as occupying a "field of bullets" and the extinction of any one species as well as the extinction of any particular family a matter of chance. In the second view, the stresses that produce extinction, whatever they may be, may be lethal for some families but not for others.

It is generally recognized that the major mass extinctions are selective. This has been documented recently for the Frasnian-Famennian event (Devonian). Pedder (4) showed that this extinction was far more severe among shallow water corals than deep water species. Similar patterns are found among other constituents of the Frasnian fauna.

The question of selectivity raises an even more fundamental issue, that of the long-term effect of extinction in the total evolutionary system. If extinction is completely nonselective, then it has no long-term constructive effect. It acts only as a perturbing factor. However, it is a basic tenet of Darwinian evolution that extinction is selective, and therefore beneficial in the long run. Except for certain tautological claims to the effect that if an organism survived it must have been better, we have very little hard evidence in favor of the traditional Darwinian view.

The situation is even more complicated, because there is a kind of selectivity which does not promote the long-term adaptation of the biota. Certain groups of organisms are, by virtue of evolutionary accident, more resistant to particular stresses than other groups. If the biota as a whole were subjected to a burst of ionizing radiation, for example, insects would survive far better than mammals, because insects are less susceptible to somatic damage from ionizing radiation. Such an irradiation might therefore cause the extinction of virtually all terrestrial mammals but might have a negligible effect or no effect at all on insects. This is a hypothetical example, of course, but it shows how an extinction may be selective without having a bearing on adaptation: the organisms involved are not normally subjected to high levels of ionizing radiation and therefore would have no way to adapt to it. This kind of selectivity may have the same sort of perturbing but nonconstructive effect on evolution as the field of bullets model. Unfortunately, nonconstructive selectivity is very difficult to distinguish from the more traditional Darwinian kind.

It is important to develop methods of analysis that can define the degree of selectivity in extinctions, particularly mass extinctions, and to develop criteria for distinguishing constructive from nonconstructive types.

DIRECTIONS FOR RESEARCH

We seem to be faced with three basic problems. First, we need far more complete data on radiations and extinctions. We need data on their taxonomic distribution, and we need as tight a time resolution as possible. The data must be tested statistically to resolve basic questions of rates of origination and extinction of species and higher taxa. Along with synoptic data on taxa, we need far more centimeter-by-centimeter information on population sizes very close to times of mass extinctions. Second, we need to develop the mathematical models necessary to evaluate origination and extinction as episodic processes. A promising approach in this context is that of Extreme Value Statistics which was developed by hydrologists for studying flood frequency (1), but which is also applicable to earthquake frequency distributions (2) and to other short-term events which have or may be considered to have random causation. A fundamental input for analysis by Extreme Value Statistics is the frequency distribution in time of extinctions of varying magnitude (6). Third, much more needs to be done with the problem of selectivity in extinction and radiation. This involves the statistical analysis of temporal associations of evolutionary activity among taxa, among organisms living in different habitats, and among organisms of different functional types. Often, the best information of this kind comes from the biota preserved on bedding planes and in cores immediately above and immediately below extinction events.

REFERENCES

(1) Gumbel, E.J. 1957. Statistics of Extremes. New York: Columbia University Press.

(2) Howell, Jr., B.F. 1979. Earthquake risk in eastern Pennsylvania. Earth Min. Sci. 48: 63-64.

(3) McLaren, D.J. 1983. Bolides and biostratigraphy. Geol. Soc. Am. Bull. 94: 313-324.

(4) Pedder, A.E.H. 1982. The rugose coral record across the Frasnian-Famennian boundary. Geol. Soc. Am. Spec. Paper 190: 485-490.

(5) Raup, D.M. 1979. Size of the Permo-Triassic bottleneck and its evolutionary implications. Science 206: 217-218.

(6) Raup, D.M. 1981. Extinction: bad genes or bad luck? Acta Geolog. Hispan. 16: 25-53.

(7) Raup, D.M., and Sepkoski, Jr., J.J. 1982. Mass extinctions in the marine fossil record. Science 215: 1501-1503.

(8) Schopf, T.J.M. 1982. A critical assessment of punctuated equilibria. I. Duration of taxa. Evolution 36: 1144-1157.

(9) Schopf, T.J.M.; Raup, D.M.; Gould, S.J.; and Simberloff, D.S. 1975. Genomic versus morphologic rates of evolution: influence of morphologic complexity. Paleobiology 1: 63-70.

(10) Sepkoski, Jr., J.J. 1981. A factor analytic description of the Phanerozoic marine fossil record. Paleobiology 7: 36-53.

(11) Stanley, S.M. 1979. Macroevolution. San Francisco: W.H. Freeman and Company.

(12) Valentine, J.W.; Foin, T.C.; and Peart, D. 1978. A provincial model of Phanerozoic marine diversity. Paleobiology 4: 55-66.

(13) Van Valen, L. 1973. A new evolutionary law. Evol. Theory 1: 1-30.

(14) Yule, G.U. 1924. A mathematical theory of evolution, based on the conclusions of Dr. J.C. Willis, FRS. Roy. Soc. Lond. Phil. Trans. (B) 213: 21-87.

Patterns of Change in Earth Evolution, eds. H.D. Holland and A.F. Trendall, pp. 15-40.
Dahlem Konferenzen 1984. Berlin, Heidelberg, New York, Tokyo: Springer-Verlag.

Large Body Impacts Through Geologic Time

E.M. Shoemaker
U.S. Geological Survey
Flagstaff, AZ 86001, USA

Abstract. The present collision rate between the Earth and asteroids $\geq$1km diameter is estimated to be $\sim$6 x 10^{-6} yr^{-1}; if intermediate albedo bodies predominate among the Earth-crossing asteroids, however, the collision rate might be as low as $\sim$3 x 10^{-6} yr^{-1}. Asteroids $\geq$10 km diameter collide with the Earth with an estimated frequency of $\sim$2 x 10^{-8} yr^{-1}. The collision rate of comets is lower than that of asteroids; it is poorly known owing to uncertainties in our knowledge of the size of comet nuclei. The supply of 1 km diameter Earth-crossing asteroids derived from the fragmentation of main belt asteroids is steady within 25%. On the other hand, a surge in the collision of 10 km asteroids, involving perhaps as many as half a dozen bodies in $\sim$5 x 10^7 yr, may have occurred during the Phanerozoic. The flux of Earth-crossing comets probably varies about 10% as the sun passes into and out of the spiral arms of the galaxy. Brief but intense comet storms probably have recurred about once every 10^8 yr, as a consequence of close passages of stars near the sun. A long-term increase in the average number of extinct comets among the Earth-crossing asteroids could have resulted from slow recovery of the Oort comet cloud after encounter of the sun with a giant molecular cloud.

INTRODUCTION

Two kinds of solid bodies large enough to be observed at the telescope share the region of space traversed by the Earth in its journey around the sun. Bodies observed only as points of light have been called asteroids (star-like); bodies with a resolved dusty atmosphere (coma), often accompanied by tails of plasma and dust, are referred to as comets (long-haired stars). The distinction, based purely on the presence or absence of an ephemeral atmosphere, is probably not fundamental. Many comets become star-like in appearance as they recede from the sun and sublimation of their surface ices diminishes and finally ceases at great distances from the sun. Moreover, some periodic comets evidently evolve into asteroids as their volatile substances are gradually lost by sublimation. A few comets with short orbital periods, for example, only rarely exhibit an observable coma; no coma has ever been detected around 944 Hidalgo which is an asteroid on an unstable orbit typical of short period comets. There are a variety of theoretical grounds to suspect that many, and perhaps a majority, of Earth-approaching asteroids are, in fact, extinct short period comets (30, 39, 52-54).

In this paper the evidence concerning the present and past rates of collision of asteroids and comets with the Earth is briefly summarized. The sources of asteroids and comets are then considered and the probable variations in the collision rate of large bodies with the Earth during the last 4 Gyr are estimated.

PRESENT POPULATION AND COLLISION RATE OF EARTH-CROSSING ASTEROIDS

The term Earth-crossing asteroid is used here for asteroids that are capable of colliding with the Earth as a result of long-range gravitational perturbations of their orbits and of perturbations of the orbit of the Earth. Three classes of Earth-crossing asteroids have been recognized: a) Aten asteroids whose orbits are smaller than the Earth's orbit and which overlap the Earth's orbit at aphelion, i.e., at their farthest distance from the Sun, b) Apollo asteroids which have orbits larger than Earth's orbit and which overlap the Earth's orbit at perihelion, i.e., at their shortest distance from the Sun, and c) Earth-crossing Amor asteroids which have orbits larger than Earth's and whose orbits do not presently overlap that of the Earth. Earth-crossing Amors overlap the Earth's orbit part of the time, however, as a result of secular changes of their orbital eccentricity and of the orbital eccentricity of the Earth. Some Apollo asteroids lose Earth overlap part of the time as a result of secular changes in their orbital eccentricity, and become Amor asteroids.

The population of Earth-crossing asteroids has been estimated from their rate of discovery in systematic surveys of the sky. Most are so small that they are too faint to be detected by the normal methods of asteroid search and discovery except when they are fairly close to the Earth. At the time of writing, fifty-three Earth crossers had been discovered, of which about forty are estimated to have diameters ≥1 km ((38), see Table 1 for recent discoveries). The discovered objects ≥1 km diameter probably represent about 2 to 5% of the total population to this size limit. From the number discovered per unit of volume of space sampled in systematic surveys, the population to a limiting absolute brightness of visual magnitude +18 is estimated to include ~100 Atens, 700 ± 300 Apollos, and ~500 Earth-crossing Amors (19, 38).

Physical properties, such as color and albedo, have been measured for less than half of these bodies, and the proportion of different physical

TABLE 1 - Recently discovered Earth-crossing asteroids.*

Provisional designation	Class	a (AU)	e	i (deg)	q (AU)	Q (AU)	V(1,0)	Diameter (km)
1982 TA	Apollo	2.30	0.769	12.1	0.53	4.07	15	~4
1982 XB	Amor	1.86	0.452	3.9	1.02	2.71	19.3	0.4
1981 ET3	Amor	1.77	0.422	22.2	1.02	2.52	15	~4
1982 YA	Amor	2.6 †	0.58 †	31	1.08	4.1 †	~17	~2

*This table supplements a list of 49 Earth-crossing asteroids given by Shoemaker (38), which is complete through May 1982.

† Poorly determined orbit.

Explanation of headings:

Class Orbital class, generally named for the first member recognized (see text).

a Semimajor axis of orbit in astronomical units. The astronomical unit (AU) is the length of the semimajor axis of the Earth's orbit.

e Eccentricity of orbit.

i Inclination of orbit.

q Perihelion distance in astronomical units.

Q Aphelion distance in astronomical units.

V(1,0) Absolute visual magnitude (mean magnitude as observed through a yellow filter and reduced to 1 AU distance from the Earth, 1 AU distance from the sun, and a position directly opposite from the sun). An increase of one magnitude corresponds to a decrease in brightness by a factor of $(100)^{1/5}$.

Diameter Diameter of a circular area equivalent to the mean cross-sectional area of the asteroid. The diameter based on measured albedo is reported to one decimal place; diameters based on an assumed geometric albedo of 0.1 are shown as approximate values.

types is not well-known. If it is assumed that half of the Earth crossers are low albedo objects (C-type) and half have moderate albedo (S-type), following Shoemaker et al. (39), then if a mean visual geometric albedo of 0.037 is taken for C-type and 0.14 for S-type asteroids, the population of Earth crossers ≥1 km is found to be ~2300. On the other hand, if the large majority of these objects has moderate albedo, as does the majority of well observed Earth crossers (24), then the population of asteroids with diameter ≥1 km is closer to 1000.

The number of discovered objects is relatively small and the known observational selection effects that influenced their discovery are large. Hence the size-frequency distribution of the population of Earth crossers cannot be obtained from the discovered asteroids alone. Large, bright objects are more easily discovered than small, faint ones, and the fraction of each size group that has been discovered certainly increases with increasing size. The largest discovered Earth crossers are the Apollo asteroids 1866 Sisyphus, with a diameter of~10 km, and 2212 Hephaistos, with an estimated diameter of 9 km (38). By extrapolating from the number of still larger known bodies that come fairly close to the Earth but are not Earth-crossing, one can estimate that the number of Earth-crossing asteroids about 10 km in diameter or larger is about eight (cf. (53)). If the cumulative frequency of objects between 1 km and 10 km in diameter is a simple power function of the diameter, the mean exponent or size index of this function is close to -2.5. The size distribution of young craters on the Moon, on the other hand, indicates that the size index for the impacting bodies between about 0.15 and 6 km diameter must be close to -1.6. It seems likely that the cumulative frequency of Earth crossers varies approximately as the -1.6 power of the diameter between 1 and 6 km diameter and then roughly follows a -5 power function between 6 and 10 km. If the size distribution above 10 km diameter can be extended on the basis of the -5 power function, it is likely that the largest object in the current population of Earth crossers has a diameter of ~5 km.

Most of the time, the orbit of an Earth-crossing asteroid does not intersect the Earth's orbit. However, as a result of precession of the major axes of the orbit of the asteroid and the orbit of the Earth, and due to secular changes in the eccentricity of both orbits, there are rare, short intervals of time during which the asteroid's orbit intersects the capture cross-section of the Earth. If an asteroid happens to pass through the capture cross-section of the Earth during one of these rare intervals, a collision occurs. The probability of such an event can be calculated by methods first presented by Öpik (30) and subsequently by Wetherill (48) and

Shoemaker et al. (39). The average probability of collision was found by Shoemaker et al. (39) to be ~2.5 x 10^{-9} yr^{-1}. When multiplied by the estimated population of 2300, this gives an average collision rate with the Earth of ~6 x 10^{-6} yr^{-1} for asteroids ≥1 km diameter. For asteroids ≥10 km diameter the collision rate is ~2 x 10^{-8} yr^{-1}.

FLUX AND COLLISION RATE OF EARTH-CROSSING COMETS

About three long period comets are observed to pass inside the orbit of the Earth each year, and the total population of long period comets comparable in size to those that have been observed is estimated to be as high as 10^{12} (46). In addition to the long period comets, ten comets with periods less than twenty years have been observed to pass inside the orbit of the Earth during the last two hundred years. The solid nuclei of almost all comets are obscured by comae when the comets are close enough for their nuclei to be observed by the normal methods of photoelectric photometry and infrared radiometry. Our rather fragmentary knowledge of the albedo, size, and other characteristics of cometary nuclei is based on observations made when the comets are distant and inactive but also faint. The combined evidence of the brightness of comets at great distance, the acceleration of short period comets by recoil from the gases and entrained particles expelled from their surfaces, and the brightness of the cometary comae suggest that the diameter of observed comets lies in the range of a fraction of a kilometer to several tens of kilometers (40, 47, 56). From the data presented by Shoemaker and Wolfe (40), an annual flux of about 300 comets greater than 1 km diameter can be estimated to pass perihelion inside the Earth's orbit. This estimate is based on the assumption that the average blue geometric albedo of inactive comet nuclei after perihelion passage is 0.03. If the albedo is as high as 0.5, then the estimated flux would have to be reduced by a factor of seventeen. Comets with periods less than twenty years contribute only a minor fraction of the total flux.

The size distribution of comet nuclei can be estimated from their distribution of brightnesses at large solar distances and from the size distribution of young impact craters on Ganymede and Callisto, the two largest satellites of Jupiter. Both lines of evidence indicate that the cumulative frequency is approximately proportional to the inverse square of the diameter (40). The annual flux through perihelion of Earth-crossing comets ≥10 km diameter is ~0.2 to ~3, depending on the assumption made about the average albedo of the nuclei.

Most Earth-crossing comets pass inside the Earth's orbit at large distances above or below the Earth's orbital plane, and hence cannot collide with the Earth. The orientation of the major axes of comet orbits is distributed essentially randomly, and we can calculate by means of Öpik's equations the fraction of comets that may be expected to collide with the Earth. The average probability of collision per perihelion passage is 3.3×10^{-9} if the distribution of orbital elements is the same as that of observed Earth-crossing long period comets. Multiplying this probability by the annual flux, we obtain collision rates of $\sim 0.06 \times 10^{-6}$ yr^{-1} to $\sim 1 \times 10^{-6}$, for comet nuclei ≥ 1 km, and $\sim 0.6 \times 10^{-9} yr^{-1}$ to $\sim 1 \times 10^{-8} yr^{-1}$, for nuclei ≥ 10 km diameter. Asteroids evidently are the dominant impacting bodies with a diameter of ca. 1 km, but among objects with a diameter of ca. 10 km active comets might be comparable in importance to asteroids if comet nuclei have a low mean albedo.

PRODUCTION OF IMPACT CRATERS AND THE GEOLOGIC RECORD OF LARGE IMPACT EVENTS

Cratering rates can be calculated from collision rates if appropriate average densities are assigned to the impacting bodies. In the case of the asteroids plausible densities can be chosen by matching asteroids to meteorites on the basis of spectral reflectance curves. Allowance should be made in the estimates of bulk density for possible void space produced by impact brecciation. Using this approach, Shoemaker et al. (39) derived mean bulk densities of 1.7 gm cm^{-3} and 2.4 gm cm^{-3} for C- and S-type asteroids, respectively. The choice of the density of comet nuclei is more uncertain, but a bulk density in the range 1.0 to 1.5 seems likely for the mixture of ices and rocky material in typical comet nuclei (40, 56).

Collision velocities can be determined accurately from the equations of celestial mechanics. The rms impact speed is 20.1 km s^{-1} for Earth-crossing asteroids (39) and 59.5 km s^{-1} for long period comets. In calculations of the average kinetic energy of bodies of a given size, the higher average impact speed of comets more than compensates for their lower density. Crater diameters can be estimated from the following equation (40):

$$d = 0.074\, c_f\, (g_e/g)^{1/6}\, (W \rho_a/\rho_t)^{1/3.4}\,, \tag{1}$$

where d is the crater diameter in km; c_f is the crater collapse factor, nominally 1.3 for initial craters on Earth larger than 4 km diameter;

g_e is the gravitational acceleration at the surface of the Earth and g the acceleration at the surface of the body on which the crater is formed; $W = \pi b^3 \delta v^2/(12 \times 4.19 \times 10^{10})$ is the kinetic energy of the impacting body in kilotons TNT equivalent; b is the diameter of the impacting body; δ its density; and v its speed, all in cgs units; $\rho_a = 1.8$ gm cm^{-3}; and ρ_t is the density of the target rock. (For a recent review of crater scaling formulas see (38)). An impacting body 1 km in diameter typically produces a crater about 20 km in diameter, and a half km body produces a 10 km crater.

Assuming that half of the Earth-crossing asteroids are C-type and half S-type, to any given limiting magnitude, Shoemaker (38) found by use of Eq. 1 that the rate of production of craters ≥10 km diameter on the continents is $\sim 2.4 \times 10^{-14}$ km^{-2} yr^{-1}. If all Earth crossers were S-type, the estimated 10 km cratering rate would drop to $\sim 1.3 \times 10^{-14}$ km^{-2}. If the craters produced by comets are added in, the predicted rate of production of craters ≥10 km increases by as much as $\sim 0.8 \times 10^{-14}$ km^{-2} yr^{-1}, if the mean albedo of comet nuclei is low. If the distribution of albedo among both asteroids and comet nuclei follows that which now seems most likely, the cratering rate to 10 km probably lies in the range 1.5 to 3.0×10^{-14} km^{-2} yr^{-1}. This range appears to be consistent with the Phanerozoic record of the formation of impact craters. Grieve and Dence (17) obtained production rates of craters ≥20 km diameter of (0.33 ± 0.2) and $(0.36 \pm 0.1) \times 10^{-14}$ km^{-2} yr^{-1} from a study of the recognized impact structures on the European and North American cartons. The corresponding average production rate for craters ≥10 km diameter is $(1.24 \pm 0.36) \times 10^{-14}$ km^{-2} yr^{-1} (38). From an assessment of a smaller set of impact structures in the central United States, Shoemaker (37) obtained a crater production rate to 10 km diameter of $(2.2 \pm 1.1) \times 10^{-14}$ km^{-2} yr^{-1}, which is essentially identical with the rate predicted from astronomical observations. The consistency between the present cratering rate derived from astronomical observations and the geological record of impact structures suggests that the average flux of impacting bodies has been constant within a factor of about 2 for the last several hundred million years.

If the cumulative frequency of Earth-crossing asteroids varies as the -1.6 power of the diameter up to 6 km diameter, and if the impact rate has been steady during the Phanerozoic, then about $25 \stackrel{\times}{\div} 2$ craters ≥100 km diameter were formed on the continents during the last 500 million years. However, no more than about one-third of the area of the continents has been mapped sufficiently to allow confident recognition of large ancient impact structures, and of this fraction only about half consists of platform or cratonic elements where most impact structures have been successfully identified. The mean age of the rocks exposed on the cratons is about 250 myr, which further reduces the frequency of exposed impact structures by about a factor of two. The number of impact structures ≥100 km diameter and younger than 500 myr that we can reasonably expect to have been discovered, therefore, is about $1/2 \times 1/2 \times 1/3 \times 25 = 2 \overset{\times}{\div} 2$. One such structure of Tertiary age (Popigai in the Soviet Union) has been recognized, and two impact structures approaching 100 km diameter (Punchezh-Katunki, of Jurassic age, in the Soviet Union, and Manicouagan, of Triassic age, in Quebec) have been identified (16, 18). Thus, within the uncertainties of observation, the late geologic record of very large impacts also appears consistent with the present flux of Earth-crossing bodies.

Impact with the Earth of about $12 \overset{\times}{\div} 2$ bodies ≥10 km diameter should have occurred during the Phanerozoic. Two thirds of these probably struck oceanic crust. Approximately 85% of the Phanerozoic impact structures in oceanic crust have been lost by subduction but a global stratigraphic record of large impact events should be preserved. A clay layer enriched in noble metals that occurs at the Cretaceous-Tertiary boundary was evidently formed by atmospheric fallout of dust produced by one such event (1, 12, 42). This layer has now been located at more than 40 sites distributed around the world (2, 32), including ten sites in nonmarine deposits. Isotopic analysis of distinctive spherules contained in the clay layer (35) support the suggestion of Emiliani et al. (7) that the impact occurred in the ocean. Another global noble metal anomaly appears to have been found associated with microtektites in deep-sea sediments at or near the Eocene/Oligocene boundary (3, 13). At least two microtektite horizons are present, however, and there might be more than one anomaly. As indicated by isotopic analysis of tektites (35) which are similar in age, a large impact that may have produced one of these anomalies occurred in relatively young continental rocks. The Popigai structure has been suggested as a source for the North American tektites (7), but the great age of the Precambrian basement beneath Popigai appears to be in conflict with this hypothesis (35). The sources are probably other very large Tertiary impact structures on the continents that remain to be recognized.

SOURCES AND PROBABLE VARIATIONS IN THE FLUX OF EARTH-CROSSING ASTEROIDS

The fate of almost all Earth-crossing asteroids is either to collide with a planet or to be ejected from the solar system. Ejection of asteroids occurs as a consequence of numerous close encounters with the Earth and with the other terrestrial planets. The eccentricity of the orbit of Earth-crossing asteroids tends to increase with time as a result of planetary encounters. In somewhat more than half the cases, their orbit becomes Jupiter-crossing before collision with a terrestrial planet. Once the dynamical evolution of an asteroid orbit comes under the control of Jupiter, the asteroid has about a 99% chance of being ejected from the solar system in less than a few million years. Typical histories of orbital evolution have been studied with Monte Carlo methods by Wetherill and Williams (54), who found that the median lifetime of Earth-crossing asteroids is about 3×10^7 years. For a steady population to be maintained, new Earth crossers of diameter ≥ 1 km must be supplied at an average rate of $\sim 2.3 \times 10^3 / 3 \times 10^7$ years $\approx 0.8 \times 10^{-4}$ yr^{-1}.

Some new Earth crossers are supplied from the main asteroid belt and others probably are newly extinct comets of very short period. The relative importance of these two sources has been the subject of some debate and is, as yet, not known. Recent dynamical investigations have revealed a variety of mechanisms by which Earth crossers can be delivered from the main asteroid belt (33, 51, 52, 55, 63). Perhaps as many as a third or more Earth crossers are of main belt origin; if so, this would explain the observed photometric similarity between many Earth crossers and asteroids found in the nearer part of the main asteroid belt (24). As will be shown, whatever the rate may be, the supply of bodies ≥ 1 km in diameter from the asteroid belt is relatively steady. Any large long-term fluctuations in the population and past flux of Earth crossers of diameter ≤ 1 km have been due almost certainly to fluctuations in the supply of objects of cometary origin. On the other hand, the supply of Earth crossers more than 10 km in diameter may have been marked by surges that were due to the breakup of large main belt asteroids.

The main asteroid belt, as measured by the semimajor axes of the asteroid orbits, extends from about 1.8 to 3.7 AU from the sun. Roughly 10^6 asteroids with diameters between 1 and 1000 km occur in this region. About 97% of these bodies are on stable orbits and are safe from encounter with any planet unless they are perturbed by collision with a smaller body or by a very close pass by one of the largest asteroids. Physical collisions by themselves cannot inject main belt asteroids or asteroid fragments from stable orbits into Earth-crossing orbits; the required changes of velocity, typically ~5 km s^{-1}, are too great. Collisional impulses of the necessary magnitude result in melting and vaporization of rocky material.

Delivery of asteroids from the main belt into Earth-crossing orbits is probably the result chiefly of the injection of collision fragments into resonances within orbital element phase space. These resonances are of two kinds: a) secular resonances, in which the rate of precession of either the apsides (perihelion and aphelion) or the nodes of an asteroid orbit is nearly equal to a principal perturbing frequency related to long-range gravitational effects of Jupiter and Saturn (57), and b) commensurabilities, where the orbital period of the asteroid is a low order simple fraction of the orbital period of Jupiter. Six major resonances that probably contribute significantly to delivery of Earth-crossing asteroids occur along surfaces that dissect the semimajor axis, eccentricity, and inclination space of main belt asteroids. Regions adjacent to these surfaces are nearly devoid of asteroids or have a much lower density of asteroids than the regions midway between the resonances (15, 57, 58).

The distribution of the resonances is such that few asteroids are remote from them. An impulse of 1 km s^{-1} or less is sufficient to displace most asteroids into the center of a resonance, and about half of all asteroids could be displaced into the margin of a resonance by means of an appropriately directed impulse on the order of 200 to 300 m s^{-1}. A significant fraction of large fragments produced by collisional disruption of asteroids receive impulses of this magnitude, as is demonstrated by the dispersion of orbital elements observed among asteroid families formed by collision (cf. 61). Collision fragments injected into the center of two of the secular resonances can become Earth-crossing solely as a result of resonant amplification of orbital eccentricity (59, 60). In the more usual case, however, a fragment is injected into the margin of a resonance and is first perturbed into a Mars-crossing orbit; encounters with Mars then lead to random walk of the orbital elements until the orbit is shifted into the center of the resonance and the body finally becomes Earth-crossing (51, 52).

The rate of collisional disruption of main belt asteroids and the resulting rate of supply of Earth-crossing fragments can be roughly evaluated as follows: The size-frequency distribution of main belt asteroids is given, with sufficient precision for the present analysis, by

$$F = kb^{\lambda}, \tag{2}$$

where F is the cumulative frequency, $k = 10^6\ km^2$, b the asteroid diameter in km, and the size index, λ, is -2. The true frequency in the diameter range of 1 to 1000 km does not deviate from that given by Eq. 2 by more than ~50%. The differential area, dA, of all asteroids of diameter b is given by

$$dA = \pi k \lambda b^{\lambda+1}\, db, \tag{3}$$

and the differential probability, dP, of collisional disruption for asteroids of diameter b is

$$dP = p\,dA, \tag{4}$$

where p is the disruption rate per unit area of asteroids of diameter b. The disruption rate, p, is itself a function of the asteroid diameter which can be approximated by

$$p = Rb^{s\lambda}, \tag{5}$$

where the coefficient R is a constant related to the probability of mutual collision among asteroids, and s is a factor composed of the following terms obtained from crated scaling

$$s = e + g, \tag{6}$$

e is an empirical energy scaling factor, and g a gravity scaling factor; from experiment, $e = 3.4/3 = 1.13$, and $g = 1/6$ (see Eq. 1); hence $s = 1.3$. Combining Eqs. 3, 4, and 5, we have

$$dP = \pi R k \lambda b^{\lambda(s+1)+1} db, \tag{7}$$

and, integrating,

$$P = \frac{\pi R k \lambda}{\lambda(s+1)+2} \left(b_{min}^{\lambda(s+1)+2} - b_{max}^{\lambda(s+1)+2}\right), \tag{8}$$

where b_{min} and b_{max} are the limits of asteroid diameter over which the probability of collisional disruption is to be integrated.

To evaluate Eq. 8, one must know the value of the collision constant R. I have evaluated the rate of production of 10 km craters on the three largest asteroids (Table 2), using Wetherill's (48) equations for collision probability, the set of orbits for numbered asteroids given by Bender (4), estimates of the distribution of asteroid types by Zellner (62), the magnitude-frequency distribution of faint asteroids from van Houten et al. (43), appropriate corrections for completeness of discovery of asteroids as a function of semimajor axis, estimates of asteroid densities based on inferences about composition from spectrophotometry, and crater scaling as given by Eq. 1, with $c_f = 1$. As shown in Table 2, the calculated production of craters ≥10 km diameter on Ceres, Pallas, and Vesta is about 2 1/2 to 7 times the present cratering rate on Earth. The 10 km cratering rate for a standard 550 km asteroid will be taken as 10^{-13} km^{-2} yr^{-1}, essentially the average of the cratering rates given in Table 2.

Next, the condition for collisional disruption is taken as the production of a crater whose diameter, as formally given by Eq. 1, is equal to the diameter of the asteroid. This condition is conservative, in the sense that disruption of small bodies may occur at somewhat lower impact energies. From gravity scaling, the collisional disruption rate of 10 km asteroids is $(55)^{2.26/6} \times 10^{-13}$ km^{-2} yr^{-1} = 4.5×10^{-13} km^{-2} yr^{-1}. From

TABLE 2 - Production of impact craters ≥10km diameter on the three largest asteroids (from unpublished calculations by E.M. Shoemaker and R.F. Wolfe).

	CERES (1000 km diameter)		PALLAS (540 km diameter)		VESTA (550 km diameter)	
Colliding small asteroids	Cratering rate (10^{-14}/km^{2}yr)	rms velocity (km/s)	Cratering rate (10^{-14}/km^{2}yr)	rms velocity (km/s)	Cratering rate (10^{-14}/km^{2}yr)	rms velocity (km/s)
C-type	4.0	4.76	9.7	11.11	2.6	4.44
S-type	1.2	4.71	3.4	11.72	1.7	4.02
"M-type"*	1.3	4.73	3.7	11.52	1.5	4.13
Total	6.5		16.8		5.8	

*All colliding asteroids other than C- or S-type have been grouped here under "M-type."
Assuming low mean albedo for comet nuclei, an average production of craters ≥10 km diameter by comet impact on the order of ~0.3×10^{-14} km^{-2}yr^{-1} should be added to the total cratering rates.

Eq. 5, the collision constant R, which is the disruption rate of 1 km diameter asteroids, is found to be 1.8 x 10^{-10} $km^{-(s\lambda+1)}$ yr^{-1}. If appropriate numerical values are used for each of the constants in Eq. 8,

$$P = 4.4 \times 10^{-4}(b_{min}^{-2.6} - b_{max}^{-2.6})\ km^{2.6}\ yr^{-1}. \qquad (9)$$

It follows that the largest asteroid to have been collisionally disrupted in the last 3 Gyr has a diameter of about 230 km. Thus, in solving P for smaller bodies, 230 km can be taken as b_{max}. Solutions for the cumulative disruption rate to various lower limiting diameters are listed in Table 3. About 10 asteroids ≥100 km diameter were probably disrupted in the last 3 Gyr, ~3600 asteroids ≥10 km were disrupted, and ~1.3 x 10^6 asteroids ≥1 km, or a number about equal to the present population were collisionally destroyed. Losses of smaller asteroids have been replaced by fragmentation of the larger bodies, and the population of small asteroids has probably been nearly steady. If the population of small bodies is in equilibrium, most asteroids of diameter close to 1 km have been formed as collision fragments during the past 3 Gyr, but only about one third of the 10 km bodies are new fragments produced in this time interval.

TABLE 3 - Rate of collisional disruption of main belt asteroids (as given by Eq. 9).

Asteroid diameters	Disruption rate	Mean interval between disruptions	Total asteroids disrupted in 3 Gyr
≥ 1 km	4.4 x 10^{-4} yr^{-1}	2.3 x 10^3 yr	1.3 x 10^6
≥ 10 km	1.2 x 10^{-6} yr^{-1}	8.3 x 10^5 yr	3.6 x 10^3
≥ 30 km	6.3 x 10^{-8} yr^{-1}	1.6 x 10^7 yr	1.9 x 10^2
≥ 50 km	1.6 x 10^{-8} yr^{-1}	6.1 x 10^7 yr	49
≥ 100 km	2.3 x 10^{-9} yr^{-1}	4.3 x 10^8 yr	7
	(3.2 x 10^{-9} yr^{-1})*	(3.1 x 10^8 yr)*	(10)*
≥ 200 km	4.5 x 10^{-10} yr^{-1}	2.2 x 10^9 yr	1

*The estimated number of main belt asteroids with diameters ≥100 km is 140 (62), 40% higher than given by Eq. 2; the disruption rate, disruption interval, and number disrupted in 3 Gyr corrected for the true estimated population is given in parentheses.

The production of new fragments can be obtained readily from Eq. 7 if the size distribution of the fragments for individual disruption events is specified. The size of asteroids in the Themis, Eos, and Koronis families (14) suggests that the cumulative size-frequency distributions of fragments have a form closely similar to Eq. 2, with $\lambda \approx -2$. Reconstruction of the hypothetical disrupted parent bodies of ten different families on the basis of the nature of the observed fragments (14) suggests that the largest surviving fragment is commonly on the order of half the diameter of the parent.

If we write the cumulative frequency of fragments, f, as

$$f = c\,D^{-2}, \tag{10}$$

where c is a constant and D the fragment diameter, then the volume of fragments, V, is

$$V = \frac{\pi}{3}c\,(D_{max} - D_{min}) = \frac{\pi}{6}b^3, \tag{11}$$

$$c = \frac{b^3}{2(D_{max} - D_{min})} \quad ,$$

where D_{max} and D_{min} are the maximum and minimum fragment diameters. Setting $D_{max} = 1/2\ b$ and $D_{min} = 0$,

$$c = b^2. \tag{12}$$

The production of fragments ≥1 km diameter is given by

$$d\Gamma = c\,dP, \tag{13}$$

where Γ is the cumulative number of fragments ≥1 km diameter; combining Eqs. 7, 12, and 13 and integrating, we obtain

$$\Gamma = \frac{\pi\ R\,k\,\lambda}{\lambda\,(s+1)+4}\quad (b_{min}^{\lambda\,(s+1)+4} - b_{max}^{\lambda\,(s+1)+4}). \tag{14}$$

Solutions to Eq. 14 are listed in Table 4.

Two thirds of the production of asteroids 1 km diameter and larger, as given by Eq. 14, comes from disruption of 2 to 10 km precursor asteroids. Only 15% are derived from precursors larger than 50 km in

TABLE 4 - production of asteroids ≥ 1 km diameter from collisional disruption of main belt asteroids (as given by Eq. 14).

Diameters of asteroids disrupted	Production of bodies >1 km	Percent of Production
2 to 10 km	$0.78 \times 10^{-3} yr^{-1}$	66
10 to 30 km	$0.23 \times 10^{-3} yr^{-1}$	19
30 to 50 km	$0.07 \times 10^{-3} yr^{-1}$	6
50 to 100 km	$0.07 \times 10^{-3} yr^{-1}$	6
100 to 200 km	$0.03 \times 10^{-3} yr^{-1}$	3
2 to 200 km	$1.18 \times 10^{-3} yr^{-1}$	100

diameter. The total production rate of bodies ≥ 1 km shown in Table 4 is 2.7 times higher than the estimated destruction rate given in Table 3. It is likely, however, that the population at 1 km diameter is actually steady and that the production of fragments ≥ 1 km has been overestimated by a factor of ~ 2.7. The actual average size distribution of fragments produced by disruption is probably weighted somewhat more heavily toward large fragments than was assumed for Eq. 10. If we correct the exponent of Eq. 10 to make the production balance the loss of asteroids at 1 km diameter, the fraction of the production from precursors smaller than 10 km diameter is even greater than shown in Table 4. The bulk of asteroid fragments of any size is produced by the disruption of precursors no more than ten times larger than the fragments.

If about half of all disruption events occur close enough to a resonance to inject fragments into the resonance, and if we assume, as an upper bound, that no more than about 10% of the fragments receive impulses of the magnitude and direction and at a time necessary to place them in the resonance, then the supply of fragments to resonances is $\leq 1/20$ of the total production rate of fragments. If the collisional production rate for fragments ≥ 1 km in the main asteroid belt is set equal to the loss rate, 4.4×10^{-4} yr^{-1}, the average yield of Earth crossers is $\leq 2 \times 10^{-5}$ yr^{-1}. This yield is $\leq 1/4$ the rate required to maintain the Earth-crossing asteroid population of 1 km diameter. It is provided, typically, by the disruption of 5 to 10 km asteroids at intervals on the order of 10^6 years. As these intervals generally are between one and two orders-of-magnitude shorter than the typical dynamical lifetime of Earth crossers, fluctuations in the number of Earth crossers ≥ 1 km derived from the main belt have usually been no greater than a few percent. About one tenth of the total yield of Earth crossers of diameters up to 1 km, however, has come from disruption events that occurred at intervals comparable to or longer than their typical dynamical lifetimes. Once every 600 million years or so, on average, disruption of an asteroid $\geq$100 km diameter may have injected a sufficient number of fragments into a resonance to have increased the total population of Earth crossers of diameter up to 1 km by as much as ~25%. Transients of this magnitude may be expected to have grown and decayed in $\sim 5 \times 10^7$ years.

Ten kilometer Earth crossers derived from the main belt have probably been delivered chiefly after disruption of 30 to 50 km precursors at intervals on the order of 30 to 100 million years. If the main belt population of 100 km diameter asteroids remained at equilibrium, the rate of supply of 10 km Earth crossers is $\lesssim(1/20 \times 1.2 \times 10^{-6} \mathrm{yr}^{-1}) \lessapprox 6$ per 10^8 years, of which one or two may be expected to collide with the Earth every 10^8 years, on average. This more or less steady supply may be close to that required to maintain the present population of 10 km Earth crossers. The disruption of a main belt asteroid $\geq$100 km in diameter at intervals on the order of the span of the Phanerozoic probably produced surges in the population of Earth-crossing 10 km bodies by a factor of about five; these surges lasted for periods on the order of 5×10^7 years. Typically, a half dozen or so bodies $\geq$10 km in diameter may have struck the Earth during these surges.

From the foregoing analysis, the chance may be as high as about 60% that a surge in the impact rate of asteroids of diameter $\geq$10 km occurred some time during the Phanerozoic. The elongate 15 x 30 km Amor asteroid 433 Eros, which approaches Earth but is not now an Earth crosser, may be an example of a body produced by the disruption of a large ($\geq$100 km?) precursor and injected into a Mars-crossing orbit, where it has become stranded in a relatively long-lived orbit by Mars encounter. There is a 20% chance that this large body will hit the Earth in the next 400 myr (53). The existence of Eros, whose dynamical lifetime is less than the span of the Phanerozoic, suggests that a major disruption event in the main belt and an accompanying surge in the impact of large bodies on Earth did, in fact, occur during the Phanerozoic. There is a slight suggestion from the statistics of the ages of terrestrial impact structures (16) that a surge in the impact rate occurred between 300 and 400 myr BP; a distinct clustering of ages is found around 300 myr (late in the Devonian).

The estimated rate of supply of Earth crossers of diameter $\geq$1 km is equal to or less than about one fourth that required to maintain the present population. This may be due to the choice of the criterion for the disruption threshold of 5 to 10 km diameter precursors. It is possible that asteroids 5 to 10 km in diameter may be disrupted when an impact crater, as given by Eq. 1, is as small as one half the diameter of the asteroid. If so, the rate of production of fragments from 5 to 10 km precursors could be four times greater than estimated above. This would remove the apparent discrepancy between the rate of injection of main belt fragments into Earth-crossing orbits and the loss rate of Earth crossers. On the other hand, the efficiency of injection of collision fragments into Earth-crossing orbits suggested earlier could be too high. Our understanding both of the disruption and the injection mechanisms is still too rough to make a closer assessment. Hence it is also necessary to consider extinct comets as the possible dominant source of Earth-crossing asteroids.

One short period comet, P/Encke, has been discovered on an orbit with aphelion sufficiently far inside the orbit of Jupiter that it is safe from close encounter with Jupiter. Apparently P/Encke has arrived in this "safe" orbit as a result of nongravitational forces associated with the blowing off of gas and dust (34). Its orbit is closely similar to that of the Earth-crossing asteroid 2212 Hephaistos. Wetherill (52) has shown that a distribution of orbits such as that observed among Earth-crossing asteroids is produced by multiple planetary encounters of bodies initially injected into Encke-like orbits. The nucleus of P/Encke has a diameter between 0.4 and 4 km as determined by radar (21) and will probably either become extinct or disintegrate in less than 10^4 years. If there is a more or less steady population of one active comet on a safe orbit such as that of P/Encke, and if the typical lifetime of activity is as long as 10^4 years, then the supply of extinct comets would be sufficient to maintain the Earth-crossing asteroid population, provided that each comet decayed to an inactive body ≥ 1 km diameter. The average active lifetime of these bodies is probably shorter than 10^4 years, and only a fraction of the comets become asteroids. The supply of extinct comet nuclei is potentially adequate, however, to provide an important fraction of Earth-crossing asteroids. The long-term stability or fluctuations in the rate of this supply depend largely on the mechanisms by which comets are delivered into short period orbits.

VARIATION IN THE FLUX OF COMETS

Long period comets arrive in the vicinity of the Earth from distances far beyond the most distant known planet. Prior to entry into the planetary region, about one third of these comets have initial orbits with semimajor axes greater than 10^4 AU (23). This distant reservoir of comets is generally referred to as the Oort sphere or Oort comet cloud. The remaining two thirds of the observed long period comets, with initial semimajor axes less than 10^4 AU, are generally thought to be Oort cloud comets that have been captured in smaller orbits during a succession of passes through the planetary region.

All but about 0.01% of the Earth-approaching long period comets are ejected from the solar system by planetary encounters, usually in times shorter than a few tens of millions of years. A flux of new comets from the Oort cloud is maintained in the vicinity of the Earth primarily by the perturbation of the cloud by passing stars (11, 20, 29). If the density of stellar material is 0.1 solar masses per cubic parsec in the solar neighborhood, if the mean encounter velocity is 30 km s^{-1}, and the estimated frequency distribution of stars near the sun is that given by Miller and Scalo (25), then on average one star with a mass greater than 0.1 solar mass has probably passed within 4 x 10^4 AU of the sun (the typical distance of an Oort cloud comet) each million years during the recent geological past. The mean mass of these stars was about 0.36 solar mass, sufficient to produce significant perturbation and shuffling of perihelion distances of Oort cloud comets. As the typical orbital periods of the comets in the cloud are about 3 to 8 myr, and as the median lifetime of a comet perturbed to Earth-crossing is about four orbital periods (20), the shuffling of perihelion distances is sufficiently frequent to maintain a fairly steady supply of comets on Earth-crossing orbits. As Oort (29) recognized and as Hills (20) has more recently shown, the present observed inner edge of the cloud at semimajor axes between 10^4 and 2 x 10^4 AU is just the threshold size of orbits beyond which the flux of comets in the vicinity of the Earth is maintained in a near steady state by stellar perturbation.

The past flux of stars near the sun has varied as the sun passed through the spiral arms of the galaxy. These arms are controlled by waves of stellar density that revolve about the galactic center at about half the rate of revolution of the stars themselves at the position of the sun (22). As there are four spiral arms (5), the sun passes through an arm about twice per galactic year (~250 Myr). The arms appear to be broken into segments and to have branches or spurs. The average frequency of encounters with regions of higher than average stellar density is on the order of 10^{-8} yr^{-1}. Amplitudes of the fluctuation in density of old stars in the density waves are thought to be on the order of 5 to 10%. In addition, massive short-lived new stars are born in the spiral arms from shock-compressed clouds of gas (41). The massive young stars are sufficiently widely spaced, however, that they make only a minor contribution to the perturbation of the comet cloud.

The quasiperiodic variation in the flux of stars near the sun has almost certainly shifted the inner edge of the region from which comets are perturbed fairly steadily into Earth-crossing orbits. During passage through spiral arms the effective inner edge of the comet source region was driven inward by the increased frequency of stellar perturbation toward a more stable region of probably higher comet density. A larger number of comets was strongly perturbed during these passages, and the flux of Earth-crossing comets was thereby increased. Conversely, when the sun was between spiral arms the Earth-crossing comet flux was lower. The expected magnitude of the oscillation of the near-Earth comet flux depends on the unknown distribution of comets inside the present inner observational edge of the Oort cloud. Calculations of the history and structure of an inner comet cloud derived from Uranus and Neptune planetesimals (Shoemaker and Wolfe, unpublished) indicate that a fluctuation on the order of 10% in the near-Earth comet flux seems likely. At the present time the sun apparently lies between the galactic spiral arms, but it is passing through a region of higher than average density of stars. Hence, the present flux of comets may be somewhat higher than the recent average flux.

Short-lived transients or bursts in the comet flux due to encounters of individual stars at distances less than 2×10^4 AU were probably superimposed on the relatively smooth, low frequency modulation of the comet flux. About once every 10^8 years, on average, a star greater than 0.1 solar mass has passed within 4×10^3 AU of the sun, near where Shoemaker and Wolfe (unpublished) find a peak in the predicted distribution of comet aphelia. As noted by Hills (20), such stellar encounters may have produced "storms" of comets near the Earth. The lifetime of such storms would have been on the order of 10^6 years, and the near-Earth flux during these storms might have been as much as one to two orders-of-magnitude greater than the mean ambient flux; the magnitude of the flux would have depended on the density and structure of the inner comet cloud.

Napier and Clube (6, 26) have suggested that comet storms were produced by encounters of the sun with giant molecular clouds (GMC's) in the galaxy. They argue that the comet cloud was stripped from the sun during each of these encounters and that a new swarm of comets was simultaneously captured. As pointed out by van den Bergh (45), however, in order for a new swarm of comets comparable to the Oort cloud to be captured, the comet density in GMC's would have to be very high. If comets are formed in GMC's, they must diffuse to interstellar space, and the space density of comets would approach that in the GMC's. As a consequence, there should be a large flux of interstellar comets in the vicinity of the Earth; this is contrary to observation. The principal effect of encounters of the sun with massive GMC's has probably been simply to deplete or strip the Oort cloud from the sun, as described by Napier and Staniucha (27). Evidently the Oort cloud is then replenished by combined stellar and planetary perturbation of the more populous and tightly bound inner reservoir of comets (45). Encounters with GMC's probably do not lead to comet storms but rather to a long-term depletion of the comet flux near the Earth, which is followed by the gradual recovery of the Oort cloud and a slow increase in the comet flux.

Short period comets are captured from the long period comet swarm by planetary encounters. It is not the comets with perihelia near the Earth that tend to be captured, however, but those with perihelia near the giant planets (9, 10). Some short period comets undoubtedly are derived from the flux of bodies arriving from the Oort cloud, but the principal source of comets delivered to very short period orbits may be the inner comet cloud, a significant fraction of which remains Neptune-crossing (Shoemaker and Wolfe, unpublished). Neptune crossers can be handed down to the control of Uranus by a succession of planetary encounters and thence to Saturn and Jupiter (10). Fluctuations in the population of Earth-crossing asteroids derived from short period comets may arise from variations in the rate of stellar perturbation of the inner comet cloud, provided that most comets in the inner cloud at any given time are not Neptune-crossing. A long-term increase in the population of Earth-crossing asteroids late in geologic time could arise from slow recovery of the Oort cloud after encounter with an unusually massive GMC.

LONG-TERM HISTORY OF THE CRATERING RATE IN THE EARTH-MOON SYSTEM

The combined evidence of the cratering record on the Earth and the Moon suggests that the average cratering rate during the Phanerozoic was somewhat higher than the long-term average cratering rate over the last 3.3 Gyr (39). The difference appears likely to be as much as a factor of two and is possibly as high as a factor of six. There is very little information at present on the possible short-term (< 1 Gyr) fluctuations in the cratering rate between the emplacement of lunar lavas at ~3.3 Gyr BP and the beginning of the Phanerozoic.

At 3.9 Gyr, the lunar cratering rate was about twenty-five times higher than the Phanerozoic cratering rate. The bombardment declined approximately exponentially between 3.9 Gyr and 3.3 Gyr (28, 36, 43). This episode has been referred to as the late heavy bombardment. Late arriving fragments of planetesimals of the Earth and Venus (50), and late arriving Uranus and Neptune planetesimals (49) probably both contributed to the late heavy bombardment. On the basis of extrapolation of an exponential function fitted to the decay of the cratering rate between 3.3 and 3.9 Gyr, most of the preserved large impact basins on the Moon probably were formed between about 3.9 and about 4.2 Gyr. The Earth almost certainly was subjected to a similar concurrent bombardment.

REFERENCES

(1) Alvarez, L.W.; Alvarez, W.; Asaro, F.; and Michel, H.V. 1980. Extraterrestrial cause for the Cretaceous-Tertiary extinction. Science 208: 1095-1108.

(2) Alvarez, W.; Alvarez, L.W.; Asaro, F.; and Michel, H.V. 1982. Major impacts and their geological consequences. In Geological Implications of Impacts of Large Asteroids and Comets on the Earth, eds. L.T. Silver and P.H. Schultz. Geol. Soc. Am. Spec. Paper 190: 305-316.

(3) Asaro, F.; Alvarez, L.W.; Alvarez, W.; and Michel, H.V. 1982. Geochemical anomalies near the Eocene/Oligocene and Permian/Triassic boundaries. In Geological Implications of Impacts of Large Asteroids and Comets on the Earth, eds. L.T. Silver and P.H. Schultz. Geol. Soc. Am. Spec. Paper 190: 517-528.

(4) Bender, D.F. 1979. Osculating orbital elements of the asteroids. In Asteroids, ed. T. Gehrels, pp. 1014-1039. Tucson: University of Arizona Press.

(5) Blitz, L.; Fich, M.; and Kulkarni, S. 1983. The New Milky Way. Science 220: 1233-1240.

(6) Clube, S.V.M., and Napier, W.M. 1982. Spiral arms, comets, and terrestrial catatrophism. Q. J. Roy. Astron. Soc. 23: 45-66.

(7) Dietz, R.S. 1977. Elgygytgyn Crater, Siberia: a probable source of Australasian tektite fields (and bediasties from Popigai). Meteoritics 12: 145-157.

(8) Emiliani, C.; Kraus, E.B.; and Shoemaker, E.M. 1981. Sudden death at the end of the Mesozoic. Earth Planet. Sci. Lett. 55: 317-334.

(9) Everhart, E. 1972. The origin of short period comets. Astrophys. Lett. 10: 131-135.

(10) Everhart, E. 1977. The evolution of comet orbits as perturbed by Uranus and Neptune. In Comets, Asteroids and Meteorites, ed. A.H. Delsemme, pp. 99-104. Toledo: University of Toledo Press.

(11) Fernandez, J.A. 1980. Evolution of comet orbits under the perturbing influence of the giant planets and nearby stars. Icarus 42: 406-421.

(12) Ganapathy, R. 1980. A major meteorite impact on the Earth 65 million years ago: Evidence from the Cretaceous/Tertiary boundary clay. Science 209: 921-923.

(13) Ganapathy, R. 1982. Evidence for a major meteorite impact on Earth 34 million years ago: Implication on the origin of North American tektites and Eocene extinction. In Geological Implications of Impacts of Large Asteroids and Comets on the Earth, eds. L.T. Silver and P.H. Schultz. Geol. Soc. Am. Spec. Paper 190: 513-516.

(14) Gradie, J.C.; Chapman, C.R.; and Williams, J.G. 1979. Families of minor planets. In The Satellites of Jupiter, ed. D. Morrison, pp. 359-390. Tucson: University of Arizona Press.

(15) Greenberg, R., and Scholl, H. 1979. Resonances in the asteroid belt. In Asteroids, ed. T. Gehrels, pp. 310-333. Tucson: University of Arizona Press.

(16) Grieve, R.A.F. 1982. The record of impact on Earth: Implications for a major Cretaceous/Tertiary impact event. In Geological Implications of Impacts of Large Asteroids and Comets on the Earth, eds. L.T. Silver and P.H. Schultz. Geol. Soc. Am. Spec. Paper 190: 25-38.

(17) Grieve, R.A.F., and Dence, M.R. 1979. The terrestrial cratering record II. The crater production rate. Icarus 38: 230-242.

(18) Grieve, R.A.F., and Robertson, P.B. 1979. The terrestrial cratering record. I. Current status of observations. Icarus 38: 212-229.

(19) Helin, E.F., and Shoemaker, E.M. 1979. Palomar planet-crossing asteroid survey, 1973-1978. Icarus 40: 321-328.

(20) Hills, J.G. 1981. Comet showers and the steady-state infall of comets from the Oort cloud. Astron. J. 86: 1730-1740.

(21) Kamoun, P.G.; Campbell, D.B.; Ostro, S.J.; Pettengill, G.H.; and Shapiro, I.I. 1981. Comet Encke: Radar detection of nucleus. Science 216: 293-295.

(22) Lin, C.C.; Yuan, C.; and Shu, F.H. 1969. On the spiral structure of disk galaxies - III. Comparison with observations. Astrophys. J. 155: 721-746.

(23) Marsden, B.G., and Roemer, E. 1982. Basic information and references. In Comets, ed. L.L. Wilkening, pp. 707-733. Tucson: University of Arizona Press.

(24) McFadden, L.A. 1983. Spectral reflectance of near-Earth asteroids: Implications for composition, origin, and evolution. Ph.D. Thesis, University of Hawaii, Honolulu.

(25) Miller, G.E., and Scalo, J.M. 1979. The initial mass function and stellar birthrate in the solar neighborhood. Astrophys. J. Supp. Series 41: 513-547.

(26) Napier, W.M., and Clube, S.V.M. 1979. A theory of terrestrial catastrophism. Nature 282: 455-459.

(27) Napier, W.M., and Staniucha, M. 1982. Interstellar planetesimals - I. Dissipation of a primordial cloud of comets by tidal encounters with massive nebulae. Mon. Not. Roy. Astron. Soc. 198: 723-735.

(28) Neukum, G.; König, B.; Fechtig, H.; and Storzer, D. 1975. Cratering in the Earth-Moon system: Consequences for age determination by crater counting. Proceedings of the VI Lunar Science Conference, vol. 2, pp. 2597-2620, Houston, Texas. Oxford: Pergamon Press.

(29) Oort, J.H. 1950. The structure of the cloud of comets surrounding the solar system, and a hypothesis concerning its origin. Bull. Astron. Inst. Netherlands 11: 91-110.

(30) Öpik, E.J. 1951. Collision probabilities with the planets and the distribution of interplanetry matter. Proc. Roy. Irish Acad. 54A: 165-199.

(31) Öpik, E.J. 1963. The stray bodies in the solar system. Part I. Survivors of cometary nuclei and asteroids. Adv. Astron. Astrophys. 2: 219-262.

(32) Pillmore, C.L.; Tschudy, R.H.; Orth, C.J.; Gilmore, J.S.; and Knight, J.D. 1982. Iridium abundance anomalies at the palynological Cretaceous/Tertiary boundary in coal beds of the Raton Formation, Raton basin, New Mexico and Colorado. Abstract. Geological Society of America Annual Meeting, 95th, New Orleans, LA.

(33) Scholl, H., and Froeschle, C. 1977. The Kirkwood gaps as an asteroidal source of meteorites. In Comets, Asteroids and Meteorites, ed. A.H. Delsemme, pp. 293-295. Toledo: University of Toledo Press.

(34) Sekanina, Z. 1971. A core-mantle model for cometary nuclei and asteroids of possible cometary origin. In Physical Studies of Minor Planets, ed. T. Gehrels, pp. 423-426. Washington, D.C.: NASA-SP 267.

(35) Shaw, H.F., and Wasserburg, G.J. 1982. Age and provenance of the target materials for tektites and possible impactites as inferred from Sm-Nd and Rb-Sr systematics. Earth Planet. Sci. Lett. 60: 155-177.

(36) Shoemaker, E.M. 1972. Cratering history and early evolution of the Moon. Lunar Sci. III: 696-698.

(37) Shoemaker, E.M. 1977. Astronomically observable crater-forming projectiles. In Impact and Explosion Cratering, eds. D.J. Roddy et al, pp. 617-628. New York: Pergamon Press, Inc.

(38) Shoemaker, E.M. 1983. Asteroid and comet bombardment of the Earth. Ann. Rev. Earth Planet. Sci. 11: 461-494.

(39) Shoemaker, E.M.; Williams, J.G.; Helin, E.F.; and Wolfe, R.F. 1979. Earth-crossing asteroids: Orbital classes, collision rates with the Earth, and origin. In Asteroids, ed. T. Gehrels, pp. 253-282. Tucson: University of Arizona.

(40) Shoemaker, E.M., and Wolfe, R.F. 1982. Cratering time scales for the Galilean satellites. In The Satellites of Jupiter, ed. D. Mossison, pp. 277-339. Tucson: University of Arizona Press.

(41) Shu, F.H.; Milione, V.; Gebel, W.; Yuan, C.; Goldsmith, D.W.; and Roberts, W.W. 1972. Galactic shocks in an interstellar medium with two stable phases. Astrophys. J. 173: 557-592.

(42) Smit, J., and Hertogen, J. 1980. An extraterrestrial event at the Cretaceous-Tertiary boundary. Nature 285: 198-200.

(43) Soderblom, L.A., and Boyce, J.M. 1972. Relative ages of some near-side and far-side Terra plains based on Apollo 16 metric photography. NASA Spec. Paper 315: 29-3 - 29-6.

(44) van Houten, C.J.; van Houten-Groenveld, I.; Herget, P.; and Gehrels, T. 1970. The Palomar-Leiden survey of faint minor planets. Astron. Astrophys. Suppl. 2: 339-448.

(45) van den Bergh, S. 1982. Giant molecular clouds and the solar system comets. J. Roy. Astron. Soc. Can. 76: 303-317.

(46) Weissman, P.R. 1982. Dynamical history of the Oort cloud. In Comets, ed. L.L. Wilkening, pp. 637-658. Tucson: University of Arizona Press.

(47) Weissman, P.R. 1982. Terrestrial impact rates for long and short-period comets. In Geological Implications of Impacts of Large Asteroids and Comets on the Earth, eds. L.T. Silver and P.H. Schultz. Geol. Soc. Am. Spec. Paper 190: 15-24.

(48) Wetherill, G.W. 1967. Collisions in the asteroid belt. J. Geophys. Res. 72: 2429-2444.

(49) Wetherill, G.W. 1975. Late heavy bombardment of the Moon and the terrestrial planets. Proceedings of the VI Lunar Science Conference, vol. 2, pp. 1539-1561, Houston, Texas. Oxford: Pergamon Press.

(50) Wetherill, G.W. 1977. Evolution of the Earth's planetesimal swarm subsequent to the formation of the Earth and Moon. Proceedings of the VIII Lunar Science Conference, vol. 8, pp. 1-16. Oxford: Pergamon Press.

(51) Wetherill, G.W. 1977. Fragmentation of asteroids and delivery of fragments to Earth. In Comets, Asteroids, Meteorites, ed. A.H. Delsemme, pp. 283-291. Toledo: University of Toledo Press.

(52) Wetherill, G.W. 1979. Steady-state populations of Apollo-Amor objects. Icarus 37: 96-112.

(53) Wetherill, G.W., and Shoemaker, E.M. 1982. Collison of astronomically observable bodies with the Earth. In Geological Implications of Impacts of Large Asteroids and Comets on the Earth, eds. L.T. Silver and P.H. Schultz. Geol. Soc. Am. Spec. Paper 190: 1-13.

(54) Wetherill, G.W., and Williams, J.G. 1968. Evaluation of the Apollo asteroids as sources of stone meteorites. J. Geophys. Res. 73: 635-648.

(55) Wetherill, G.W., and Williams, J.G. 1979. Origin of differentiated meteorites. In Origin and Abundance of the Elements, Second Symposium, ed. L.H. Ahrens, pp. 19-31. Oxford: Pergamon Press.

(56) Whipple, F.L. 1978. Comets. In Cosmic Dust, ed. J.A.M. McDonnell, pp. 1-73. New York: John Wiley and Sons.

(57) Williams, J.G. 1969. Secular perturbations in the solar system. Ph.D. Thesis, Los Angeles, University of California.

(58) Williams, J.G. 1971. Proper elements, families, and belt boundaries. In Physical Studies of Minor Planets, ed. T. Gehrels, pp. 177-181. Washington, D.C.: NASA-SP 267.

(59) Williams, J.G. 1973. Meteorites from the asteroid belt? Abstract. Eos. Trans. Am. Geophys. Union 54: 233.

(60) Williams, J.G. 1973. Secular resonances. Abstract. Bull. Am. Astron. Soc. 5: 363.

(61) Williams, J.G. 1979. Proper elements and family memberships of the asteroids. In Asteroids, ed. T. Gehrels, pp. 1040-1063. Tucson: University of Arizona Press.

(62) Zellner, B. 1979. Asteroid taxonomy and the distribution of the compositional types. In Asteroids, ed. T. Gehrels, pp. 783-806. Tucson: University of Arizona Press.

(63) Zimmerman, P.D., and Wetherill, G.W. 1973. Asteroidal source of meteorites. Science 182: 51-53.

Patterns of Change in Earth Evolution, eds. H.D. Holland and A.F. Trendall, pp. 41-61.
Dahlem Konferenzen 1984. Berlin, Heidelberg, New York, Tokyo: Springer-Verlag.

Sudden Changes in Atmospheric Composition and Climate

O.B Toon
Ames Research Center, NASA
Moffett Field, CA 94035, USA

Abstract. Volcanic eruptions, collisions with galactic dust lanes, and asteroid impacts could all alter the composition of the atmosphere in significant ways. Asteroid impacts may be the major cause of large, sudden changes in atmospheric composition and climate by injection of dust and nitrogen oxides. Volcanic eruptions produce much less dust than large asteroid impacts but may still cause significant climatic changes. Galactic dust lane collisions probably do not affect the Earth's climate unless very dense dust lanes are encountered. Numerous questions remain in studies of these phenomena.

INTRODUCTION

The Earth's climate is very sensitive to the composition of the atmosphere. The composition of the atmosphere as well as climate have varied greatly throughout geologic history. Many changes, such as the rise of oxygen and ozone levels, the decline of CO_2 levels, and the initiation of ice ages, were probably gradual requiring millions or at least thousands of years. Much attention has been given to understanding these changes. By contrast some events may have occurred which have produced dramatic changes in atmospheric composition and climate over a few years or centuries. Much less study has been devoted to these changes and much less is known about them. Among the events and processes which may quickly alter atmospheric composition and climate, volcanic eruptions, solar system collisions with galactic dust lanes, and terrestrial collisions with asteroids and comets are perhaps the most important.

The mass of the atmosphere is $5.2x10^{21}$ gms, which is equivalent to the mass of a 10 meter thick layer of density 1 gm cm^{-3} covering the Earth. The quantity of oxygen and nitrogen in the atmosphere is so large that it is not likely to be altered significantly except by the degassing of a magma reservoir or impacting body of volume larger than $5x10^6$ km^3.

Even the concentration of carbon dioxide, which currently comprises about 0.03% of the atmosphere, is difficult to alter. Humans are now producing about $1.5x10^{16}$ gms CO_2 yr^{-1} largely by burning fossil fuels; of this quantity about half is removed from the atmosphere within a few years of release. Volcanoes worldwide (excluding lava extrusion along oceanic ridges) produce about 1 km^3 of lava each year (4). If this lava contains 2% of CO_2 (see (2) for discussion), then ca. $6x10^{13}$ gm CO_2 yr^{-1} are released to the atmosphere from this source. Only the largest volcanic lava production of historic times, the 10^3 km^3 of the 934 A.D. Eldgjer eruption (11) could have yielded a quantity of CO_2 equal to the quantity added by fossil fuel burning in a single year. The Deccan Trap volcanism of 65 million years ago produced 10^6 km^3 of lava (15), but the average rate of lava production was probably only ca. 1 km^3 yr^{-1}. To have doubled the Earth's atmospheric CO_2, more than 10% of the Deccan Trap lava would had to have been produced in ca. 100 years. Although it is possible for volcanism or impacts of meteorites to affect the CO_2 budget significantly on short time scales, such changes cannot be common phenomena.

It is much easier to upset the budget of trace components in the atmosphere. For example, a large fraction of the ozone in the atmosphere could be destroyed by the injection of relatively small amounts of nitrogen oxides (about $5x10^{13}$ gms); these gases can be produced by heating air to high temperatures (e.g., (32)). The injection into the atmosphere of 10^{13} gms of particulate matter or of gases that can be converted into particulates, can lead to a reduction of sunlight that is large enough to produce small changes in climate. The significance of such effects increases exponentially with the mass of the injected particles.

The following sections will deal with the ability of various processes to affect the concentration of trace species in the atmosphere and the Earth's climate.

VOLCANOES

The effects of volcanoes on climate are reasonably well–known (e.g., (16)). For a few years after an eruption vivid twilight and other optical phenomena provide evidence of enhanced levels of stratospheric aerosols. Most of the long-lived debris consists of sulfuric acid particles that are generated by the photochemical oxidation of sulfur dioxide gas vented into the stratosphere. The silicate ash injected by volcanoes is removed from the stratosphere within a few months due to the fairly large size and rapid sedimentation velocity of most ash particles. Figure 1 illustrates the evolution of the dust mass following the 1963 Mt. Agung eruption.

The mass of debris in the stratosphere may be determined from observations of the optical depth following volcanic eruptions (Fig. 1). The volume of particulate debris in the stratosphere some months after

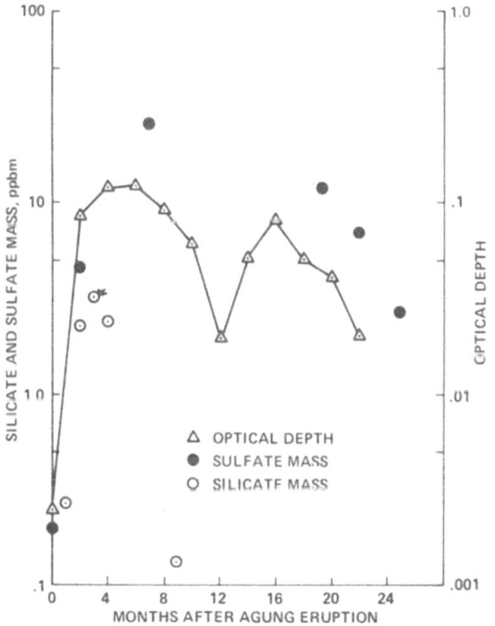

FIG. 1 - The mass of volcanic silicate ash, the mass of sulfate, and the optical depth as a function of time following the Mt. Agung eruption of 1963 (30). The time evolution of the sulfate mass matches that of the optical depth showing that sulfuric acid particles are primarily responsible for the "dust" veils seen after volcanic eruptions. The ash falls quickly from the stratosphere due to the relatively large size of the injected particles.

the eruption of El Chichon (1982), Krakatoa (1883), Katmai (1912), and Agung (1963) was only about 10^{-2} km^3, and most of this debris probably consisted of sulfuric acid rather than volcanic ash (6, 30).

Following large, single volcanic eruptions, the mean temperature at the Earth's surface is observed to decrease on average by 0.5°C or less for a period of a year or two. Even after the most violent eruption of modern history, that of Tambora in 1815, the world average temperature only dropped by 1°C (16). Localized effects were somewhat greater. For example, during the summer of 1816 record low temperatures were observed in New England. Even these temperatures were only about 2.5°C lower than normal; this represents about 2.5 standard deviations from the normal (17).

There are about 500 active volcanoes in the world, some of these will probably erupt at nearly the same time. Table 1 provides some estimates of the probability of achieving a given temperature perturbation over a fixed time period due to multiple eruptions.

We assume that during periods of high volcanic activity volcanic eruptions have a Gaussian probability distribution. If so, the waiting time, t_w, between epochs of length t, with N eruptions, is given by $t_w = t\ (N/n)^2$, where n is the mean number of eruptions per interval t during the period t_w (3). In order to determine the desired number of eruptions N, we first pick the temperature change we desire to obtain. Then, following

TABLE 1 - The waiting time, t_w, between epochs of length, t, with designated optical depth and temperature change, ΔT. We have omitted t_w for $N > 500$ because it exceeds the number of presently active volcanoes.

Optical depth	T(°C)	N decade total	N century total	t(years) n=4/ decade t=decade	t(years) n=4/ century t=century	t(years) n=2.5/ decade t=decade	t(years) n=25/ century t=century
0.15	-1	15	150	$1.4x10^4$	$1.4x10^5$	$3.6x10^2$	$3.6x10^3$
0.45	-3	45	450	$1.3x10^5$	$1.3x10^6$	$3.3x10^3$	$3.3x10^4$
0.6	-4	60	600	$2.3x10^5$	-	$5.8x10^3$	-
1.0	-6	100	1000	$6.3x10^5$	-	$1.6x10^4$	-
2.0	-10	200	2000	$2.5x10^6$	-	$6.4x10^4$	-

Pollack et al. (25, 26), we assume that each eruption yields an annual mean optical depth of 0.1 and relate the temperature to optical depth using the curves of Pollack et al. Their calculations include the effects of volcanic debris both on sunlight and on terrestrial thermal radiation. The calculations agree well with the observed temperature changes after single eruptions. However, the physics of the dense sulfuric acid clouds which would result from multiple explosions might lead to larger particle sizes, shorter-lived clouds, and to smaller climatic effects than the calculations suggest. We assume that N cannot exceed the approximately 500 active volcanoes in the world during any short period (19). Finally we obtain two choices of n from Lamb's (16) estimate of four major eruptions and twenty-five significant but lesser eruptions per century during the period from 1500 to 1960. Similar numbers of eruptions occurred for 10^4 years during the Late Wisconsin (10). The frequency and intensity (n) of periods of high volcanic activity, which may not be Gaussian, deserve further investigation.

Table 1 shows that if volcanically active periods only last 300-400 years, as did the period of 1500-1900, then even with twenty-five eruptions per century only one decade will occur in which the temperature perturbation is as large as 1°C. On the other hand, if such periods often last 10^4 years, then once during such a period a decrease of 6°C might occur for a decade. Of the 500 active volcanoes, only a small fraction are of the type that produces explosive volcanic eruptions. If 20% are explosive, Table 1 indicates that even if all of the explosive volcanoes in the world were to erupt during a century, the mean temperature during that century would only be 1°C cooler than normal. Of course, a great increase in snow cover might amplify such a temperature drop. If all of the explosive volcanoes erupted in 10 years, a cooling of about 6° might occur. It is clear that a large number of Krakatoa-sized eruptions are needed to produce large temperature changes.

A number of investigators have suggested that some single large volcanic eruptions could produce exceptionally dense volcanic clouds. For example, the Krakatoa eruption is thought to have blown 6 to 18 km^3 of ash into the sky (16). The Tambora eruption of 1815 is estimated to have ejected 100 to 300 km^3 of material (16). The volume of the Toba caldera implies an injection of close to $2x10^3$ km^3, 75,000 years ago (14). By comparison, the layer of meteoritic and crustal material which covered the Earth at the Cretaceous-Tertiary boundary is thought to have had a volume of approximately $5x10^3$ km^3 (1). On the basis of these numbers, Kent

(14) suggested that large volcanic eruptions should be just as important for extinction events and sudden changes to planetary climate as the impact of large meteorites. Unfortunately, direct comparisons between these numbers are very misleading. Much of the asteroid material was globally distributed and presumably resided in the atmosphere for at least several months. However, most volcanic ejecta fall very close to the erupting volcano in a matter of hours. We have already pointed out that the stratospheric cloud following the Krakatoa eruption contained only 10^{-2} km^3 of material, much of which was probably sulfuric acid.

The energy contained in a one megaton explosion is sufficient to lift about 10^{-2} km^3 of rock to an altitude of 20 km working only against gravity. The mechanical energy of the Krakatoa eruption was equivalent to about 100 megatons (5), about 1 km^3 of rock could therefore have been lifted into the stratosphere. Almost all of the rock would have been in the form of large particles which could not have remained in the atmosphere for more than a few hours or days. Turco et al. (34) have shown that even the one-day-old material in the May 1980 Mt. St. Helens cloud contained only 20% by mass of ash particles small enough to remain aloft for more than a few weeks. Murrow et al. (22) showed on the basis of grain size analyses that at most 0.8% of the ejecta grains are small enough ($< 2\ \mu m$) to remain aloft. If 1% of the total debris produced during the eruption consisted of such small grains, then not more than 10^{-2} km^3 of ash could be expected to remain in the Krakatoa cloud after a few weeks.

If 10% of the Toba material reached the stratosphere and 1% were in small grains, then 2 km^3 of debris might have been injected into the stratosphere. Alternatively, if the ejected rock mass scales as the caldera volume and if roughly 10^{-2} km^3 of debris were injected by Krakatoa, then the Toba volcano might have injected about 1 km^3 of ash. Both values are much less than the dust injected by the asteroid impact studied by Alvarez et al. (1). It is also possible that volcanic events, which are so much less powerful than asteroid impacts, do not lead to a distribution of fine dust over a wide area. Large, local dust loadings produce density flows on the flanks of volcanoes, thereby removing much of the injected material from the atmosphere.

Scaling arguments based upon comparisons of volcanic caldera volume and dispersed meteoritic dust volume are not very reliable. It might be possible to scale the impact energy of the asteroid and the mechanical energy of the eruption. The kinetic energy of the asteroid discussed by Alvarez et al. (1) is equivalent to 10^8 megatons of TNT. This energy is 10^7 times larger than the mechanical energy of the May 18, 1980, Mt. St. Helens eruptions or 10^6 times larger than that of the 1883 Krakatoa eruptions (5). The ratio of fine debris injected into the stratosphere by the asteroid and by the Krakatoa eruption also seems to be about 10^6. On this basis the effects of volcanoes are probably insignificant compared to those of large asteroid impacts.

If large volcanic eruptions really can inject 1 km^3 of fine debris into the stratosphere, their climatic impact might be significant. The calculations of Gerstl and Zardecki (9) and of Toon et al. (31) both show that 1 km^3 of dust would reduce light levels to about 5 to 10% of normal. Plants would probably continue to photosynthesize, but with some difficulty at such low light levels. Very large temperature changes might well occur for these optical depths, although I am not aware of calculations that apply specifically to this case.

As discussed previously, much of the material in typical volcanic clouds is sulfuric acid rather than silicate ash. Unfortunately, the quantity of sulfur dioxide vented to the stratosphere is not closely related either to caldera size or eruption energy. For example, the relatively small Agung eruption produced nearly as great a stratospheric cloud as did the Krakatoa eruption (27). Studies of volcanic sulfate in polar ice cores suggest that the Tambora eruption produced a cloud containing 150×10^{12} gm sulfate, or about .07 km^3. If so, this exceeded the sulfate content of the Krakatoa cloud by a factor of about three (11). If the Toba eruption produced ten times as much, it would have injected about 1 km^3 of sulfate. Sulfate particles can grow very quickly, and it is likely that large sulfate injections will simply lead to growth of large particles and thence to rapid particle removal. This process has not been simulated numerically. However, Rampino and Self (27) have suggested on empirical grounds that the critical point occurs on the scale of Krakatoa and Tambora eruptions.

In summary, volcanic eruptions certainly can affect climate. Generally such effects are limited to mean temperature depressions of perhaps 1°C for time periods of years to centuries. Extremely large single eruptions do occur but are probably not able to inject more than about 1 km^3 of long-lived material into the stratosphere; such quantities of injecta might be enough to produce a cooling of many °C for periods of a few years. If a large fraction of the Earth's volcanoes were to erupt within a few years of each other, very large (5°C) temperature changes could occur for time periods as long as a decade; light levels would not be reduced enough to eliminate photosynthesis.

COLLISIONS WITH GALACTIC DUST LANES

Recently consideration has been given to the possibility that climate changes may accompany the passage of the solar system through galactic dust lanes (e.g., (29)). If the Earth accreted dust without any gravitational focusing, then the accreted mass per cm^{-2} would be

$$M = 1/4\ V\ M_H\ n_H X,$$

where V is the Earth's velocity through the cloud, M_H is the mass of an H_2 molecule, n_H is the number of hydrogen molecules per cm^3, and X is the mass ratio of dust to hydrogen. The value V is thought to be about 20 km sec^{-1}, X is thought to be about 10^{-2} (29). Hence

$$M = 1.6 \times 10^{-20} g\ cm^{-2}\ sec^{-1}\ n_H\ .$$

Number densities of H_2 on the order of 5×10^3 are needed to obtain a flux of particulate mass comparable to that due to interplanetary dust grains reaching Earth. Of course, gravitational focusing of dust might increase the dust flux for a given H_2 abundance by factors as large as 10. According to Talbot et al. (29), the Earth may have passed through as many as five clouds with $n_H \gtrsim 10^3\ cm^{-3}$ during its history. Such clouds would not even have doubled the normal influx of dust particles.

Hunten et al. (12) have reviewed the process by which debris enters the Earth's atmosphere under normal circumstances. Each day about forty metric tons of meteoritic debris, equivalent to 10^{-16} gm cm^{-2} sec^{-1}, enters the atmosphere. The mass median weight is about 10 μgm which corresponds to a radius of about 100 μm for a 2.0 gm cm^{-3} body. Two-thirds of this mass is between 10^{-6} and 10^{-3} μgm (50 and 500 μm). Particles smaller than about 50 μm are the classical micrometeorites and do not ablate completely on entering the atmosphere. Due to their relatively large size, these particles quickly fall from the atmosphere and so are of little consequence for the properties of the atmosphere. Very large particles do not ablate completely either. Except for extremely large meteorites with radii on the order of a km, these are of little consequence as far as placing material into the atmosphere. However, two thirds of the mass of incoming meteoritic debris is in the size range which completely ablates in the atmosphere. This ablated smoke recondenses in the atmosphere into submicronized particles which coagulate to form larger particles. Due to the relatively small size of these particles, they can accumulate in the atmosphere. Hunten et al. (12) suggested that under ambient conditions there may be 50-100 meteoritic particles cm^{-3} with a size of less than about 100°A at 30 km altitude.

At the current influx of 10^{-16} gm cm^{-2} sec^{-1}, it would require $5x10^6$ years to account for the iridium levels discovered by Alvarez et al. (1). A meteorite influx level equivalent to 10^6 times the current rate would cover the Earth with the observed quantity of iridium in only five years. In this case the duration of the event would be limited by the time during which debris fell on the Earth rather than by the time it takes particles to coagulate to a diameter of several microns and then to fall out of the atmosphere.

We have evaluated numerically the amount of debris which would remain in the atmosphere given a steady influx rate. The models which we have used are similar to those of Hunten et al. (12) and Toon et al. (31). Figure 2 illustrates calculations of the optical depth of the atmosphere for various meteoritic influx values in which a relatively steady influx occurs. In all cases we followed Hunten et al. (12) in assuming that the meteorite mass ablates between 80 and 95 km altitude and then recondenses into smoke particles that are typically 10°A in size. These particles then coagulate to form larger particles. (If in fact the size distribution of dust in galactic dust lanes varies from that of the current dust, our results will probably overestimate the importance of dust since we assume the longest possible residence time for the dust.) The calculations in Fig. 2 present the optical depths after five years of input, by which time the concentration of dust has reached steady state.

The results of these calculations show that an influx of dust 10^4 or 10^5 times the normal rate produces optical depths on the order of unity. Such optical depths would influence climate to the same extent as very large amounts of volcanic activity (e.g., (25, 26)). To achieve an optical depth of 10^{-1} requires $n_H = 10^6$ cm^{-3}, and an optical depth of 1 requires $n_H = 10^7$ cm^{-3}. It does not seem likely that Earth has encountered such dense clouds, although McCrea (20) argues that they exist.

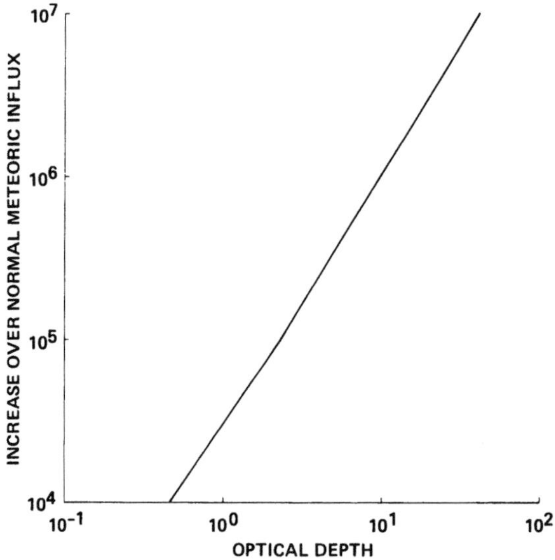

FIG. 2 - The optical depth due to small meteoritic particles whose flux exceeds that of current particles.

Galactic dust lanes also modify the solar luminosity and the solar wind. Apparently even clouds which are not very dense ($n_H = 10^3$) are capable of pushing the solar wind inside the Earth's orbit (29). Until recently it was believed that a statistical connection had been established between the Earth's climate and the solar wind (see (29)). However, this proposed relationship has subsequently been disproved (28). Although numerous statistically significant relationships seem to relate solar activity and weather, none of these has been shown to be based on a plausible physical connection. Nevertheless, the magnitude of solar flares and solar u.v. fluxes clearly vary and clearly affect the ionosphere and the chemistry of the upper atmosphere, and perhaps relations with the lower atmosphere will someday be found.

Solar luminosity changes, if they occurred, would affect the climate. McCrea (20) argues that collisions with clouds having $n_H = 10^3$ cm^{-3} would increase solar luminosity <1%, which would create temperature changes of <1°K. (Few meteorologists would support McCrea's contention that an increase in solar luminosity will cause ice ages.) Should one encounter clouds of 10^5 to 10^7 H_2 cm^{-3}, solar luminosity increases from

1 to 100% could occur, and some of these would indeed have dramatic effects on climate. The probability of encountering such clouds does not seem to have been explored thoroughly, but Talbot et al. (29) infer that the probability is not very high, and Dennison and Mansfield (7) have shown that no such cloud is as nearby as it should be if the last glacial epoch were triggered in the manner suggested by McCrea (20).

Yet another approach to galactic dust lanes has been taken by McKay and Thomas (21). They argue that the H_2 in the galactic dust lanes would affect the chemistry in the upper atmosphere even if $n_H = 10^3$ cm^{-3}. They find that the concentration of ozone in the mesosphere will be decreased due to chemical reactions; this in turn will lead to cooler temperatures and an expanded number of noctilucent clouds. Although McKay and Thomas' (21) chemistry may be plausible, their inference that noctilucent clouds could affect climate is not. If the total abundance of water above 70 km altitude is 3 ppm of H_2O, a cloud with an optical depth of only .02 will be produced even if all the water condenses. Such a cloud could produce a temperature decrease no greater than that of very low level volcanic activity. Perhaps such clouds do occur, but if so they probably do not have a strong influence on climate.

In summary, the terrestrial effects of passage through a galactic dust lane with $n_H = 10^3$ would probably be minimal. Even densities of this magnitude are only likely to be encountered a few times in Earth history. Some changes in solar wind, upper atmospheric chemistry, and noctilucent clouds might occur, but at this time they do not seem likely to have significant implications for the Earth's surface. The probability of encountering a cloud with $n_H = 10^5$ to 10^7 cm^{-3} seems very remote but does not seem to have been considered very explicitly in the literature. Such clouds could inject large numbers of particles into the Earth's atmosphere and could have a significant effect on solar luminosity.

COLLISIONS WITH METEORITES

Bodies of every size from submicron to several kilometers in diameter and with a variety of compositions have struck the Earth at speeds in excess of 10 km sec^{-1}. The impact of large bodies may affect surface climate in two ways. On entering the atmosphere the ablational heating of meteorites can produce significant quantities of nitrogen oxides which among other things can destroy ozone. Second, the larger bodies impacting the surface may inject large quantities of dust into the atmosphere and thereby alter the Earth's radiation budget.

Turco et al. (32, 33) examined the effects of the 1908 Tunguska meteorite impact which they believed might have deposited 10^3 megatons of energy in the upper atmosphere and created 30×10^6 metric tons of NO in the stratosphere. Such an NO injection could have depleted 45% of the ozone in the Northern Hemisphere, large ozone reductions may have persisted for several years after the meteorite impact. Data on ozone amounts in 1909-12 are consistent with a 30 $\pm$15% reduction, although the number of observations of ozone are too small to allow accurate mean ozone concentrations to be calculated. The climatic effect of the ozone perturbation was calculated to be small (0.1°C). A 45% ozone depletion could result in a tripling of erythemal active ultraviolet radiation at the ground (30°N, noontime, spring).

Turco et al. (32, 33) showed that the principal uncertainty in calculation of the amount of NO produced is the uncertainty in the quantity of energy of ablation which is transmitted to the atmosphere as heating. About 1.3×10^{12} ergs are required to produce one gram of NO and one megaton of energy released should produce about 3×10^{10} gms of NO. Bodies of the radius of Tunguska ($\simeq$75 meters) probably strike Earth once every 1000- 20000 years. The Tunguska meteor, however, seems to have coupled its energy very efficiently ($\simeq$100%) to the atmosphere by breaking apart at high altitude. After disintegrating, the body spread out slightly, and this increased the mass of atmosphere traversed.

The energy of an incoming body of mass, m, and radius, R, which is coupled to the atmosphere can be estimated crudely as

$$E = e \frac{mV^2}{2} = (.15) \left(\frac{2\pi (yR)^2 \, Psfc}{g \quad \cos\theta \, m} \right) \frac{mV^2}{2} = (yR)^2 \, 7.5 \times 10^{15} \text{ erg} ,$$

where we assume V=40 km sec^{-1}, its largest plausible value. The angle of entry, measured from the zenith, is given by θ.

For grazing entry the spherical shape of the Earth would need to be taken into account. The mass of NO produced is about $(yR)^2 \, 6 \times 10^3$ gms. Here the coupling coefficient, e, is taken to be 15% of the ratio of the mass of atmosphere swept through by the meteor to the mass of the meteor. Momentum conservation requires that the meteor pass through at least its own mass to transfer momentum, and about a factor of .15 is required to transfer all the momentum to the atmosphere. The mass of the atmosphere per unit area is given by the ratio of the surface pressure,

Psfc, to the gravitational acceleration, g. The factor y is a spreading factor which represents the possible expansion of the meteorite due to breakup in the atmosphere or to debris at the ground that are sent back into the atmosphere at very high speed.

The equation implies that for bodies which do not break up in the atmosphere (y=1), radii on the order of 750 meters are needed to produce as much NO as Tunguska might have produced. A 5 km radius body such as the one investigated by Alvarez et al. (1) might have produced $1.5x10^{15}$ gms of NO (e = $1.5x10^{-4}$), enough to destroy 90% of the ozone layer as discussed by Turco et al. (33). Since the total fixed nitrogen produced each year is about $2x10^{14}$ gms, the meteor may have provided the biota with about four times the normal fixed nitrogen.

Lewis et al. (18) have also considered the impact of a large meteor on the NO budget. O'Keefe and Ahrens (23) suggest that about 40% of the energy of the large meteor would be transmitted to the atmosphere, primarily through the crater ejecta. (Contrast .4 to $1.5x10^{-4}$ if the ejecta are not considered.) For a 5 km radius bolide, about $5x10^{18}$ gms of NO, which is about 0.1% of the atmospheric mass, should be produced. As Lewis et al. point out, the opacity of the large quantity of NO_2, which would result from the NO, would be very great at visible wavelengths, but it would probably not affect photosynthesis due to the narrowness of the NO_2 absorption band. However, such high NO_X levels are toxic to many plants and animals and could also affect the pH of the surface layers of the ocean.

The effects of a large meteorite impact on atmospheric dust levels have been discussed in detail by Toon et al. (31) and Pollack et al. (24). Alvarez et al. (1) discovered that the Earth's surface is covered worldwide with about 1 gm cm^{-2} of dust, from the impact of a large meteorite 65 million years ago. Toon et al. (31) discussed the effects of dust loadings of .01, .1, 1, and 10 gm cm^{-2}. These dust loadings all exceed a critical value such that small particles will coagulate to produce micron-sized particles resulting in a cloud removal time which is nearly independent of cloud density. As illustrated in Fig. 3, such massive clouds maintain optical depths greater than 10 for periods of a few months. A cloud of 0.01 gm cm^{-2} has a volume of about 20 km^3 and is thus much larger than the largest conceivable volcanic dust cloud.

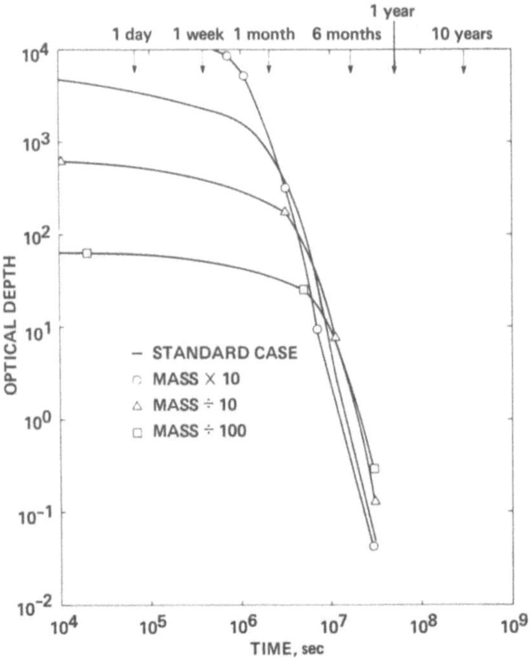

FIG. 3 - Toon et al. (31) considered the duration of large dust loadings for a number of different cases. The standard mass is 1 gm cm^{-2} spread uniformly over the Earth. The duration of large optical depths is not very sensitive to the injected mass.

Figure 4 illustrates the light levels at the surface for several values of dust loadings, while Figure 5 shows the change in the Earth's surface temperatures that might result. Clearly such low light levels and low continental temperatures, even for a few months period, could have an important impact on life.

There are several questions that have not been fully addressed in these calculations. These questions include the frequency of the injections of large amounts of dust, the particle size distribution of the dust, the means by which the dust was spread through the atmosphere, and the effects of dust on the coupled ocean-atmosphere climate.

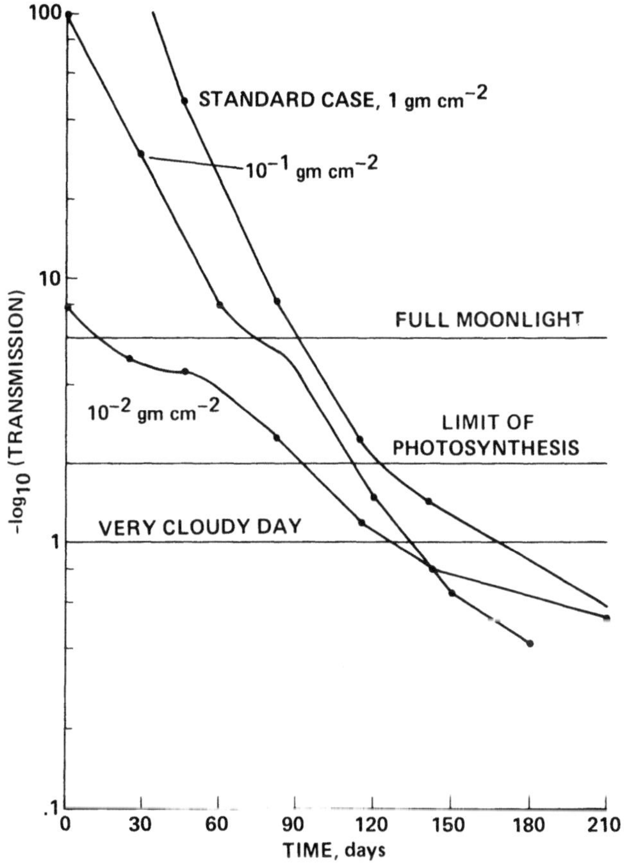

FIG. 4 - The logarithm (base 10) of the transmission of visible light is illustrated as a function of time for various mass loadings. The transmission levels equivalent to three natural phenomena are indicated. For the larger two mass loadings, not enough light reaches the surface to allow vision for several months. Photosynthesis could not occur for any of these cases for about three months.

The frequency of impacts of bodies of the size that struck the Earth 65 million years ago is about 1 per 100 million years (8). If we assume that the mass raised in an impact is proportional to the energy, E, of the impact, we can use the lunar crater size distribution for which N

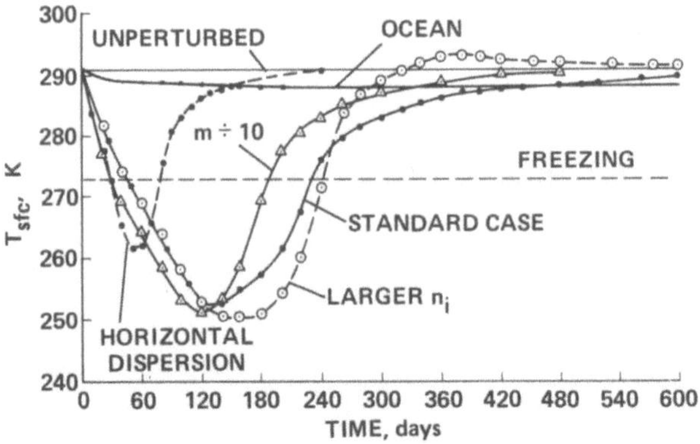

FIG. 5 - The surface temperature is illustrated for several possible dust loadings. The standard case has 1 gm cm^{-2} of dust. The m/10 case is for 0.1 gm cm^{-2} mass loadings. The larger n_i case is for more absorbing dust which decreases the Earth's albedo. In the horizontal dispersion case the dust takes two weeks to cover the Earth. We estimate the cooling will be two or three times less than shown within a few hundred kilometers of a coastline. In continental interiors atmospheric heat transport from the oceans might reduce the cooling by 20% to 30%.

is proportional to $D^{-1.7}$, and the approximation that the diameter D is proportional to the energy to the 1/3.4 power, to infer that the number of meteorites injecting at least 10^{-2} of the mass of the K-T bolide must be about $N \alpha E^{1/2}$ = 10 per 100 million years. This number is larger than what one might feel is in accord with the apparent lack of large-scale extinctions in the last 65 million years. Perhaps the number of impacts has been overestimated, their ability to cause extinctions is overestimated, or smaller impacts do not yield globally distributed dust clouds whose density is simply proportional to their impact energy. Or perhaps significant extinctions have occurred but have simply not yet been recognized.

The particle size distribution of dust produced by an impact is not known. The nature of dust in volcanic eruptions suggests that <1% of the dust ejected from impact craters is submicron in size. If the observed 1 gm cm^{-2} of dust is indeed submicron in size, then the crater volume must have been on the order of $2x10^5$ km^3. If the crater was 5 km deep with a radius of 75 km as expected by Emiliani et al. (8), it would have the required volume of $3.5x10^5$ km^3. Hence, it is reasonable that the debris layer found by Alvarez et al. (1) could consist of fine dust. However, it has been suggested by Jones and Kodis (13) that the energy of the impact was sufficient to hurl material worldwide. Thus, a significant fraction of the debris layer could have been made of fairly large particles. Particles larger than 5 μm will fall out of the atmosphere in a short period of time (31). If all of the debris were larger than 5 μm, then no significant effects on light levels or climate would be expected. If only a fraction (1 to 10%) were smaller than 1 μm, then the cloud mass would be lower than simulated as a standard case in Figs. 3-5. This would not affect the likelihood that light levels and continental temperatures changed, but it would explain why extinction events are not more common. The event 65 million years ago would then be at the low mass end of events that could be significant. Attempts to obtain the original size of the dust grains in the boundary layer clay would be of great interest.

The manner in which dust spreads through the atmosphere also needs considerable attention. Jones and Kodis (13) suggest that the initial impact of a large asteroid is capable of spreading dust and debris ballistically over most of the Earth's surface. Such a process would instantly cover all regions of Earth with a dust layer, insuring a worldwide extinction event. However, much of this debris might consist of rather large particles which would not remain long in the atmosphere. If the impact only distributed debris over a very localized area, the weight of the debris would greatly exceed the mass of the atmosphere. A debris flow similar to an avalanche or to the density flows seen on the flanks of volcanoes could quickly deposit the material at the surface. Toon et al. (31) suggested that a cloud as massive as the one studied by Alvarez et al. (1) must cover at least 1% of the Earth's surface to avoid such density flows. Although volcanic dust clouds spread over the Earth during a period of many months, a dense cloud might well induce a strong flow in the stratosphere which would distribute the cloud globally in a few weeks. Such a process accounts for the rapid dispersal of Martian dust storms.

Figure 5 illustrates the changes in temperature which might occur over the ocean and continents following a large injection of dust. Very little happens over the oceans because there is such a large heat reservoir in the upper 100m of the water column; the continents, however, cool very rapidly. These two systems will interact. Coastal regions would not experience as large a temperature decrease as would the continents. However, the winter conditions experienced annually show that atmospheric transport of heat from the warm oceans cannot prevent low temperatures on the continents fairly close to coastal regions. In a darkened atmosphere convection tends to be suppressed, so that precipitation patterns would be strongly modified. The details of climate during a meteoritic "blackout" clearly remain to be worked out; it is certain, however, that there would be considerable climatic differences between the various regions of the Earth.

CONCLUSIONS

Fortunately for life, the Earth seems to be remarkably safe from events which might quickly alter the composition of its atmosphere. The major gases, oxygen, nitrogen, and carbon dioxide, have such great abundances compared with the volumes of lava which is typically degassed or with the volume of asteroidal bodies, that their concentrations are not likely to be altered on short time scales. Volcanic activity is probably not able to inject large enough quantities of debris into the atmosphere to block enough sunlight to prevent photosynthesis. However, single, very large explosions, or simultaneous explosions of many of the world's volcanoes might produce dense enough dust veils to decrease continental temperatures substantially. Collisions between the solar system and galactic dust lanes probably do not have a significant effect on surface climate. However, if the sun ever encountered a very dense cloud containing 10^5 to 10^7 H_2 cm^{-3}, solar luminosity might increase greatly. The probability of encountering such a cloud does not seem to be high, although I am not aware of detailed probability estimates. Asteroid or comet collisions do pose a hazard to Earth. Even relatively small bodies may produce significant quantities of NO_X during ablation and thereby alter the Earth's ozone budget. Collisions with large asteroids or comets may occasionally inject immense quantities of dust into the atmosphere. These materials block sunlight, drop continental temperatures, shut off photosynthesis, and perhaps prove toxic to many life forms. Although much further work needs to be done on the effects of meteorite collisions, it appears at this time that such impacts are the major cause of large, sudden changes in the global environment.

Although the phenomena discussed above are those I believe most likely to have affected the Earth catastrophically in the past, it should not be overlooked that human beings are now altering the Earth and its atmosphere in major ways and on a short time scale. The rise of CO_2 due to fossil fuel burning during the next century may well produce one of the most rapid, large alterations to climate in geologic history.

Acknowledgements. I thank NASA's climate program and R. Schiffer for their support. E. Shoemaker made valuable suggestions for improving the manuscript. M. Gomes' help in its preparation is also appreciated.

REFERENCES

(1) Alvarez, L.W.; Alvarez, W.; Asaro, F.; and Michael, H.V. 1980. Extraterrestrial cause for the Cretaceous-Tertiary extinctions. Science 208: 1095-1108.

(2) Anderson, A.T. 1975. Some basaltic and andesitic gases. Rev. Geophys. Space Phys. 13: 37-55.

(3) Budyko, M. 1969. Climate change. Soviet Geog. 10: 429-456.

(4) Cadle, R.D. 1975. Volcanic emissions of halides and sulfur compounds to the troposphere and stratosphere. J. Geophys. Res. 80: 1650-1652.

(5) Colgate, S.A., and Sigurgursson, T. 1973. Dynamic mixing of water and lava. Nature 244: 552-555.

(6) Deirmendjian, D. 1973. On volcanic and other particulate turbidity anomalies. Adv. Geophys. 16: 267-296.

(7) Dennison, B., and Mansfield, V.N. 1976. Glaciations and dense interstellar clouds. Nature 261: 32-34.

(8) Emiliani, C.; Kraus, E.B.; and Shoemaker, E.M. 1981. Sudden death at the end of the Mesozoic. Earth Planet. Sci. Lett. 55: 317-334.

(9) Gerstl, S.A.W., and Zardecki, A. 1983. Reduction of photosynthetically active radiation under extreme stratospheric aerosol loads. Geol. Soc. Am. Spec. Paper 190: 201-210.

(10) Gow, A., and Williamson, T. 1971. Volcanic ash in the Antarctic ice sheet and its possible climatic implications. Earth Planet. Sci. Lett. 13: 210-218.

(11) Hammer, C.U.; Clausen, H.B.; and Dansgaard, W. 1980. Greenland ice sheet evidence of post-glacial volcanism and its climatic impact. Nature 288: 230-235.

(12) Hunten, D.M.; Turco, R.P.; and Toon, O.B. 1980. Smoke and dust particles of meteoric origin in the mesosphere and stratosphere. J. Atmos. Sci. 37: 1342-1357.

(13) Jones, E.M., and Kodis, J.W. 1982. Atmospheric effects of large-body impacts: the first few minutes. Geol. Soc. Am. Spec. Paper 190: 175-180.

(14) Kent, D.V. 1981. Asteroid extinction hypotheses. Science 211: 649-650.

(15) KTEC. Cretaceous-Tertiary extinctions and possible terrestrial and extraterrestial causes. Syllogeus 39: 1-51.

(16) Lamb, H. 1970. Volcanic dust in the atmosphere: With a chronology and assessment of its meteorological significance. Phil. Trans. Roy. Soc. 226: 425-533.

(17) Landsberg, H.E., and Albert, J.M. 1974. The summer of 1816 and volcanism. Weatherwise 27: 63-66.

(18) Lewis, J.S.; Watkins, G.H.; Hartman, H.; and Prinn, R.G. 1982. Chemical consequences of major impact events on Earth. Geol. Soc. Am. Spec. Paper 190: 215-222.

(19) Macdonald, G.A. 1972. Volcanoes. New Jersey: Prentice Hall.

(20) McCrea, W.H. 1975. Ice ages and the galaxy. Nature 255: 607-609.

(21) McKay, P. and Thomas, G.E. 1978. Consequences of past encounter of the Earth with an interstellar cloud. Geophys. Res. Lett. 5: 215-218.

(22) Murrow, P.J.; Rose, W.I.; and Self, S. 1980. Determination of the total grain size distribution in a volcanian eruption column, and its implications to stratospheric aerosol perturbation. Geophys. Res. Lett. 7: 893-897.

(23) O'Keefe, J.D., and Ahrens, J.J. 1982. The interaction of the Cretaceous-Tertiary extinction bolide with the atmosphere, ocean, and solid Earth. Geol. Soc. Am. Spec. Paper 190: 103-120.

(24) Pollack, J.B.; Toon, O.B.; Ackerman, T.P.; McKay, C.P.; and Turco, R.P. 1983. Environmental effects of an impact-generated dust cloud: Implications for the Cretaceous-Tertiary extinctions. Science 219: 287-289.

(25) Pollack, J.; Toon, O.; Sagan, C.; Summers, A.; Baldwin, B.; and Van Camp, W. 1976. Volcanic explosions and climatic change: A theoretical assessment. J. Geoph. Res. 81: 1071-1083.

(26) Pollack, J.; Toon, O.; Summers, A.; Baldwin, B.; Sagan, C.; and Van Camp, W. 1976. Stratospheric aerosols and climatic change. Nature 263: 551-555.

(27) Rampino, M.R., and Self, S. 1982. Historic eruptions of Tambora (1815), Krakatoa (1883) and Agung (1963): Their stratospheric aerosols and climate impact. Quatern. Res. 18: 127-143.

(28) Shapiro, R. 1979. An examination of certain proposed sun-weather connections. J. Atmos. Sci. 36: 32-34.

(29) Talbot, Jr., R.J.; Butler, D.M.; and Newman, M.S. 1976. Climatic effects during passage of the solar system through interstellar clouds. Nature 262: 561-563.

(30) Toon, O.B., and Pollack, J.B. 1982. Stratospheric aerosols and climate. In The Stratospheric Aerosol Layer: Topics in Current Physics, ed. R.C. Whitten, vol. 28, pp. 121-147. Berlin: Springer Verlag.

(31) Toon, O.B.; Pollack, J.B.; Ackerman, T.P.; Turco, R.P.; McKay, C.P.; and Liu, M.S. 1982. Evolution of an impact-generated dust cloud and its effect on the atmosphere. Geol. Soc. Am. Spec. Paper 190: 187-200.

(32) Turco, R.P.; Toon, O.B.; Park, C.; Whitten, R.C.; Pollack, J.B.; and Noerdlinger, P. 1981. Tunguska meteor fall of 1908: Effects on stratospheric ozone. Science 214: 19-23.

(33) Turco, R.P.; Toon, O.B.; Park, C.; Whitten, R.C.; Pollack, J.B.; and Noerdlinger, P. 1982. An analysis of the physical, chemical, optical and historical impacts of the 1908 Tunguska meteor fall. Icarus 50: 1-52.

(34) Turco, R.P.; Toon, O.B.; Whitten, R.C.; Hamill, P.; and Keese, R.G. 1983. The 1980 eruption of Mount St. Helens: Physical and chemical processes in the stratospheric clouds. J. Geophys. Res. 88: 5299-5319.

Patterns of Change in Earth Evolution, eds. H.D. Holland and A.F. Trendall, pp. 63-74.
Dahlem Konferenzen 1984. Berlin, Heidelberg, New York, Tokyo: Springer-Verlag.

Geochemical Markers of Impacts and of Their Effects on Environments

K.J. Hsü
Geological Institute, Swiss Federal Institute of Technology
8092 Zurich, Switzerland

Abstract. Geochemical markers include anomalies in trace element abundances, in stable-isotope ratios, and abrupt changes in mineralogic and chemical composition. Enrichment of siderophile and depletion of rare-earth elements indicate extraterrestrial impact. Certain changes in oxygen and carbon isotope ratios signify environmental changes consequent upon an impact. The worldwide geochemical records across Cretaceous/Tertiary boundary sections all point to mass mortality caused by a large body impact which also triggered environmental catastrophes and mass extinction.

TRACE ELEMENTS AS GEOCHEMICAL MARKERS

Mass extinction and sedimentological anomalies were suggestive of the occurrence of a cosmic event at the end of Cretaceous. The enrichment of the siderophile elements in sediments at, or very near, the paleontologically-determined Cretaceous/Tertiary (CT) provided the first positive confirmation of this suggestion (3).

The strength of the evidence from trace-element anomalies in the boundary clay rests on: a) their magnitude and sharpness, b) the proportional enrichment of noble elements as in meteorites, c) the concomitant depletion of the terrestrial rare-earth elements, and d) the global distribution of the anomaly.

The iridium content of sediments at the terminal Cretaceous anomaly is two or three orders-of-magnitude greater than the background concentration of this element (3). In fact, the extraterrestrial components seem to be excessive in the boundary clays of Denmark and Spain (23). Unusual concentrations of trace elements are not uncommon in deep-sea sediments deposited at very slow rate; an example of this is the enrichments of heavy metals in manganese nodules. However, the iridium-rich boundary clays were deposited in a time interval of less than 10^5 years (21), and very probably in 10^3 or even 10^2 years (20, 30). We know of no processes on Earth which could cause a global enrichment of the siderophile elements in sediments to this extent in such a short time.

Ganapathy measured nine trace elements in the boundary clay of Denmark, including several noble elements, and found an abundance-pattern which he interpreted as that of type 1 carbonaceous chondrites (11). Asaro later demonstrated that the ratios of noble elements in the Danish boundary clays have values very similar to those of C-1 chondrites (4). We know of no natural processes other than impact which can concentrate the noble elements in sediments with such abundance ratios. A very slow sedimentation rate does not adequately explain the enrichment of siderophile elements in the terminal Cretaceous boundary clays, but such a possibility cannot be completely excluded. The depletion of rare-earth elements (REE), however, has to be explained in terms of an unusually high rate of extraterrestrial input, which diluted the REE of the terrestrial detritus (32). We know of no processes other than impact which could result in a concomitant enrichment of siderophiles and depletion of REE in a sediment.

The hypothesis of a terminal Cretaceous impact predicts the occurrence of an iridium anomaly globally at or near the paleontologically dated C/T boundary, regardless of the depositional environment of the host sediments. As Alvarez has noted, this expectation has been fulfilled in 36 of the 37 areas investigated: the iridium anomaly was found in terminal Cretaceous sediments deposited in continental, in open marine, and in deep marine environments (2). At the one locality (out of two) in Montana where the iridium anomaly was not found, the sediments are fluviatile, and there may be an obscure unconformity at the boundary. It seems fair to say that so far the predictive power of the impact theory has been phenomenal.

Arguments against the theory of a terminal Cretaceous large body impact are as follows: a) the timing of the iridium anomaly and that of the terminal Cretaceous extinction are not exactly identical everywhere;

b) the clays in the boundary sediments are terrestrial and not very different from those in adjacent sediments; c) anomalous enrichments of iridium have been found at stratigraphic horizons other than the end of the Cretaceous; d) the anomalous concentration of iridium can be explained by other mechanisms (volcanism, submarine diagenesis).

The merit of the first argument is questionable, because of the paleontological uncertainties in placing the boundary. That the last dinosaur bones were not found at exactly the same level as that of the iridium anomaly could be due to incomplete sampling of randomly preserved fossils (2).

The second argument is neither particularly relevant nor factually accurate. The composition of the clay minerals in the boundary sediments at many places, such as at the Atlantic deep-sea drilling Site 524 investigated by us, are indeed no different from those above and below, because contributions from detrital sources predominated throughout (20). Where the iridium anomaly indicated the presence of a large extraterrestrial component, such as in Denmark or at the Pacific deep-sea drilling Site 465A, the clays at the boundary do indeed have compositions anomalous compared to those of the adjacent sediments ((4), and Kastner, personal communication). Kastner showed me an X-ray diffraction pattern of the boundary clay at Stevnsklint, Denmark; the pattern is typical of pure smectite, and I agree with her interpretation that the clay is either an altered volcanic ash, or altered impact-ejecta fallout.

The third argument is not irrelevant. Iridium enrichment at other stratigraphic levels may indicate impacts at other times. Even if these anomalous occurrences turn out to be products of terrestrial processes, they do not constitute proof against the hypothesis that the excess iridium in terminal Cretaceous sediments is extraterrestrial. The fourth argument is weak, because none of the alternative ideas have led to useful predictions. They represent at best ad hoc explanations of single phenomena and have been largely discredited by a synthesis of all known facts (28).

In conclusion, the evidence seems compelling to me that a large extraterrestrial body hit the Earth at the end of Cretaceous. The timing of the impact event is so close to the C/T boundary that the suggestion that the impact triggered a catastrophic extinction is quite reasonable.

The next step is to attempt to answer the following three questions: a) what was the nature of the bolide? b) where was the site(s) of impact? and c) how did the impact bring about mass extinction?

ASTEROID OR COMET?

While I agreed with the Alvarez team that a meteor hit, I am less convinced that the bolide was an asteroid. Their evidence for an asteroid impact was based upon their interpretation of the trace-element chemistry: the noble element ratios agree "very well with expectations for C-1 chondrites and somewhat less well for other chondrites. Terrestrial and iron meteoritic materials usually have considerably different values" (4). Kyte pointed out, however, that the Ni/Ir ratio of the DSDP Site 465A sample is 7,000, compared to the value of 23,000 for C-1 chondrites (22). Furthermore, the nonvolatile components of comets also have a chemical composition similar to those of carbonaceous chondrites, as spectral analyses of cometary fragments and chemical analyses of extraterrestrial dust in the stratosphere have indicated (15). The trace-element chemistry of the boundary clay may therefore not be able to discriminate an asteroid from cometary fallout.

There is indirect evidence to suggest that a comet rather than an asteroid was the terminal Cretaceous bolide. Clube and Napier favored the idea of cometary impact, which is an integral part of their astronomical theory that comets grown in molecular clouds are captured by the Sun as it passes through the spiral arms of the Galaxy (7). Alvarez presented several critical arguments against this astronomical theory (1). Kyte favored a cometary source, because the computed extraterrestrial component in the Danish boundary clay seemed too high; he believed that a cometary object disintegrated in the atmosphere in the manner of the Tunguska event (14). Such extraterrestrial fallout, undiluted by crater ejecta, might best explain the extraordinary iridium abundance in the clay (23). However, this argument does not necessarily refute the asteroid hypothesis, because the terrestrial component in the finest fractions of an ejecta fallout could have been small enough to account for the trace-element chemistry of the Danish boundary clay (13). Smit favored a cometary scenario because of the anomalous concentrations of the more volatile trace metals (K, As, Se, Sb, Zn) in the Spanish boundary clays (31).

I proposed a cometary impact when it occurred to me that cyanide in the nucleus of a comet could have provided the poison to kill off oceanic planktons (17, 18). I have since been told, however, that the cometary cyanide would have been decomposed and therefore detoxified by the intense shock pressures after a cometary impact (24). Meanwhile, alternative scenarios (e.g., heavy-metal poisoning, suppression of photosynthesis) have been proposed to account for the mass extinction; these mechanisms do not depend on the nature of the bolide, i.e., whether it was a comet or an asteroid.

A cometary impact may explain why the known craters of probable terminal Cretaceous age are all smaller than predicted (12). A low density comet should excavate a crater much smaller than that by an asteroid of the same mass (19). However, we are not certain if any of the "candidate" craters are the impact site of the postulated event; the crater size argument may therefore be inapplicable. The question whether the bolide was an asteroid or a comet is clearly unanswered, although some fragmentary evidence seems to favor the latter alternative.

WHERE DID THE BOLIDE HIT?

I postulated an oceanic impact in 1980, because I knew of no terminal Cretaceous craters. McCone called my attention to two craters in southern Russia: the Kamensk crater, which has a diameter of 25 km, and the Gusev crater of 3 km. They were apparently formed by fallen fragments of a single cosmic body that had broken apart during its fall (24). Even the larger of the two, however, is an order-of-magnitude smaller than that originally postulated by Alvarez and others (3, 17). On the other hand, it would seem too much of a coincidence to expect that such large craters of the right age should be excavated by bolides which had nothing to do with the terminal Cretaceous impact event.

Chemical evidence regarding the location of the impact site has been provided by investigations of sanidine spherules from the iridium-rich boundary clay in Spain (29, 31). The spherules are probably recondensed particles from impact melt which have been subsequently altered by diagenesis, although a volcanic origin cannot be ruled out (29). The oxygen isotope evidence ($\delta^{18}O_{smow} = +27.5$) is compatible with an interpretation of diagenesis by low-temperature reactions with seawater (9). The Nd and Sr isotope chemistry of the spherules suggests that they cannot have been derived from an old continent; an oceanic impact was more likely. The ejectas, however, cannot have come solely from an ocean crust ($\epsilon^{Nd} \cong +8.0$, $\epsilon^{Sr} \cong -25.0$), but should have included significant proportions of ocean water and/or marine sediments, with higher values of ϵ^{Sr} and lower values of ϵ^{Nd} (29). Smit and ten Kate came to the same conclusion on the basis of their study of REE-depletion in the Danish boundary clay (32). The ejectas cannot have been derived from an REE-enriched continental crust; the observed REE-depletion suggests derivation from oceanic crust and/or from pelagic sediments.

Sanidine spherules are very abundant in the Spanish boundary clay. They range in size from a few to a few hundred microns (32). Another occurrence of silicate spherules in a terminal Cretaceous boundary clay was reported from northern Italy (6). Larger pieces of molten ejectas should have been projected and lofted backwards along the initial bolide trajectory; they would have been cooled by radiation and conduction, and would have reentered the Earth's atmosphere as microtektites or tektites (26).

The paleontological record could provide some information regarding the target site. Hickey has noted a peculiar pattern of terminal Cretaceous plant extinction: the Aquilapollenites floras of Siberia and western North America became almost wholly extinct, while the tropical floras suffered relatively little damage (16). This fact led Emiliani and others to propose a North Pacific splashdown (8), although a Siberian impact should have led to a similar result.

Still another approach to search for the target site involves the assumption that the large body impact triggered volcanism. The hot spot volcanism along the Emperor-Hawaii Seamount Chain started too early and volcanism in Iceland too late to be linked to the terminal Cretaceous impact event. Only the volcanism that produced the Deccan trap seemed to have started at about the "right time" (25). If the two events are linked, one might search for a giant crater in the Indian Ocean.

In conclusion I have to admit that we have no clear indications as to where the target site was. Furthermore, the chance of finding such an impact is small if the bolide fell into the ocean (17).

SCENARIOS FOR EXTINCTIONS AFTER IMPACT

A large body impact should result in extinctions by mass mortality, or the extinctions could be the consequences of environmental stresses induced by the crash. The paleontological record is not clear-cut. The Cretaceous/Tertiary boundary has been assumed to signify the synchronous extinction of the "Cretaceous taxa," and the first appearances of the

"Tertiary taxa." In fact, a transition interval, in which the "Cretaceous" and the "Tertiary" fossils coexist, has ben noted in many localities (34). Commonly the first appearance datum of the "Tertiary taxa" is designated as the contact. The terminal Cretaceous iridium-anomaly is found almost everywhere at this C/T contact. The presence of the "Cretaceous taxa" above the contact (i.e., above the Ir-anomaly) has been variously explained. Thierstein believed that the "Cretaceous taxa" had become extinct, and that the fossil skeletons were reworked into the earliest Tertiary sediments by resedimentation and/or bioturbation (34). PerchNielsen found, however, in two instances, at least, that the fossil skeletons of the "Cretaceous taxa" in the transition zone give geochemical signals typical of the earliest Tertiary; they were thus considered survivors and descendants of survivors of the terminal Cretaceous catastrophe on their way to extinction because of an environmental crisis after the impact event (20, 27). Geochemical studies have given considerable evidence of environmental perturbations of very significant magnitude. The presence of boundary clay almost devoid of calcareous fossils, for example, is an indication of low productivity or of increased dissolution in waters that were more than normally acid (18).

A carbon isotope anomaly, with a change in $\delta^{13}C$ of minus 1 to 3 per mil, is found in planktic fossils across the C/T boundary (5, 8, 20). Normally a surface-to-bottom carbon isotope gradient is present in a steady-state ocean, because the utilization of light carbon by photosynthetic organisms leads to a depletion of ^{12}C in dissolved carbonate in ocean water; planktic skeletons thus have a $\delta^{13}C$ about 2 per mil more positive than the benthic tests. The terminal Cretaceous event eliminated this gradient, suggesting that the utilization of light carbon for organic production had been reduced to nearly zero. The sharpness of the perturbation indicates that the catastrophe occurred within a very short time interval of about 10^3 years or less (5, 20). In other words, the ocean was almost sterile after the impact event, and this catastrophic state may have lasted for thousands of years, before plankton-production and light-carbon utilization again reached a sufficient degree to reestablish the surface-to-bottom carbon isotope gradient.

A second carbon isotope anomaly has been found at several localities: at a horizon some 10^4 years younger than the iridium anomaly, and at the horizon when the fossils belonging to the "Cretaceous taxa" all disappeared (20, 35). McKenzie believed that this may represent a second productivity crisis at the time of the final extinction of the "Cretaceous taxa," when the "Tertiary" organisms were not yet completely established (20).

What caused mass mortality in the oceans? Of the various scenarios presented, two are still viable, namely, suppression of photosynthesis because of solar insulation by impact ejecta in the stratosphere (3, 20) and chemical pollution of the oceans (17, 20). We have little chemical evidence to evaluate the two. Some of the anomalously enriched trace elements in the boundary clay (Os, Ni, As, etc.) are poisonous. Their quantity is, however, so small that they would have been harmless if they had been homogeneously distributed in the oceans (10). On the other hand, very poisonous heavy metals, such as osmium, could have reached toxic levels if their distribution had been restricted to some surface currents (20).

Combination of atmospheric oxygen and nitrogen to form NO_x and the consequent destruction of the Earth's ozone layer can also cause mass mortality (19, 24). Like cyanide poisoning, serious pollution of the Earth's oceans by nitrogen compounds should leave a signal in the isotopic composition of nitrogen-bearing fossils. We are planning to investigate $^{15}N/^{14}N$ ratios of fossil dinoflagellates across the boundary but have not been able to obtain enough sample material for our measurements.

Shallow marine benthic organisms also suffered sudden extinction (33). Those from the Tethyan province and those with pelagic larvae seem to have been particularly severely affected. I have suggested chemical pollution of equatorial surface currents to explain these observations, but hard evidence is lacking (18, 20).

The cause of dinosaur extinction on land is a controversial subject. The evidence seems to favor extinction under thermal stress (8, 17, 20, 25). Paleotemperature data based on measurements of the isotopic composition of oxygen in the tests of marine fossils have not yielded systematic results. There seemed to be cooling across the boundary at one place and warming at another (5). Our results from South Atlantic DSDP Site 524 suggest that there were temperature oscillations which reached a maximum of about 10°C increase some 30,000 years after the impact event (20). However, the evidence is still too contradictory, and the role of thermal stress can only be presented as a provocative hypothesis to encourage further work.

SUMMARY

Geochemical investigations have served to indicate that a large body impact took place at the end of Cretaceous. The bolide was either an asteroid or a comet. Preliminary chemical data suggests an oceanic impact in the northern hemisphere; two small craters of terminal Cretaceous age in southern Russia could have been "splashdown" sites. The impact triggered mass mortality within 10^3 years after the event; the cause appears to have been solar insulation and/or chemical pollution. Mass extinctions followed during the next 30,000 years; environmental stresses eliminated most of the marine plankton, many of the shallow marine benthic animals, the Aquilapollenites flora of the high northern latitude, and all land animals weighing more than 25 kg.

Acknowledgements. I am grateful to W. Alvarez, J. Smit, and others for discussions which led to the improvement of this manuscript. I regret that the first draft was written before publication of the Special Paper of the Geological Society of America on Implications of Extraterrestrial Impacts, so that some more up-to-date references were not cited.

REFERENCES

(1) Alvarez, L. 1982. Critique of the Clube-Napier hypothesis on the origin of comets. Syllogeus 39: 76-77.

(2) Alvarez, L. 1982. Experimental evidence that an asteroid impact led to extinction of many species 65 million years ago. Proc. Natl. Acad. Sci., in press.

(3) Alvarez, L.; Alvarez, W.; Asaro, F.; and Michel, H.V. 1980. Extraterrestrial cause for the Cretaceous-Tertiary extinction. Science 208: 1095-1108.

(4) Asaro, F. 1982. Abundance ratios of noble elements at Cretaceous-Tertiary boundary. Syllogeus 39: 6-9.

(5) Boersma, A., and Schackleton, N.J. 1981. Oxygen- and carbon-isotope variations and planktonic-foraminifer depth habitats, late Cretaceous to Paleocene, Central Pacific, Deep Sea Drilling Project Sites 463 and 465. Initial Reports. Deep Sea Drilling Project 57: 513-526.

(6) Castellarin, A.; Del Monte, M.; and Frascani, F. 1974. Cosmic fallout in the "hard grounds" of the Venetian region. Giorn. di geol. 39: 333-346.

(7) Clube, S.V.M., and Napier, W.M. 1982. Spiral arms, comets and terrestrial catastrophism. Q. J. Roy. Astro. Soc. 23: 45-66.

(8) Emiliani, C.; Kraus, E.B.; and Shoemaker, E.M. 1981. Sudden death at the end of Mesozoic. Earth Planet. Sci. Lett. 55: 318-334.

(9) Epstein, S. 1982. The $\delta^{18}O$ of the sanidine spherules at the Cretaceous-Tertiary boundary. Proceedings of the 13th Lunar and Planetary Science Conference, p. 205. Houston: Lunar and Planetary Institute.

(10) Feldman, P. 1982. Chemical poisons from comets. Syllogeus 39: 71-72.

(11) Ganapathy, R. 1980. A major meteorite impact on earth 65 million years ago: evidence from the Cretaceous-Tertiary boundary clay. Science 209: 921-923.

(12) Grieve, R.A.F. 1981. The record of impact on Earth: Implications for a major Cretaceous/Tertiary boundary event. Geol. Soc. Am. Spec. Paper 190: 25-38.

(13) Grieve, R.A.F. 1982. Iridium content in impact melts and ejecta dust. Syllogeus 39: 32-34.

(14) Halliday, I. 1982. Looking back on the Tunguska Event. Syllogeus 39: 138-139.

(15) Halliday, I., and Kyte, F. 1982. Comet compositions Syllogeus 39: 12-13.

(16) Hickey, L. 1981. Land plant evidence compatible with gradual, not catastrophic change at the end of the Cretaceous. Science 292: 523-531.

(17) Hsü, K.J. 1980. Terrestrial catastrophe caused by cometary impact at the end of Cretaceous. Nature 285: 201-203.

(18) Hsü, K.J. 1981. Origin of geochemical anomalies at Cretaceous-Tertiary boundary. Asteroid or cometary impact. Oceanologica Acta No. SP: 129-133.

(19) Hsü, K.J. 1982. A scenario for terminal Cretaceous Event. Initial Reports. Deep Sea Drilling Project 73, in press.

(20) Hsü, K.J.; He, Q.; McKenzie, J.; et al. 1982. Mass mortality and its environmental and evolutionary consequences. Science 216: 249-256.

(21) Kent, D.V. 1977. An estimate of the duration of the faunal change at the Cretaceous-Tertiary boundary. Geology 5: 769-771.

(22) Kyte, F. 1982. Abundance ratios of noble elements at Site 465A. Syllogeus 39: 10-11.

(23) Kyte, F.T.; Zhou, Z.; and Wasson, J.T. 1980. Siderophile-enriched sediments from the Cretaceous-Tertiary boundary. Nature 288: 651-656.

(24) Lewis, J.S.; Watkins, G.H.; Hartman, H.; and Prinn, R.G. 1982. Chemical consequences of major impact events on Earth. Geol. Soc. Am. Spec. Paper 190: 215-222.

(25) McLean, D.M. 1982. Deccan volcanism and the Cretaceous-Tertiary transition scenario. Syllogeus 39: 143-144.

(26) O'Keefe, J.D., and Ahrens, T.J. 1982. Impact mechanics of the Cretaceous-Tertiary extinction bolide. Nature 298: 123-127.

(27) Perch-Nielsen, K. 1982. Maastrichtian coccoliths in the Danian: Survivors or reworked "dead bodies". International Association of Sedimentologists, 3rd European Meeting, Copenhagen, Abstracts, p. 23.

(28) Russell, D.A., and Rice, G. 1982. Cretaceous-Tertiary extinctions and possible terrestrial and extraterrestrial causes. Syllogeus, no. 39. Ottowa: National Museum of Canada.

(29) Shaw, H.F., and Wasserburg, G.J. 1982. Age and provenance of the target material for tektites and possible impactites as inferred from Sm-Nd and Rb-Sr systematics. Earth Planet. Sci. Lett. 60: 155-177.

(30) Smit, J., and Hertogen, J. 1980. An extraterrestrial event at the Cretaceous-Tertiary boundary. Nature 285: 198-200.

(31) Smit, J., and Klaver, G. 1982. Sanidine spherules at the Cretaceous-Tertiary boundary; cometary material? Nature 292: 47-49.

(32) Smit, J., and ten Kate, W.G.H.Z. 1981. Trace element patterns at the Cretaceous-Tertiary boundary. In Cretaceous Research, Ph.D. Dissertation, J. Smit, vol. 3, pp. 307-332, University of Amsterdam.

(33) Surlyk, F., and Johansen, M.B. 1982. Extinction pattern of late Cretaceous brachiopods compatible with catastrophic change of the marine calcareous shelled biota, Abstracts, p. 49. American Association for the Advancement of Science, 148th National Meeting, Washington, D.C.

(34) Thierstein, H.R. 1981. Late Cretaceous calcareous nannoplankton and the change at the Cretaceous-Tertiary boundary. Soc. Econ. Paleont. Min. Spec. Publ. 32: 355-394.

(35) Williams, D.F.; Wealy-Williams, N.; Thunell, R.C.; and Leventer, A. 1982. Detailed stable isotope and carbonate records from the late Maastrichtian-early Paleocene section of Site 516F. Initial Reports. Deep Sea Drilling Project 72, in press.

Standing, left to right:
Walter Alvarez, Dieter Fütterer, Andreas Wetzel, Brian Toon, Kevin Padian, Gene Shoemaker, Digby McLaren.

Seated, left to right:
David Raup, Jan Smit, Tove Birkelund, Ken Hsü, Jere Lipps.

Patterns of Change in Earth Evolution, eds. H.D. Holland and A.F. Trendall, pp. 77-102.
Dahlem Konferenzen 1984. Berlin, Heidelberg, New York, Tokyo: Springer-Verlag.

The Possible Influences of Sudden Events on Biological Radiations and Extinctions
Group Report

K.Padian, Rapporteur
W. Alvarez
T. Birkelund
D.K. Fütterer
K.J. Hsü
J.H. Lipps
D.J. McLaren
D.M. Raup
E.M. Shoemaker
J. Smit
O.B. Toon
A. Wetzel

INTRODUCTION

This report, and the Workshop discussions leading to it, stem from the growing realization that physical events of short duration may have had frequent, significant, and lasting effects on the Earth's biota. For the purposes of the report, physical events are considered relevant only if they are: a) relatively sudden, b) have a global or nearly global biologic effect, and c) are amenable to analysis through a combination of physical prediction and empirical observation. The time durations of the physical events themselves cannot be strictly defined, but most are short enough to be considered instantaneous on a geologic time scale. This generally means a duration of 10^4 years or less, with a definite bias toward truly instantaneous events. The geographic scale of events and their geochemical, sedimentologic, and biologic effects can be determined from theoretical models and from the geologic record; there are, however, many uncertainties attached to the theoretical predictions and the incompleteness of the geologic record prevents a full understanding both of their cause and their effects.

The composition of the Workshop group was biased toward one particular segment of geologic time, the Cretaceous-Tertiary or K-T boundary, and a degree of partisanship for certain physical and biologic factors. However, we attempted to consider a wide variety of phenomena and situations and to avoid the temptation to "solve" or judge the relative merits of competing hypotheses to explain events at the K-T boundary. Our strategy was to use the data base, methodologies, and theoretical (components) treatments associated with a variety of specific events in Earth history in order to derive valid generalizations concerning the major problems and potential solutions in the rapidly expanding field of physical-biological interactions. Our "conclusions," such as they are, relate specifically to unsolved problems about these interactions and include recommendations on directions for future research.

Our discussions were greatly facilitated by the papers prepared by Shoemaker, Raup, Hsü, and Toon (all this volume). These are necessarily incomplete and somewhat controversial, but they provided excellent summaries of most of the major problems. They are concerned with: a) the principal physical events that have a potential for significant effects on evolution, b) the evolutionary patterns that need explanation in biological and physical terms, and c) the range of approaches for testing hypotheses for specific evolutionary effects of particular physical events. Throughout, we tried to focus on methods of detection, discrimination, and resolution as well as on the need for operational definitions of various phenomena in order to facilitate testing of hypotheses and to promote mutual understanding between members of the widely different scientific fields that impinge on these problems.

We took as a given that the Earth and its biota have been subjected to episodic physical stresses of short duration, that the effects of these phenomena can be modeled theoretically as well as observed to some extent in the geologic record, and that large-scale patterns of evolution can be recognized empirically and analyzed in terms of the tempo and mode of biologic change. We also started with the notion that rates of geologic and biologic change have been subject to significant short-term fluctuations and excursions during certain intervals in Earth history. This departure from strict "substantive uniformitarianism" (sensu Gould) is required by the increasing number of documented cases where rates of change seen in the geologic record depart from the norms set by observations in the present-day environments. We can only expect to succeed in learning anything new about historical processes and patterns if we keep our minds open to significant deviations from normal rates. However, we also regard with great caution assertions of the operation of special causes in the absence of sound criteria by which such assertions can be tested and measured.

This report is not an attempt to chronicle the Workshop discussions, although we did proceed generally along the lines given here. The text is divided into three parts; these deal with physical phenomena, biological phenomena, and questions for future research; emphasis is placed on those issues which we believe to be of prime importance.

SHORT-TERM PHYSICAL EVENTS WITH POSSIBLE BIOLOGICAL CONSEQUENCES

Many classes of physical events influence the world's biota. In the space available it would be impossible to catalog and discuss all of them. At the Workshop we tried to make this task practicable by considering only those classes of events that a) have global consequences, b) have significant and to some degree predictable biological effects, and c) can be detected by physical and/or biological methods. We then tried to define the relative importance of these classes of events through geologic time.

Classification and Evaluation of Physical Phenomena

The phenomena that concern us here could be organized in several ways. They could be viewed as manifestations of three distinct kinds of physical processes that are ultimately responsible for other patterns (see Fig. 1). Or, they could be viewed as proximal causes of certain biological phenomena. We found it useful to consider those phenomena which have biological effects that fall within the purview of our study.

1. Short-term local phenomena. These include fires, floods, earthquakes, small impacts and other astrophysical agents, and certain volcanic eruptions. Although they may have significant local biologic effects, it is highly improbable that they are the cause of mass extinctions. Their biologic effects are generally limited to "ecologic time," the range observable through human history (0-10^4 years), and thus of little long-term importance. Although we regard this class of phenomena as of great interest in other contexts, we do not consider it to be within our purview.

2. Potentially important physical phenomena with undetectable biological effects. Some phenomena have been proposed as potentially important influences on the Earth's biota but, for one reason or another, the proposed connection has not produced viable theories. For example, fluctuations

IMMEDIATE CAUSES		ULTIMATE CAUSES		
		MANTLE CONVECTION	SOLAR AND ORBITAL CHANGES	IMPACTS AND SUPERNOVAE
SEA-LEVEL CHANGES		X		
CLIMATIC CHANGES	LONG TERM	X	X	
	SHORT TERM		X	X
RADIATION INTENSITY CHANGES	LONG TERM		X	
	SHORT TERM	X	X	X
POISONING		X		X
SHOCK EFFECTS				X

FIG. 1 - Causes of extinctions: some possible immediate causes plotted against ultimate causes. An X indicates that an immediate cause could have been initiated by the one or more ultimate causes indicated (from (1)).

in solar activity have potentially important effects, because the sun has a prime influence on the biota. Unfortunately, scientifically useful observations of the sun's activity are limited to the past few hundred years, and attempts to extrapolate these observations to the geologic past are risky. There is no reason to postulate major excursions in the sun's luminosity during Phanerozoic time. However, some astronomical phenomena, such as passage of the solar system through cosmic dust clouds, might pulse the sun's luminosity on a temporary basis. It is unclear how much change in the solar constant could be expected, and measuring such effects through time is highly problematic.

Fluctuations in the orbital patterns of the Earth, such as those that are the basis of the Milankovitch theory of ice ages, may have direct or indirect biological results. For example, the Triassic-Jurassic rift lakes of the east coast of North America experienced cycles of filling and draining that have been estimated from varve counts to be very close to 21,000 years in duration. During each cycle, the lakes became populated with organisms, notably fishes, that show rapid evolutionary activity in the form of morphologic variability and speciation. New biological cycles began with each refilling of the lakes. The cycles were, however, short and the potential for long-term effects on evolution remain unexplored.

Supernovae have received much attention as potential causes of biologic perturbations. Supernovae are large enough and some occurred close enough to the Earth to be the potential source of biologic effects, but the means of detecting them are virtually nonexistent. A supernova would probably have to have burst within half a light year from Earth in order to produce the anomalous iridium concentrations seen at the K-T boundary. The relationship between supernova size, frequency, and distance from the Earth and the biological effects of supernova explosions are still uncertain.

Geomagnetic reversals have been proposed as potential causes of biologic change, notably extinction. Apparently the synchroneity between geomagnetic reversals and biologic change is not as close as had originally been thought; what we seem to be seeing are two sets of high frequency events whose correlation is not statistically significant. Most extinction occurs in the absence of magnetic reversals, and most magnetic reversals have occurred without detectable mortality or extinction in the biologic reord. Several microtektite horizons in the geologic record have been detected close to intervals of magnetic reversal, and this raises the possibility that extraterrestrial objects may perturb the geomagnetic field, or "push it over the edge" when in a weakened state. However, there is no magnetic reversal near the K-T boundary, where there is the strongest evidence so far for a large bolide impact. Furthermore, it is now thought that the weakening of the Earth's magnetic field during a reversal would only modify the cosmic ray flux to a degree that is roughly comparable in biologic terms to an organism moving to a zone of high geomagnetic latitude.

3. Short-term phenomena with possibly significant, global biotic effects. Volcanism is normally regarded as a local phenomenon, although it is controlled and initiated by deep-seated geophysical processes that are global in extent. Volcanic activity at any given time may be relatively local (up to the extent of the arc of a continental plate), but the effects of volcanic eruptions are both local and worldwide. The immediate effects of volcanism are local mass mortality and habitat destruction. But long-term changes in the environment of a region and its biota may also be brought about by long periods of volcanic activity. Observations and models of ballistic effects indicate that regardless of size, no eruption is likely to suspend a significant quantity of material in the stratosphere for periods longer than a few years. The maximum projected effects of the most intense worldwide volcanic activity are still limited to lowering the ambient temperature by a few degrees and reducing light levels by up to 50% for a period of a few months.

Changes in ocean chemistry and circulation may have a variety of causes and may be relatively rapid; their hydrologic and biologic effects may be either local or worldwide. The mechanisms of change and their potential effects are only beginning to be appreciated and understood. Ocean waters are variously mixed, and patterns of circulation are directly tied to patterns of the organisms that inhabit them. The El Niño Southern Oscillation (ENSO) is an example of a local phenomenon which has the potential of affecting large parts of the ocean. Recent research suggests that the current cycle represents a considerable excursion from the normal range of El Niño. We are led to wonder whether certain geophysical changes might be able to cause continued excursion from its normal cyclical levels, or perhaps to neutralize the oscillation completely. If so, we can expect to see considerable effects on ocean mixing and attendant biologic effects. Turnover of deep anoxic water masses, an effect with many possible oceanic and geological causes, is a mechanism that could trigger extensive poisoning of oceanic organisms. The relative rapidity of these changes and the relative extent of their biologic consequences are probably complex, highly variable, and difficult to determine. The supply of bottom water is sensitive to conditions in the polar regions as well as to the formation of very saline waters. Large fluctuations in the temperature and salinity of ocean bottom waters probably occur.

4. Gradual short-term phenomena with gradual biotic effects. Glaciation, transgression and regression of the oceans, and continental plate movement are members of this class. In some cases their causes may be synergistic. In all cases, significant biotic effects are observed and expected, even though they may be neither "catastrophic" on a short-term scale nor globally correlatable in the geologic record. Admittedly, these processes are on the long end of the temporal scale that we considered and are not sudden in the strict sense used here. However, they may be relatively rapid in some cases, and they are evidently correlated with biotic effects. At some point in the late Tertiary, for instance, a land bridge formed between North and South America. Whether the exchange of species was effectively instantaneous, and whether the speciation and extinction that followed was short-lived, remain problematic.

5. Extraterrestrial phenomena with measurable physical effects. Supernovae were considered above as impractically rare and undetectable, but other extraterrestrial agents may be more amenable to study. We have mentioned the possible small effects of cosmic dust clouds on the sun's luminosity, and their potential direct effect on the Earth as dust in the upper atmosphere. However, the density of the cloud would have to be very high to produce effects that are significant. It is not clear whether gravity-focusing could enhance the effects.

The influence of the impact of extraterrestrial objects (asteroids and comets) are of great interest, because they can be detected and because their effects on the Earth's biota are potentially large. In our discussions, we proceeded from the physical and biological data to inferences about their interactions, methods of detection and evaluation, and the relative importance of such impacts. The data have been summarized in many recent publications, and research in this field is burgeoning. We did not attempt to evaluate all the competing scenarios, nor the accuracy of the various data bases, but to establish what lines of evidence were secure or uncertain, concordant or discordant.

It is now clear that the Earth has been bombarded throughout its history by extraterrestrial objects of various sources, sizes, and compositions. Shoemaker's review in this volume summarizes the theoretical and empirical evidence for the occurrence of impacts and the variation of their rate during Earth history. At present there appears to be no criterion for distinguishing with confidence between the impact effects of an asteroid and a comet. It is presumed that their effects on the atmosphere are more or less similar, but there is as yet no consensus regarding the sequence of events during entry and the nature of the physical and atmospheric shocks.

The frequency of terrestrial impacts has been calculated from direct observations of Earth-crossing asteroids and comets, from the geologic record on the moon and Earth, and from calculations of the probability of impacts. Several intriguing problems remain unsolved. The record of impacts is better preserved on the moon than on the Earth; the reasons for this include the lower rate of erosion and the absence of tectonic activity, organic ground cover, and an atmosphere on the moon. It has been estimated that stony objects must have a diameter ≥150 m to penetrate the Earth's atmosphere. Bolides with a diameter above this threshold generally form craters larger than 3 km in diameter. Although the terrestrial impact record is much less complete than the lunar impact record, our knowledge of the chronology of Phanerozoic craters on the Earth is far more reliable.

The effects of extraterrestrial bombardment on the Earth depend on the infall frequency of objects of various sizes. Has the infall of such objects been random through time? Have there been significant excursions from the normal flux during the Phanerozoic? What agents might have caused such excursions? The available record is incomplete. Approximately 100 impact structures 3 km in diameter or larger are known, but only about half of these are well dated. Although the available data are rather sparse, they suggest that variations in the flux during the last several hundred million years have been minimal. On the other hand, frequency of impact could have varied widely as the sun has passed through the spiral arms of the galaxy.

The diameter of the extraterrestrial object that is thought to have struck the Earth at the end of the Cretaceous period was estimated at about 10 km, through extrapolation of the estimated amount of iridium distributed worldwide. But it is possible that an object much smaller than the hypothesized K-T bolide could have produced similar effects. If so, our thinking regarding the flux of ET objects of sufficient size to have a large effect on the biota could change dramatically, because the Earth has received a large number of impacts of such smaller objects within the last 10^8 years.

If this turns out to be the case, we will need to ask why we do not see many mass extinctions comparable to that of the K-T boundary. It may be that most impacts of this type do not affect biologic diversity. It should be possible in time to examine the total biologic record of extinctions and to define the nature of the processes that have produced such adverse effects on the biota.

EVOLUTIONARY PATTERNS THAT MAY BE AFFECTED BY SHORT-TERM PHYSICAL PHENOMENA

Concepts and Definitions in the Analysis of Biologic Diversity

The intricacies of evolutionary patterns and processes, ecological and evolutionary concepts, and the ability to recognize and resolve these in the historical record are terra incognita to many scientists working on the problem of sudden physical/biotic interactions. We therefore identified a series of ecological and evolutionary concepts that may be useful as a common ground for future discussions.

Species of sexually reproducing organisms are generally distributed as interbreeding populations spread over a variable geographic area. It is usually quite difficult to measure the exact limits and extent of gene exchange in living populations, and fossil populations are almost impossible to define with certainty. Fossil species are defined purely on morphologic criteria. Thus, paleontologists use the concept of the assemblage (or death assemblage as opposed to life assemblage) to describe collections of members of a fossil species without inferring that they all lived at the same time, died at the same time, or interbred. A typical slab of Devonian shale may be covered with brachiopod shells, yet it is virtually impossible to know whether these animals lived together or died together. All we know is that they have been preserved together. They may also be preserved with other organisms as a part of repeatable associations of two or more taxa in the fossil record. Associations differ from communities, a term used in modern ecology, because communities are composed of organisms that live together, presumably with a variety of species interactions such as predation and competition. These can be discerned only rarely in the fossil record; for similar reasons the concept of the ecosystem, which calibrates energy flow in a biota, is equally elusive to paleontology.

In the fossil record, then, we are dealing largely with the remains of organisms about which we know only that they were preserved together. Species diversity (corresponding to the term "species richness" of the ecologist) is the number of species represented in a given unit of rock. Abundance is the number of individuals present in the unit. Biomass is the total amount of organic matter present in a unit measure (sometimes expressed as numbers of individuals) and is difficult to assess in geologic situations for reasons that should be evident from the above discussions. Two other ecological concepts, productivity (the rate of fixation of organic carbon per unit measure) and standing crop (the number of living individuals per unit of area or volume at one time) are also difficult to assess from fossils. We can, however, examine the distribution of relative abundance of individuals belonging to sympatric species (i.e., those preserved together) through the concept of equitability, an index of this distribution. As a simple example, one hundred organisms can be distributed among ten species of ten individuals each (high equitability), or among six species of which one species comprises ninety-five individuals and the other five have one each (low equitability). The concept of equitability is of some use when considering the possible pace of biotic change under short-term stress.

It is important to distinguish between two kinds of death in the fossil record. Mortality is the death of a single individual or individuals; extinction implies the elimination of the last member of a species and is, of course, "forever." The issue is complicated somewhat by the recognition of two kinds of extinction (see Raup, this volume): one in which the species dies out without issue, the other where a species is transformed into another species differing slightly in morphology but recognizably distinct from its predecessor. The latter, "pseudoextinction," does not imply the cutoff of a continuously breeding population through time and therefore does not qualify as a proper extinction. Inclusion of such events may distort "true" patterns of evolutionary change, but each "pseudoextinction" event also implies a new "pseudoorigination" event, and such events may be more or less equally common in faunas through time, thereby assuring that the bias will be averaged out. A mass extinction, as traditionally (if informally) recognized, involves a significant decline in species diversity or change in species composition over a relatively short time and over a relatively large geographic area. By contrast, mass mortality represents the death of a large number of individuals in a relatively short time, irrespective of geographic extent. In mass mortality, relative equitability of abundance among species may in part determine the extent of extinction. Following the example of one hundred individuals given above (for equitability), if 90% of the individuals were eliminated, in the first case nearly all species would be expected to persist, but in the second case it would be highly unlikely that one would find more than two or three represented among the ten survivors (all other factors being equal). For reasons discussed above, mass mortality is extremely difficult to demonstrate in the fossil record, because it cannot be assumed that organisms preserved together actually lived or died together. Thus, while it should be possible to have mass mortality without mass extinction, this subtlety is virtually lost in the fossil record, and the relative instantaneousness of death cannot be established in the vast majority of cases.

A number of rapid species replacement can qualify as mass extinction phenomena (Fig. 2). In the first, nearly all species are eliminated without issue and without replacement at the hypothetical boundary (dotted line). In the second, all species become extinct, but they are quickly replaced

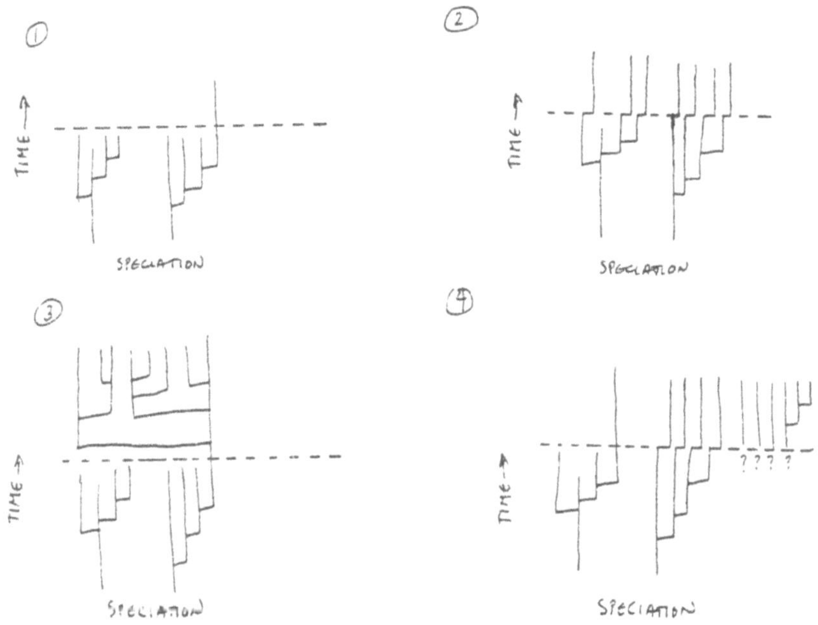

FIG. 2 - Four possible models of mass extinction and subsequent replacement. For explanation see text.

by close relatives, either from previous speciation events or by migration from elsewhere (or both). In the third case, one or a few lineages survive and quickly diversify by speciation. The fourth case represents a combination of the other three plus incursions of previously unrepresented taxa, a pattern that is quite commonly observed. In all four cases, very rapid and extensive extinctions occur. Obviously, there are other possibilities and combinations of circumstances. The point here is that the patterns of extinction at taxonomic levels other than the species level (which will be equally high in all three cases) are quite different. In the first case, nearly all higher and lower taxonomic categories become extinct and species diversity crashes. In the second case, there is virtually no extinction at higher taxonomic levels and species diversity remains constant. In the third case, there is almost complete extinction of higher taxonomic levels but species diversity remains constant. In the fourth case, some higher taxa become extinct and species diversity shows a net increase over time. This indicates the importance of considering a variety of evolutionary scenarios and of considering changes in several taxonomic levels.

Originations and extinctions of taxa can be expressed as numbers or as rates. Number of originations and extinctions is usually used to describe particular geologic sections; origination and extinction rates are calibrated per unit time (usually averaged per million years). The difference between these parameters is often significant. For example, in a given deposit, five out of seven species in a clade (evolutionary group) may become extinct during the deposition of a bed or at a boundary within it. Without precise dating of these sediments it is impossible to compare the changes with extinctions of other taxa at other times. For a given time interval it may be completely normal for such a high proportion of species to become extinct. This underscores the value of studying characteristic rates of evolution for taxa in the fossil record in order to form a baseline of data for evolutionary research. Studies of such rates have shown that there are considerable variations in the average duration of species in different taxonomic groups; however, when taxonomic survivorship curves are compared among a wide variety of groups, the shape of the curves is remarkably similar. This has been interpreted by some to imply that the probability of a species surviving from one interval of time to the next may be stochastically constant for its taxonomic group; other workers have disputed this interpretation or considered it spurious. We cannot specify how long a particular species <u>should</u> last, but we have a reasonable idea of the typical range of rates and of significant excursions from those rates. Nevertheless, much more work is needed to make the data useful for evolutionary studies of high resolution.

In any given time interval, origination and extinction rates may be high or low. If both are high, there will be no net change in the "standing crop" of species diversity (as contrasted with abundance). But this will also be the case if both rates are low. To distinguish between these two cases, which have the same net effect, the concept of <u>turnover</u> is applied. In the first case, turnover of species is high, and in the second it is low.

Table 1 shows that different kinds of changes in evolutionary rates may have the same net effects. If the origination rate increases, species diversity will increase as long as the extinction rate does not also increase. But note also that a mass extinction can result either from an increase in extinction rate or a drop in origination rate. No doubt different episodes of high net extinction have been caused by both mechanisms, and it is quite possible that during the same extinction "event," different species or taxonomic groups may have experienced severe net extinctions for

TABLE 1 - Relationships of the concepts of origination rate (α), extinction rate (Ω), turnover rate, and standing diversity. Arrows do not indicate the absolute magnitude of rates, only their synergistic effects (double arrows).

α	Ω	Turnover Rate	Standing Diversity
↑	↑	↑↑	-
↓	↓	↓↓	-
↑	-	↑	↑
↓	-	↓	↓
-	↑	↑	↓
-	↓	↓	↑
↑	↓	-	↑↑
↓	↑	-	↓↓

different statistical reasons, which may or may not reflect heterogeneous biological causes.

Biological Factors and Rates Related to Evolution: Specific Cases

Origination and extinction are common "facts of life" in evolutionary history, and it is obviously impossible to find causes for the extinction of every individual species through time. Evolutionary patterns, however, may be discerned after careful analyses of taxonomic groups and faunas through time. In effect, there are three levels of analysis. The data can be analyzed initially in terms of the distribution of species and clades through time; this includes measurements of fluctuations in relative numbers, at differing taxonomic levels. Second, the basic data may be segregated with respect to habitat and facies preservation, as, for example, for different subtidal marine facies. Paleobiologists have analyzed associations from intertidal, nearshore, and offshore facies from the Paleozoic through the Tertiary and described patterns of variation through time. Intertidal faunas have remained low in diversity, the diversity of nearshore faunas rose slightly in the Mesozoic and Cenozoic, and the diversity of offshore faunas virtually exploded in the Tertiary after a significant increase in the Mesozoic. The third level of analysis, which follows from the other two, is to generalize, if possible, from the presumed functional and ecologic patterns emergent from considerations of evolutionary history. This third level is a synthesis of taxonomic and evolutionary change and may hold the key to the relative importance of suggested causal mechanisms where significant deviations from "normal" evolutionary rates have occurred.

Several specific examples from a wide variety of taxa, habitats, and ages were briefly considered in our group discussions to illustrate the data bases, the analytic effects, the problems of stratigraphic and temporal resolution, and the ecologic and evolutionary patterns that require explanation or that present difficulties. Some basic points of the presentations are repeated here in order to give some idea of the practical side of evolutionary analysis.

A question that occurs over and over in the analysis of extinction patterns is whether biological common denominators of surviving vs. non-surviving taxa can be identified. Consideration of the biology of present-day planktonic foraminifera, and comparison with their cognates across the K-T boundary, is a case in point. Modern species in polar regions live in more or less homogeneous water columns and are characterized by relatively simple morphology. By contrast, subtropical and equatorial species are quite diverse and complex morphologically, and some migrate vertically in the water column during growth and reproduction. In the short term, immediately after the K-T boundary event, only one species of planktonic foram survived; extinction of species occurred whether they were cosmopolitan or provincial, complex or simple. The apparent descendants of the one surviving species appeared suddenly in a burst of speciation immediately after the K-T boundary; they all have a simple morphology inherited from the sole surviving species. The burst of speciation slowed after some time, and species remained simple and widely distributed. Complex provincial equatorial forms, as well as species, genera, and families of cosmopolitan species, disappeared and were replaced by cosmopolitan species of simpler morphology in low species diversity. Radiolarians, diatoms, and other planktonic forms seem to show similar, if weaker, patterns. Species inferred to have migrated vertically in the water column were absent. Organisms that required a heterogeneous water column were not present in the lower Paleocene; this suggests a change in oceanic circulation. These observations suggest the occurrence of an instantaneous ecological event, perhaps triggered by a bolide impact that exerted a profound effect on pelagic ecosystems, and followed by a long period of oceanographic recovery in what are inferred to be rather homogeneous oceans. This may indicate that the K-T event had both short- and long-term effects, and that these produced a major biological reorganization. An ecologic state comparable to that prior to the K-T boundary did not reappear for millions of years.

Megainvertebrates in a relatively complete shallow marine geologic section show taxonomic changes across the K-T boundary that suggest rapid evolutionary extinction and replacement. In the last twenty-five million years of the Cretaceous, the White Chalk fauna developed slowly into a major soft-bottom system. At the boundary the facies show the effects of a slight regression, but no patterns appear to be explained by this event. Most benthonic forms (bryozoans, brachiopods, bivalves, and echinoderms) pass through the boundary without great change on the generic level, but with extensive extinction and turnover on the species level. Environmental stress is suggested by extensive coccolith blooms. The first 1.5 meters above the K-T boundary contain very few species, but subsequently a new fauna appeared, composed of species closely related to those that disappeared at the K-T boundary. The extinction in this section evidently did not affect taxonomic levels higher than species for almost all groups; congeneric species mostly replace Cretaceous species in the Tertiary. This suggests that members of these genera did not become extinct, but lived elsewhere during the barren period just above the K-T boundary. The thickness of the barren strata suggests that environmental stress continued for 50,000-100,000 years. This duration could, of course, be due to a short-term event with long-term effects, or due to a long term process or processes. The extinct Cretaceous organisms that did not reappear in the Tertiary were apparently those that needed to live on chalk bottom. If their habitat was destroyed, as the facies record suggests, at least local extinction would have occurred. Unfortunately, the Danish section appears to be one of the few relatively complete sections where these changes can be observed.

Analysis of long-term evolutionary trends in fossil vertebrates indicates that the record is plagued with lacunae that are often substantial and with relatively small samples, as well as with pervasive biases in the environments of preservation. For example, during all periods from the Permian through the Cretaceous we have virtually no middle period records for terrestrial faunas; before the Permian the record of terrestrial environments virtually assumes the character of Lagerstätten (fossil bonanzas of preservation). Nonetheless, in certain geologic sections it is possible to ask many detailed questions about the pace and timing of evolutionary change. In any attempt to account for the extinction of the dinosaurs at the K-T boundary, the presence of several complicating factors should be understood. First, almost all genera of dinosaurs are monotypic, or should be regarded as such, and so are many families. Thus, the genus level is effectively equivalent to the species level in considering diversity and is far more appropriate and reliable for representing the fossil record. Second, if taxonomic assignments based on overall similarity, size, or the vagaries of preservation are discounted, virtually no genus of dinosaur, pterosaur, or crocodile (the three Jurassic-Cretaceous archosaurian groups) survived its own formation (including appearances in other formations considered contemporaneous). In many or most cases, this applies even to parts of formations. It must be stressed that, because the terrestrial vertebrate record is relatively spotty and discontinuous, this pattern could be an artifact of sampling. Nevertheless, if the data are taken at face value, the extinction rates per temporal equivalent of "formations" are effectively 100%.

What is unusual about the disappearance of the dinosaurs at the end of the last recognized Maastrichtian formations is not that they all became extinct - this was usual for them - but that they failed to be replaced by other members of their clades. No long-term hiatus in deposition has been proposed in the most complete terrestrial sections spanning the K-T boundary, and so we must regard these extinctions as both real and relatively rapid. However, the temporal resolution of this event is unsettled in both absolute and stochastic terms. Latest Maastrichtian faunas varied in composition both geographically and environmentally, and different clades and faunas became extinct at different times in different areas and environments; we must therefore be prepared to accept the probability of multiple, synergistic mechanisms for these terrestrial extinctions.

It is apparent that all vertebrates above 25 kg body weight became extinct; however, smaller members of their clades, and, of course, their young, also died out. Most of the non-archosaurian components of these faunas (turtles, crocodiles, and champsosaurs) crossed the K-T boundary without serious effect but appear to have suffered extensive extinctions in the mid-Paleocene. Large or specialized terrestrial vertebrates are usually more affected by extensive extinctions, in spite of the notion that large animals perceive their environments as relatively "fine-grained." The ephemerality of large Tertiary mammal species suggests that dinosaurs, with their enormous body size, held on for surprisingly long periods of time in the Mesozoic Era. This may have been possible largely because of equable, stable Mesozoic climates. Toward the end of this time their environment changed. Angiosperms became dominant and climatic conditions deteriorated. By analogy with the ecology of elephants, it is not unlikely that such environmental threshold effects might well have resulted in relatively rapid extinction; nonetheless it is surprising that the generic diversity of dinosaurs in the Campanian and Maastrichtian was comparable within a factor of two. "Old" clades (such as the carnivorous dinosaurs) as well as "new" clades (such as the ceratopsians) appear to have been either holding their own or diversifying in the latest Cretaceous. Unfortunately, adequate terrestrial exposures during this time interval are few. Only continued, fine-scale analysis of stratigraphic sections across the K-T boundary can help us to gain further insight into these patterns. Single sections may reveal very little about the pace and character of a process that was evidently a combination of very heterogeneous changes.

The fossil record of terrestrial plants across the K-T boundary is complicated by several biological factors peculiar to plants. Plants are notoriously sensitive to climate and climatic change, and plant composition in a given environment may vary greatly along a 10-meter-long moisture gradient. This indicates that in any given locality the stratigraphic plant profile may vary drastically as much from very local changes in moisture conditions (for example, from a meandering stream) as from any ecological disturbance or evolutionary crises. It is therefore not safe to assume that correlation of facies type reflects precise temporal correlation. (This is shown, for example, by coal deposits that appear to cross magnetostratigraphic boundaries over their geographic range, implying diachronous deposition akin to a transgressing shoreline.)

It is inferred in studies of plant species abundance and diversity that centimeter-by-centimeter compilations of taxonomic change may indicate mass mortality; yet it cannot be inferred that this pattern automatically translates into mass extinction. This is because plants, unlike most animals, can leave their next generation in the form of dormant seeds: local catastrophes are not necessarily final for plants, which are otherwise at a disadvantage due to their inability to walk away from adverse conditions. As a result, some postulated effects of short-term catastrophes, such as a severe reduction in light and sudden temperature drops, may not have caused extinctions in high latitude plants. On the other hand, toxins produced or spread by a catastrophic event may well have had catastrophic effects on plant life. As we learn more about the abundance, distribution, and paleoecology of plants, they may serve to check the sensitivity and to evaluate the relative plausibility of models for biologic effects of physical crises.

At present, plants seem to have had a heterogeneous distribution at the K-T boundary, and they were apparently quite differently affected in various geographic settings. Different plant provinces were affected to different degrees. Unfortunately, much of the work to date is based on the study of pollen, which cannot always be tied to their mother plants on the species level. Plant morphology shows mosaic evolution, i.e., not all parts of a given plant evolve at the same rate. Pollen may be quite conservative in many morphologically and ecologically variable groups, so that ecologic inferences are best not based on pollen alone. The plant evolution during the Cretaceous is complicated by the appearance and radiation of angiosperms in the middle and late Cretaceous. The high rates of evolution of this group continued across the K-T boundary. Species diversity shows a net increase across the boundary, though some species and higher taxa were affected adversely; as mentioned above, there was a great deal of geographic heterogeneity. These factors must be taken into consideration in the interpretation of patterns of botanical evolution.

The above examples are only the barest sketches of some very complex problems. Such problems are not unique to the K-T interval; they have been well studied there because the record is relatively good, the taxonomic interest is great, and the events relatively recent compared to other mass extinctions. In the Cambrian, for example, trilobite biomeres (megazones) were replaced completely five times within a 35-my period. Each biomere has a history of rapid expansion in abundance and diversity, followed by a period of more normal evolutionary rates, and finally by a sudden elimination of nearly all members of the biomere; no visible decline preceded these eliminations. The extinct species were replaced partly by close relatives and partly by taxa new to the area. This apparently cyclical pattern is simpler than events at the K-T boundary, because it involved almost entirely trilobites, and because there is no apparent biologic or geologic "noise" in these systems. No unifying mechanism has been proposed that explains this overall evolutionary pattern.

The Frasnian-Famennian extinction in the late Devonian is an example of a more severe extinction. Organisms that were seriously affected included the shallow-water benthos in many of the continental areas of the world within the tropical and subtropical zones. Animals that disappeared include 142 out of 148 species of shallow-water corals, virtually all stromotoporoids, four superfamilies of brachiopods, and most families of trilobites. The extinction affected a huge number of animals that may have represented as much as 80% of the animal biomass in the affected environments. Succeeding Famennian faunas are very different. The event appears to have taken place within one conodont subzone; its duration was less than 1/2 million years, perhaps very much less. In each section examined, the change is found to have taken place across one bedding plane. It is never transitional.

The organisms affected were largely filter feeders, probably with planktonic larvae. Suggested mechanisms of extinction include fresh water, bad water, cold water, and turbid water. The most probable of these would appear to be turbid water, and an oceanic bolide impact has been suggested as a source of such turbidity.

The most extensive extinction in the Phanerozoic, that at the Permo-Triassic boundary, is less well-known, because good boundary sequences are rare or absent. Some ecological selectivity is evident, but the faunal data are still too sketchy and anecdotal to provide a basis for synthesis.

TESTING HYPOTHESES OF CAUSAL CONNECTIONS BETWEEN PHYSICAL AND BIOLOGICAL EVENTS

This was the main goal of our group. Can connections be established or improved between data and patterns from the physical and the biological sciences? What must scientists from diverse disciplines do to make their fields understandable and approachable to others? How can common ground be established, and how can competing or disjunct hypotheses and complex scenarios be evaluated?

The papers by Shoemaker, Raup, Hsü, and Toon (all this volume) have tried to organize and to clarify some of these questions and issues. The following synthesis, though by no means complete, is meant to be a first step toward further rapprochement between the physical and biological sciences. Our group discussed and developed these questions with the viewpoint that scientists have a right to understand each other's data, to know the limits of the data and resolution of hypotheses, and to expect conditions under which hypotheses could be rejected or provisionally accepted as robust. The questions that follow are loosely organized in the realm of physical events, biological events, and possible connections between the two.

Physical Hypotheses and Scenarios

1. Operational definitions of geochemical anomalies, detection of anomalies, and evaluation of their source are all required. Anomalies are the primary evidence for individual, large bolide impacts on Earth and for scenarios related to possible biotic effects of such collisions. It is therefore of primary importance that agreement on methods and criteria for the identification of such anomalies be established in the fields of geochemistry and geophysics.

At present, several kinds of signals interpretable by physical scientists indicate anomalies. Most emphasis has been on the stable isotopes of carbon and oxygen and on iridium and associated trace elements. Analyses of iridium are, in nearly all cases, standardized against the background abundance of iridium in sediments. The measured values of this background flux (generally ≤ 10 parts per trillion) do not vary significantly from zero, but individual values may fluctuate. Usually, an anomaly can be recognized because it is at least ten times higher than the background values in a particular section. Expression of this value has been in terms of concentration in specific samples (ppb, ppm, ng/g) or of vertically integrated abundance (ng/cm^2) at a recognized stratigraphic anomaly. It would be helpful to standardize the definition and recognition of anomalies, as well as their expression relative to "normal" levels (perhaps against a background noise of cosmic dust levels).

Anomalies in the concentration of the noble metals can be terrestrial or extraterrestrial in origin. What are the methods for determining such sources, how do their occurrences vary, and what are the confidence levels for establishing sources? Anomalies of terrestrial and extraterrestrial sources should have different associations and co-occurrences with other elements and may perhaps be accompanied by other geological indicators (for example, microtektites or other residues of ejecta). Can these be established and differentiated with reasonable confidence? What will be the accepted norms and ranges? Should they be expected to vary under different conditions of sedimentation, facies type, or age? Where will the major uncertainties lie?

2. Relatively little is known about the geochemical behavior of many rare earth elements, noble elements, etc., in sediments, principally because they are so rare, because until recently data have been sparse, and because no one has asked the important questions about their abundance, distribution, and source. Iridium anomalies, for example, have been found in many kinds of sediments; at the K-T boundary they are widely

distributed in coals and clays. Why should this be? Is such preservation primarily correlated with the geochemistry of the sediments, the energy under which they were deposited, the type of sediment formed with the Ir-producing event, post-depositional diagenetic effects, or some other factors? To what extent does mixing reorganize the distribution of iridium and other such elements, and is there anything unusual about these patterns? The sedimentologic, geochemical, and geophysical properties of elements that are used to detect sudden physical events are a largely uncharted and fascinating field of physical and biologic interactions.

3. Iridium and other anomalies are evidently more widely distributed in sediments than had been previously realized. Many of these anomalies may turn out to be extraterrestrial in origin. One of the principal properties of extraterrestrial anomalies is that they are temporally instantaneous. Such globally distributed anomalies can therefore act as time markers and allow us to ask questions about the worldwide distribution of environments, organisms, and climates at particular instants of Earth history. With such markers we could study environmental patterns at a level of resolution heretofore attainable only in the very recent past (e.g., the CLIMAP project).

The isotopic composition of oxygen in marine fossils may provide information on paleotemperatures, global ice volume, paleosalinity, etc.; the isotopic composition of their carbon is related to ocean productivity and/or ocean chemistry. Anomalous isotopic perturbations have been found in sediments synchronous and/or slightly younger than those with an iridium anomaly. Explanations of these anomalies have been suggested (see Hsü, this volume). Considerable work on amplified sections is, however, necessary before convincing conclusions can be drawn from such data.

In order to ask such questions and elucidate such patterns, stratigraphers must have criteria for correlating anomalies that allow reasonable confidence in the correlations. Because few anomalies are likely to be as large as that at the K-T boundary, resolution of these events is liable to decrease with decreasing size of the anomaly. Even at the K-T boundary the signal level shows wide fluctuation both geographically and between environments of deposition. What methods and confidence intervals will stratigraphers be able to use in attempting to correlate anomalies? Will these differ from those used in other kinds of "event stratigraphy," such as ash beds and basalt flows? How will our ability to use them in correlation change as the signals become weaker?

Biological Patterns and Processes

1. So far, the principal interface proposed between physical and biological events has centered around mass extinctions. The concept of mass extinctions is widely used in the literature and is generally understood by most evolutionary biologists. But how exactly can mass extinctions be operationally defined and quantified? Is there a certain threshold level of extinctions above which we can identify episodes of mass extinction? Is such identification a question of counting taxa, counting specimens, calculating rates, elucidating causes, or some combination of these?

In the course of our discussions we realized rapidly the need for clarification of these evolutionary concepts. We suggest that mass extinctions may be defined operationally in the context of calculated "normal" taxonomic rates and the occasional excursions from these rates. At certain points in time, and within and among faunas, high numbers of apparently synchronous extinctions occur. The entire fossil record of the clades (evolutionary groups) affected at a particular boundary has to be examined with the aim of assessing normal rates and variations for origination and extinction of taxa at various hierarchical levels. If, at a given boundary, these rates are significantly higher than usual for a large number of taxonomic groups, then by definition a mass extinction can be said to have occurred. There will probably never be a way to define "mass extinction" in absolute terms, because biological history is non-repeatable, and taxonomic diversity is highly variable through time. However, in relative terms such operational definitions ought to be heuristically useful and may lead to better understanding of evolutionary rates.

2. The fundamental question just expressed suggests that a great deal of additional analysis of taxonomic rates of evolution is needed in order to answer questions related to biotic pattterns and their perturbations throughout geologic time. Paleontologists have, and probably always have had to some extent, a good intuitive understanding of the taxonomic patterns, tempos, and histories of the groups which they study. These groups could be examined statistically to determine "normal" rates of taxonomic evolution for individual groups, how these rates fluctuate and whether there are intervals of Phanerozoic time when these rates changed drastically for many or all of the members of these groups. Patterns within subgroups of these larger groups might be of great evolutionary interest, and the same could be said of analyses within particular habitats and faunas.

3. In a similar vein, ecological studies of many groups through time should be undertaken to determine the biological and functional factors that may underlie certain rates of taxonomic evolution. Why are some extinctions so selective? Answers to this question are of particular interest to nonbiologists, because they will be of greatest help in identifying predicted effects of physical changes or perturbations.

The three questions listed above have been given in the order of their increasing remoteness from the data base. The distribution of species and their changes through time are the most basic evolutionary phenomena that can be studied. In attempts to study faunas and habitats, certain assumptions are made about mode of life, associations in life, and environment. Studies of ecological and biological factors of extinct organisms are even more inferential. This level of investigation is the most difficult to approach; the conclusions that are reached tend to inspire the least confidence but are perhaps the most interesting of the three. However, work in this area cannot be undertaken without studies at the other two levels, and broad generalizations or simplistic explanations should be avoided.

Possible Connections Between Physical and Biological Factors

Most of the connections discussed in our group related to hypotheses of impact events and their projected physical and biological consequences.

1. The basic need in discussing questions of synchroneity of physical and biologic effects is for precision in stratigraphic and temporal control, and for understanding of the episodic nature of sediment accumulation. Thicknesses of sediment do not easily translate into elapsed time, and the stochastic preservation of depositional environments, compounded by factors such as compaction, bioturbation, and diagenesis, complicates the problem further.

Among the stratigraphic data for physical-biological interactions, standardized estimates of stratigraphic and temporal completeness are particularly important. Historical studies of these intervals are already attempting to develop the finest taxonomic and sedimentologic resolution possible by concentrating on the best available geologic sections. Limits of confidence in measurements taken from these studies, both on the physical and biological sides, should be developed and made explicit.

2. Scenarios are very complex statements about historical events. They require elucidation and testing in historical, biological, and physical terms. The various sets of data used in these scenarios must be kept as independent lines of evidence. The logical structure of a scenario, which is very difficult to disprove, is such that rejection of one component hypothesis of the scenario does not necessarily lead to rejection of the entire scenario. Instead, modification of certain details of the overall scenario occurs. Competing scenarios may not be entirely incompatible, but this can only be demonstrated if their logical structures, including component hypotheses and their data bases, are made explicit. Attempts should be made to give conditions under which scenarios and their component hypotheses would have to be greatly modified or rejected, and which lines of evidence could be interpreted in alternative ways.

There are no absolutes in evolutionary history, and most major "events" or intervals in evolutionary history appear to be quite complex. Therefore, oversimplified hypotheses and sweeping generalizations must be avoided. Data that do not fit certain hypotheses or scenarios should be regarded as unresolved, not necessarily irrelevant. They cannot be swept under the rug of vast generalizations, nor assumed to be part of patterns that they do not fit. Otherwise, inclusions of such data will magnify effects artificially and distort baseline data. As a result, the true character of evolutionary change and the possible influences of physical phenomena will never be illuminated.

3. The possibility that short-term physical effects have affected biological evolution, and the predicted and empirical evidence of the frequency of such stimuli, suggest the possibility of widespread influence within the available fossil record. It now seems reasonable to ask whether, if bolide impacts are frequent and can affect evolution, many stage and zone boundaries might be related to sudden extinctions that are perhaps due to external causes. In asking this question, it must be kept in mind that many zones and stages are properly defined by first (not last) appearances. It remains probable that many first appearances are made possible by previous extinctions. However, because this question has never been asked, the answer is certainly not known.

4. In a similar vein, further research is needed on the threshold level of ballistic effects of impacts that are sufficient to cause the extensive changes in temperature and light projected for the hypothesized K-T bolide. It appears that an asteroid much smaller than that thought to

have collided with the Earth at the K-T boundary may have been able to create comparable effects. If so, then we might expect there to have been many such events, since the frequency of impact probability increases as the size of the bolide decreases. As Toon points out (this volume), "apparently the number of impacts is overestimated, their ability to cause extinctions is overestimated, or smaller impacts do not yield globally distributed dust clouds in a simple proportion to their impact energy." This point deserves further attention from both physical and biological scientists and may be closely linked to the factors outlined in point 3 above.

5. In the recent past, specific physical effects of extraterrestrial phenomena have been modeled and studied empirically. Biological consequences have also been predicted from these physical effects, largely by physical scientists. Biological scientists hold the key to many of these problems and to the understanding of patterns seen in the organisms they study. Physical and biological hypotheses are logically and empirically separate, but dialogue between the two groups of practitioners is the only way to elucidate patterns and to accept or reject the validity of processes that have been proposed to underlie them. Although patterns of changing diversity in marine microorganisms are paleobiological matters, these changes may be governed by geochemical and geophysical processes (e.g., ocean circulation, mixing, and poisoning). How severe and how fast can such changes be? What indicators can we use to detect them, and how fine can our resolution of these mechanisms be? These are questions for physical scientists. Conversely, projected effects and even empirical evidence of significant geophysical or geochemical perturbations are not sufficient to establish biotic effects; these must show up in the biologic record if they are to be demonstrated conclusively. Under the stimulus of the Dahlem Workshop, we hope to have taken some steps toward an understanding of these interactions. Although we have not solved any empirical problems, we hope to have avoided the continuation or proliferation of some problems that are due to lack of mutual understanding or appreciation of the data, the methods, and the issues in the pertinent fields of scientific endeavor.

Acknowledgements. The members of our group extend heartfelt thanks and affection to Silke Bernhard and her staff for creating the Dahlem Workshop format and environment, for bringing together a great range of scientific experience and viewpoint, and for world-class organization, help, hospitality, and good will. H.D. Holland and A.F. Trendall spearheaded and facilitated the advances made at this meeting. A.H. Knoll helped enormously to clarify the botanical and paleobotanical aspects of our discussions.

As rapporteur I would like to thank the members of the group for clear discussions, stimulating ideas, help in codifying the final report, and for putting aside some substantial differences of fact and interpretation in order to advance the larger questions of method and theory. An inch was frequently relinquished to gain a mile, and for such efforts I am deeply appreciative. D.M. Raup was instrumental in turning the preliminary manuscript into a semblance of English, and the writing and editing help of all was most useful.

REFERENCES

The group decided to omit extensive bibliographic citations in the report. Much of the current data and interpretations on interactions between physical and biological events are cited in the papers by Shoemaker, Raup, Hsü, and Toon (this volume) or can be found in a recent publication that grew out of a 1981 conference in Snowbird, Utah (2). This review centered on current understanding of the K-T boundary and other extinctions with possible physical influences. Our discussions tended to focus on issues and methods dealing with a wide range of data and problems, and less on the data themselves, which were used mainly to explain the nature of inquiry, its complexities, and the difficulties encountered in various fields of importance to the problems at hand.

(1) McLaren, D.J. 1983. Bolides and biostratigraphy. Geol. Soc. Am. Bull. 94: 313-324.

(2) Silver, L.T., and Schultz, P.H., eds. 1982. Geological implications of impacts of large asteroids and comets on the earth. Spec. Paper 190. Boulder, CO: Geological Society of America.

Patterns of Change in Earth Evolution, eds. H.D. Holland and A.F. Trendall, pp. 103-121.
Dahlem Konferenzen 1984. Berlin, Heidelberg, New York, Tokyo: Springer-Verlag.

Changes in Sea Level

M. Steckler
Lamont-Doherty Geological Observatory of Columbia University
Palisades, NY 10964, USA

Abstract. A number of methods have been developed for determining sea level changes. While they give similar results for the large-scale variations (>10-20 m.y.), there is disagreement over the magnitude of eustatic changes and over whether shorter-term variations are smooth or spasmodic.

INTRODUCTION

Early geologists called on vast sea level changes to explain the evolution of the Earth. Neptunian theories presumed the oceans to have covered all of the present mountain belts during the early history of the Earth. With the development of modern geology, there still remained the question of the origin of flat-lying marine deposits now located well above present sea level. Suess (35) pointed out that this called for either a higher sea level in the past of a rising of the land. He introduced the use of the term "eustatic" to refer to global changes of sea level. Numerous geologists have noted the synchronous periods of continental submergence and exposure of the land surface in the geologic record. Various authors have ascribed these differences primarily to either tectonic movements of the land, eustatic changes of the sea, or associated movements of both (tectono-eustacy). Others have rejected the idea that there was any temporal variation in the rate of tectonic activity or in the position of sea level.

Fairbridge (7) has given an historical summary of the development of the various concepts involved with eustacy. The debate still continues regarding the relative influence of eustatic changes in sea level and tectonics in explaining the temporal changes in facies distribution and sedimentation patterns in the geologic record. During this century geologists and geophysicists have applied various techniques in an effort to map changes in the position of sea level. A better understanding of vertical and horizontal motions of the Earth's lithospheric plates and recent advances in technology have helped address the sea level problem. However, estimates of the amplitude of sea level changes still vary by nearly an order-of-magnitude and rates from $1\text{-}10^4$ m/m.y.

Thus, while there is a better understanding of the causes of sea level changes and their effects in the geologic record, there is still a great deal of controversy concerning the shape of the eustatic curve. This is due to the complex interaction of sea level changes with the numerous geologic processes active on the Earth's surface. The purpose of this paper is to review current ideas about eustacy, to indicate the current points of consensus, and to discuss the points of contention. In this, I shall concentrate on the long-term (>1 m.y.) eustatic variations in the Phanerozoic record, particularly those during the last 200 m.y.

METHODS OF ESTIMATING SEA LEVEL CHANGES

Sea level and its rate of change leave their imprint in a number of ways; these include the distribution of sedimentary and biofacies, the areal extent of the sea, and rates of sediment accumulation. The difficulty is that the Earth does not record sea level directly but only its convoluted effects. None of these effects, however, are uniquely due to sea level variations. Hence, in attempting to determine the stand of sea level through geologic time, one must always keep in mind the effects of local and/or global tectonic episodes and variations in sediment supply. Sea level affects a number of different aspects of the geologic record; geologists and geophysicists have therefore approached the problem of changes in sea level from differing points of view and have developed a variety of different techniques for detecting eustatic changes. Since the geologic record is complex and the effects of sea level variations on other processes and of tectonic activity on sea level are incompletely understood, the several approaches have yielded different answers.

One of the most widely used methods of determining sea level changes is to map the succession of transgressive and regressive facies. An important caveat in studying the stratigraphic record in this way is that transgressions and regressions are not synonomous with absolute sea level rises and falls. The change in the depth of deposition of sediments is dependent on a balance of three factors: sea level changes, local tectonic subsidence or uplift, and sediment supply. An overwhelming supply of sediments can produce a regression even during rapid sea level rises, and rapid subsidence of the land surface can result in a transgression during times of sea level fall (5). The interchangeability of subsidence and sea level has resulted in the use of the term relative rise (or fall) in sea level. Hancock (12) attaches relatively little importance to tectonics and views eustatic sea level variations as exercising the major control over transgressions and regressions. Jeletzky (17) has attempted to demonstrate the validity of the opposite view, i.e., that local tectonics control the sedimentary regime of the continents, that worldwide changes of sea level have exerted a minor effect at best, and that the effects of eustatic fluctuations are only preserved during infrequent quiescent periods. These are extreme views that are biased by the geologic setting in which the authors have worked.

Due to the difficulty of separating eustacy and tectonics, the stable interiors of continents and epeiric seas where subsidence and uplift rates are low have been the locus of much work on sea level. Shallow water environments which are particularly susceptible to sea level changes are common there. Relative changes of sea level are indicated by variations in the facies and movement of the strand line. In order to determine worldwide sea level changes it is necessary to correlate facies changes over large areas (10,11), preferably located in regions of different tectonic regimes.

Another method commonly used to define changes in sea level has involved the areal distribution of marine sediments on the continents, rather than facies changes along cross-sections or in isolated locales. Attempts to reconstruct continent-wide marine incursions have been somewhat limited by the subsequent erosion of shallow water facies and the obliteration of evidence for the location of shorelines. A number of paleographic maps and atlases suitable for this purpose have been compiled. These atlases need to be constructed with finely divided time intervals, otherwise the maximum extent of the seas will be overestimated by the summing of varying distributions (3, 48). For quantitative estimates of sea level changes hypsometric curves that define the areas flooded for given eustatic rises (19, 48) must be available. The slope of hypsometric curves for low elevations is crucial in making quantitative estimates. Yet, this slope will vary with time due to tectonics and changes in sea level.

An alternative to using cratonic sedimentary sequences is to use the stratigraphic succession in a region where the tectonics are well understood. For example, tectonic subsidence at Atlantic-type continental margins is thermal in origin; hence, by correcting borehole subsidence records for compaction, paleobathymetric variations, and sediment loading, Watts and Steckler (45) were able to separate the effects of tectonic subsidence from those of eustatic variations. Wood (46) and Hardenbol et al. (14) modified Watts and Steckler's (45) technique slightly by using explicit thermal models for margin subsidence rather than exponential curves.

Vail et al. (40) have developed an interesting and sophisticated technique based on the use of seismic stratigraphy for determining relative changes in sea level. They recognized onlap and offlap patterns within seismic sequences and used these to contruct regional charts of the landward extent of coastal onlap (Fig. 1). They then modally averaged a large number of regional onlap charts to estimate changes in global sea level. Since his initial papers, Vail has modified the direct relation proposed between coastal onlap and sea level (for a more detailed discussion of the use of seismic data to estimate sea level see Vail et al. ((40), Parts 3 and 4) and Vail and Todd (41)). A basic assumption made in this technique is that seismic reflections follow bedding planes and are synchronous. Because the data used by Vail et al. (40) are confidential, considerable effort still needs to be put into demonstrating whether sequence boundaries are synchronous and globally correlative (6). If the underlying assumptions prove to be correct, the work of Vail et al. (40) adds a significant new set of data for evaluating global sea level changes.

An entirely different approach to estimating sea level changes does not use the observational record of its effects, but calculates its magnitude from the known causes of sea level changes. The idea that changes in the volume of the ocean basins was the main cause of eustatic variations goes back to Suess (35). Pitman (26) and Donovan and Jones (6) examined the possible causes of sea level changes and concluded that except for glaciation, which has been important only during restricted intervals in the geologic past, only variations in mid-ocean ridge volumes were capable of producing the observed variations of sea level. Desiccation of small ocean basins could produce rapid changes but only of ~15 m (6). Crustal shortening, sediment influx, and mid-plate volcanism are estimated by Pitman (26) to be at least a factor of three slower than mid-ocean ridge volume changes and of smaller magnitude. On this basis Hays and

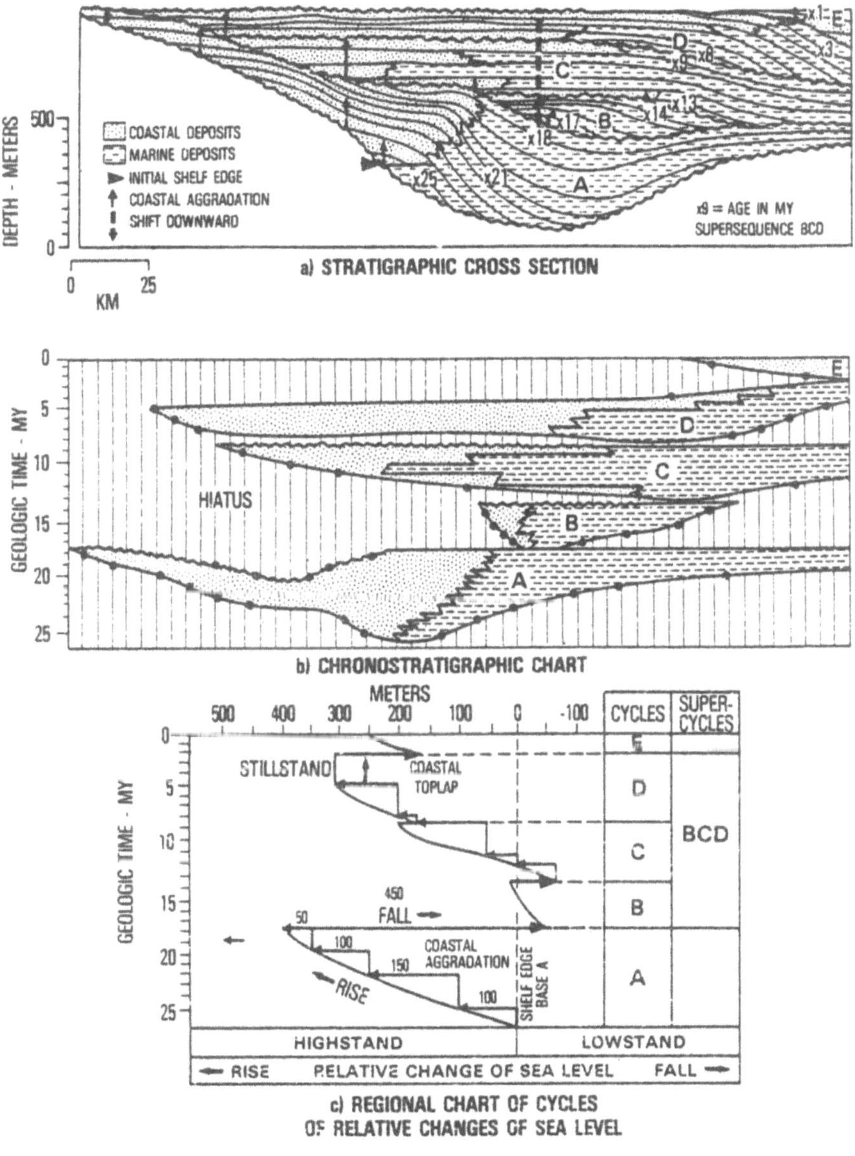

FIG. 1 - Procedure for constructing regional chart of cycles of relative changes of sea level (40).

Pitman (16), Pitman (26), and Kominz (18) calculated the stand of sea level during the past 85 m.y. by estimating the changes due to changes in the volume of MOR's since mid-Cretaceous time.

The idea that sea level changes are global and synchronous has been challenged by Mörner (21-23). Fairbridge (7) pointed out that rapid changes in the geoid could produce local sea level changes because the hydrosphere would readjust much more quickly than the land. This effect is used in glacial rebound studies to aid in estimating the viscosity profile of the Earth. Readjustment occurs within 10-20,000 yrs. Mörner (22, 23), however, has suggested that geoidal changes in the time scale of 10^6-10^7 yrs are effective in producing regional eustatic changes. His theories imply that the solid Earth either does not respond or responds in a lesser fashion to these long-term changes, because concomitant adjustment of the solid Earth and hydrosphere would result in an absence of relative sea level changes. Thus, while geoidal eustacy is operative for geoidal changes occurring more rapidly than 10-20,000 years, I find it doubtful that it influences longer-term sea level changes.

AMPLITUDE

Despite the variety of methods applied to determining sea level changes, consensus regarding their amplitude remains elusive. The geologic record shows changes due to rising or falling sea level, but the magnitude of the variations are difficult to estimate. Sediment loading can amplify an apparent rise, and rapid tectonic subsidence can mask it. Thus, there is more agreement on the timing of sea level change than on their amplitude.

The difficulties in making amplitude estimates can best be illustrated by an example. Estimates of the stand of sea level at the time of the Late Cretaceous sea level maximum, one of the largest marine incursions during the Phanerozoic, range from 0 (17) to 650 (13) m above present sea level. Most estimates, however, either fall in the range of 150-200 m or 300-350 m (Fig. 2). It is encouraging that different methods give results that cluster about similar values, but the fact that they group around two rather different values underscores the difficulty of designing techniques to measure the absolute stand of sea level.

The "high" estimates of sea level have generally come from attempts to estimate the stand of sea level from the present elevation of exposed marine beds and from calculations based on volume changes of mid-ocean ridge. These have generally been supported by modeling of seismic

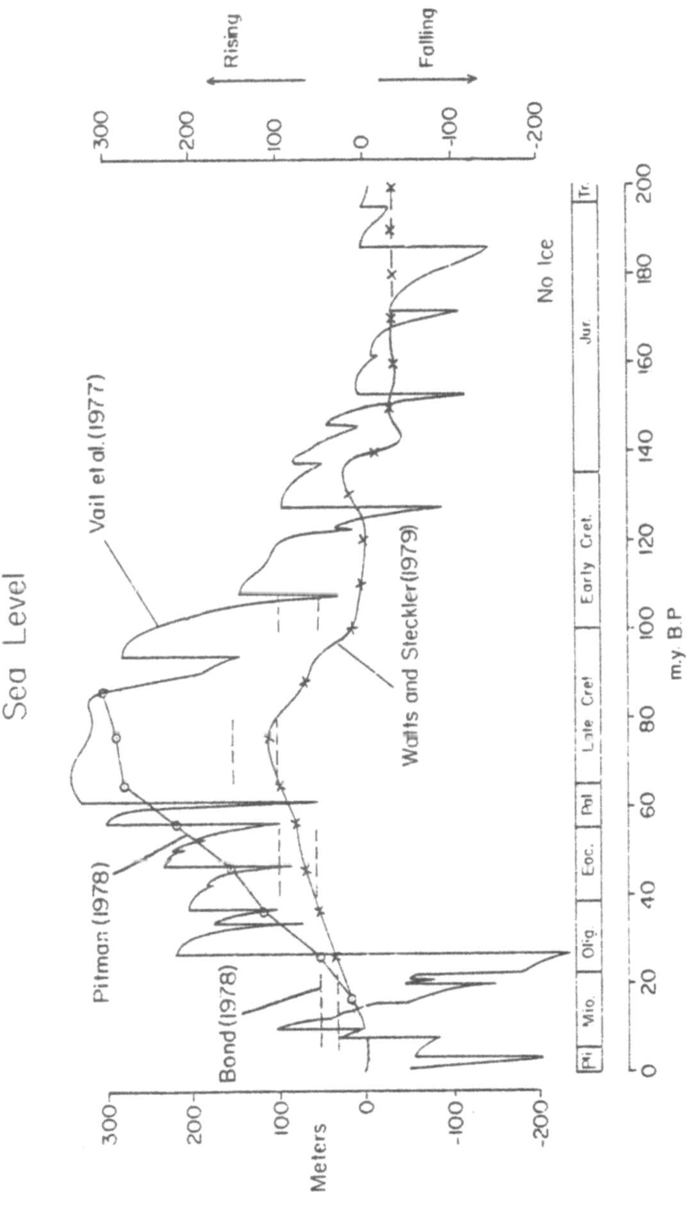

FIG. 2 – Comparison of various recently proposed sea level curves.

stratigraphy. Lower estimates of sea level stand during the Late Cretaceous come from studies of continental flooding and continental margins.

In order to estimate past values of sea level from the present elevation of older beds, it is necessary to correct for changing water depths, sediment deposition, and isostatic and tectonic movements. Sleep (26) chose a location in Minnesota at a craton location away from tectonic activity and estimated that there had been a sea level drop of approximately 300 m between the Coniacian and today. Results based on the study of isolated locations must be viewed with caution. Bond (2) noted that comparisons of paleogeographic maps and present topography suggest broad epeirogenic movements which may not be detectable by studying individual locations. Hancock and Kauffman's (13) estimate of a 650 m sea level rise during the Late Cretaceous illustrates many of the sources of error that need to be considered in this type of study. For example, by neglecting to correct their estimate for isostatic adjustment due to sediment and water loading, they introduce errors in excess of 200 m. They use only data from stable massifs, yet the elevation of their Albian reference surface varies by over 400 m and over 200 m after correcting for sediment loading.

Calculation of sea level from reconstruction of variations in mid-ocean ridge volumes has also yielded large estimates for the Cretaceous height of sea level. The validity of this method is limited by uncertainties in our knowledge of seafloor spreading rates and ridge lengths. Many of the ridges which need to be included in the modeling have since been subducted. Hays and Pitman (16) obtained a height of 521 m above present-day for sea level at 85 m.y.b.p. Pitman (26), using an improved data set, revised this estimate to 350 m for the same period (Fig. 2), in good agreement with Sleep (29). Parsons (25), however, analyzed the oceanic area-age distribution and variations in the rate of seafloor generation and could produce only a 150 m variation in sea level. He considers the occurrence of the much larger changes needed to produce sea level changes of 300-350 m to be unlikely (Parsons, personal communication). Kominz (18) has recently redone Pitman's (26) calculations incorporating new global reconstruction and time scales and has found a sea level drop of 230 ± 100 m since the Late Cretaceous. Factors other than mid-ocean ridge volumes may be significant in these calculations. Hager (9) has suggested that sea level will also be strongly affected by the location of subduction, i.e., whether subduction occurs beneath island arcs or continents. Harrison et al. (15) estimated that variable sedimentation rates and orogeny contribute a 60 m correction to the ridge volume estimates, while Schlanger et al. (28) have suggested that extensive mid-plate volcanism requires an additional 40-80 m of sea level rise.

Lower sea level estimates for the Late Cretaceous come from numerous studies of continental flooding and continental margins. Estimates on this basis generally range from ca. 180 to 200 m (19, 48). The precision of quantitative estimates is limited by the need to know how the hypsometric curve has varied with time. Bond (1, 2) developed a method for identifying changes of hypsometry in individual continents. He used modal averages of sea level estimates to identify continents whose shape has been modified by tectonic activity and for which the present hypsometric curve gave anomalous results, and concluded that the maximum sea level stand was ~+170 m. Harrison et al. (15) developed an empirical relationship between the hypsometric curve and continent size to extrapolate continental hypsometry into the past, but this curve is not very sensitive to the elevations where flooding occurs (0-300 m).

Watts and Steckler (45) derived a sea level curve based on the history of the subsiding continental margin of eastern North America (Fig. 2). However, by fitting a decaying exponential they probably removed some of the long wavelength components of sea level changes. Their curve has a shape similar to that of Pitman (26) or Bond (1), but with an amplitude of only 109 m. Attempts to apply this method to the North Sea (46, 47) found that the use of no sea level correction at all was better in this area than one with a high stand of sea level in the Late Cretaceous. This is probably due to the fact that the tectonic subsidence of the North Sea was more complex, particularly during the Paleocene (27), than had been assumed. Hardenbol et al. (14) applied the technique to North Africa, where most of the Tertiary section is missing and where recent uplift has occurred; they estimated that sea level has fallen 281 m since Late Cretaceous time.

Recently, Watts and Thorne (in preparation) have modeled the subsidence of the Baltimore Canyon using a thermal model which includes lateral heat flow and flexure (32, 33), compaction with variable lithologic constants, and erosion of exposed beds. They have produced synthetic stratigraphic sections using several different sea level curves. When they included Vail et al's. (40) first-order curve (calibrated to Pitman (26)) the result was a margin without Tertiary sediment accumulation. This occurred because the 350 m sea level fall between the Late Cretaceous and the present is as large as the 300-400 m of tectonic subsidence which occurs on the continental shelf during this period. Thus, there is little to no relative movement of the basement relative to sea level and hence no sediment accumulation.

On the other hand, the use of a very low amplitude sea level curve from Watts and Steckler (45) also failed to reproduce the stratigraphy correctly. With only 109 m of sea level fall, the amplitude was too low to produce the coastal onlap pattern and the pinchout of sedimentary layers on the margin. All of the sedimentary horizons extended across the shelf and coastal plain, nearly to the fall line. In order to reproduce the large changes in landward extent of coastal onlap, the maximum rates of sea level fall need to be similar to the rates of tectonic subsidence, but the long-term rates must be lower; otherwise no sediment accumulates. Watts and Thorne (in preparation) estimate that a sea level fall of 145 m between the Late Cretaceous and the beginning of the Miocene (83-25 m.y.b.p.) agrees best with the observed stratigraphy. It should be noted, however, that in Watts and Thorne's (in preparation) model the bathymetry of the shelf and the location of the shoreline are kept constant. Variations in water depth are one of the major effects of eustatic changes and may affect the results, although the estimated depths of water in the Late Cretaceous are similar to those observed at present.

CYCLICITY OF SEA LEVEL

The variation of sea level with time has also generated its share of debate. Estimates of the number of rises and falls, their timing, and the rate of sea level change all vary from author to author. Vail et al. (40) subdivided sea level changes into a hierarchy of cycles of relative rises and falls. On the largest scale, they identify two first-order cycles during Phanerozoic time. The first-order cycles begin during the Late Precambrian-Cambrian and during the Triassic-Jurassic; they coincide with times of major continental plate breakup, in agreement with the hypothesis of plate tectonic control of the stand of sea level. Vail et al.'s (40) supercycles range in duration from 10-80 m.y. A number of authors have suggested a cyclicity in the Earth's history on the order of 60 m.y. (36, 43) or 32 m.y. (8). Vail et al.'s (40) supercycles closely match Sloss' (30) sequences for the Paleozoic.

Sloss and Speed (31) have suggested that Sloss' (30) sequences are not sea level cycles, but a consequence of global tectonic cycles. This again raises the question of separating absolute sea level changes from tectonics. Models of flexure at continental margins (32-34) predict coastal onlap following rifting. Watts (44) showed that several of Vail et al.'s (40) supercycles coincide with major episodes in the breakup of Gondwanaland, suggesting that part of the record of coastal onlap at the supercycles scale is due to tectonics and not sea level.

At a finer scale the time transgressive nature of facies changes has at times caused difficulties in correlation and has resulted in the proliferation of recognized sea level events. Thus the number of transgressive episodes observed in the Cretaceous varies from one (49) to fifteen (4). Cooper (4) assumes that transgressions are extremely rapid. According to Hancock and Kauffman (13) this accounts for Cooper's identification of the same transgressive sequence as three events by observing it in different locations. Hancock and Kauffman (13) find clear evidence for only five Cretaceous transgressions.

Vail et al. (40) consider seismic sequence boundaries to be synchronous. Whether or not this is the case has yet to be demonstrated conclusively. Certainly there is a good worldwide correlation of Vail's major unconformities (e.g., (20)). The frequency of sequence boundaries in many cases, however, appears to approach or exceed the resolution of worldwide biostratigraphic data for dating and correlation. Hallam (10,11) derived a eustatic curve for the Jurassic using facies analysis and estimates of areal flooding. His curve shows many similarities to Vail's (39), both in timing and magnitude, although differences do exist. Olsson et al. (24) have proposed that the major eustatic fall in the Tertiary occurred in the Late Eocene to Early Oligocene in the eastern U.S. coastal plain, rather than during the Late Oligocene as found by Vail et al. (40); the time discrepancy between the two interpretations is ~7 m.y.

Initially, Vail et al. (40) equated their coastal onlap curve with sea level (Fig. 3), implying that eustatic variations commonly consist of slow rises followed by rapid falls. This proposal brought about much discussion in the geologic community (6, 10) and has caused a revision of their eustatic curve (38, 39), by incorporating other concepts (10, 26, 42). Vail and Todd (41) subdivided unconformities into three types and used corrected subsidence curves to determine long-term changes in sea level. This has resulted in a "rounding-off" of the shape of their eustatic curves relative to their coastal onlap curve and has made the rates of sea level rise and fall more similar (Fig. 3). Some of their eustatic sea level falls, such as the one in the Late Oligocene, are still extremely large and rapid, and cannot be explained by any known mechanism. Their use of a single subsidence curve, in areas where the mapped onlap patterns extend for tens of kilometers across strike with large differential subsidence, is

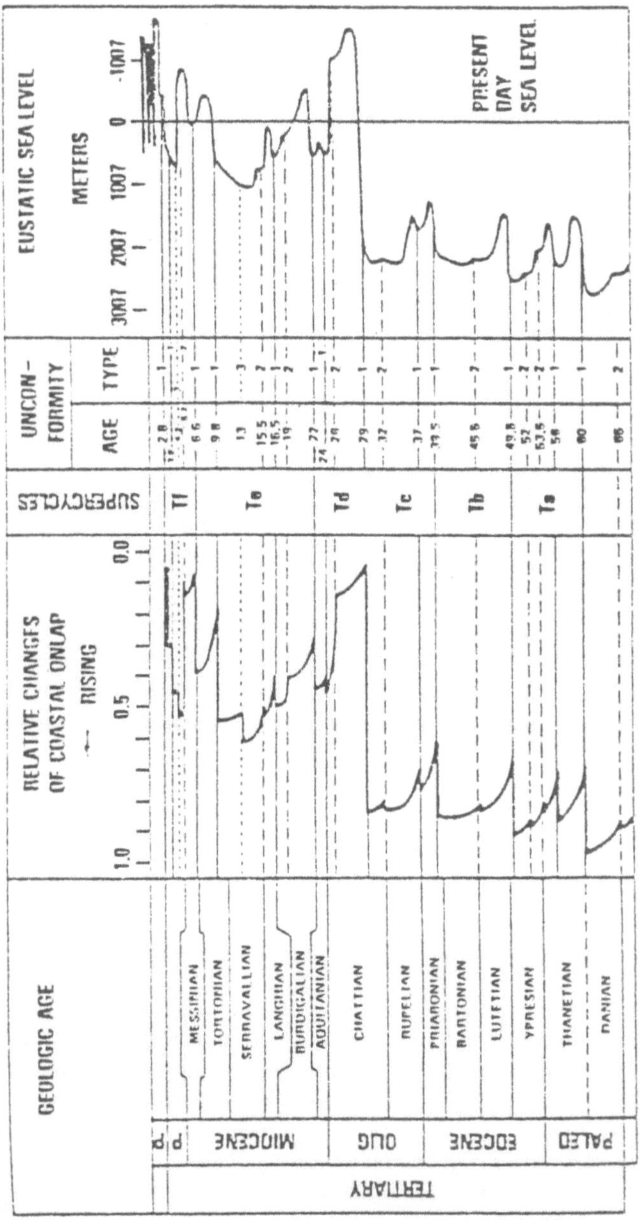

FIG. 3 – Comparison of global cycles of relative coastal onlap for Cenozoic (40) with interpreted eustatic changes of sea level (37).

probably inadequate for determining absolute sea level. Vail and Hardenbol (38) still show large, rapid falls of sea level (Fig. 3). These are probably too large to be explained by any known mechanism outside of glaciation (6). The large magnitude of the short-term variations gives the Vail curves their spasmodic appearance. The rapid Late Oligocene fall is equal in magnitude (350 m) to the entire rise that occurred during the Cretaceous. If rapid changes, other than glacial change, prove to be as large as this, then sea level variations are indeed highly spasmodic.

Pitman ((26) and personal communication) interprets the coastal onlap and strand line observations differently. He interprets variations in the position of the shoreline as being due to changes in rates of relative sea level change (26). In subsequent modeling he allowed for the deposition of nonmarine coastal plain sediments. Coastal plain sediments were found to onlap even during a relative fall in sea level as long as the rate of fall was increasing. Vail et al. (40) noted that they commonly observed regressions during the later part of a cycle to coastal onlap. The results of an exercise in modeling making use of these observations are shown in Fig. 4 (Pitman, personal communication). Using a continously falling sea level curve with sharp variations in the rate of change of sea level, Pitman was able to reproduce a pattern of transgressions, regressions, and coastal onlap similar to Vail et al.'s (40). The sea level curve used is intended simply as an experiment to investigate the possibility of a different interpretation of the Vail et al. (40) data. It demonstrates that the observed global coastal onlap patterns could be due to much smoother sea level variations than has been suggested. Turcotte and Kenyon (37) have modeled the stratigraphy produced by sea level changes and found that coastal onlap could be produced either by slow rises of sea level followed by rapid fall or by rapid rises followed by slow falls. They concluded that the rate of erosion rather than the shape of the sea level curve was the dominant factor.

Clearly, further work is needed to determine whether sea level variations are smooth or spasmodic. Many of the available methods for determining sea level have neither the time nor the absolute amplitude resolution to yield the answer. Detailed facies mapping and the observational record of coastal onlap through seismic stratigraphy do seem to have the necessary resolution, but more work is needed on the relationship between coastal onlap and eustatic sea level.

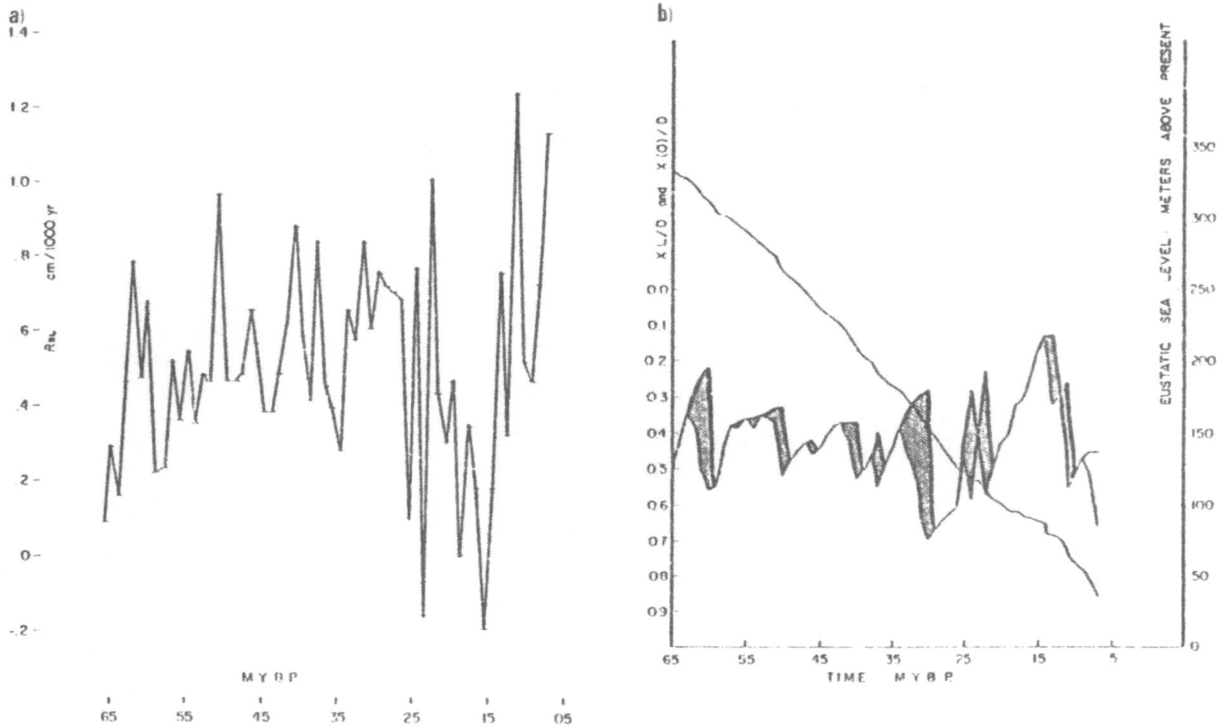

FIG. 4 - a) Rate of sea level change estimated from Vail et al. (40) assuming regression at the end of each coastal onlap cycle. b) Sea level curve, and coastal onlap and shoreline position calculated from rate of sea level change in a). Hatched regions are coastal sediments bounded by maximum extent of coastal onlap (top) and the shoreline (bottom) (Pitman, personal communication).

SUMMARY

While the last decade has seen advances in both theoretical and observational techniques, global agreement on a eustatic curve remains elusive. Attempts to quantify sea level changes have resulted in a number of disparate estimates; most of these fall between 150-200 m or between 300-350 m. Uncertainties in each technique introduce a large amount of scatter, and we appear to be deficient in our knowledge both of the interactions of sea level and of the geologic record. A major problem is the difficulty of separating tectonics and eustacy.

There is now widespread agreement regarding the long-term (200-300 m.y.) variations of sea level. On shorter time scales this agreement breaks down. Some transgressive and regressive episodes appear to be rapid and spasmodic, but it is unclear whether this is also true of the causal eustatic changes. Seismic stratigraphy suggests that rapid eustatic variations have occurred, but modeling of continental margin stratigraphy suggests that relatively smooth variations of sea level can reproduce the observed spasmodic patterns of onlap and offlap. Thus, there is a pressing need to understand the response of the stratigraphic record to sea level changes in the presence of a dynamic Earth.

Acknowledgements. I would like to thank G. Bond, K. Deffeyes, G. Karner, J. Thorne, and A.B. Watts for critical reviews of the manuscript and M. Kominz, W. Pitman, J. Thorne, and A.B. Watts for allowing me to use unpublished material. This work was supported by Lamont-Doherty Geological Observatory Institutional Funds and a grant from Arco. L-DGO Contribution Number 3491.

REFERENCES

(1) Bond, G. 1978. Speculations on real sea-level changes, and vertical motions of continents at selected times in the Cretaceous and Tertiary periods. Geology 6: 247-250.

(2) Bond, G.C. 1979. Evidence for some uplifts of large magnitudes in continental platforms. Tectonophys. 61: 285-305.

(3) Cogley, J.G. 1981. Late Phanerozoic extent of dry land. Nature 291: 56-58.

(4) Cooper, M.R. 1977. Eustacy during the Cretaceous: its implications and importance. Peleogeog. Paleoclimat. Paleoecol. 22: 1-60.

(5) Curray, J.R. 1964. Transgressions and regressions. In Papers in Marine Geology, ed. R.L. Miller, Shepard Commemorative Volume, pp. 175-203. New York: Macmillan Co.

(6) Donovan, D.T., and Jones, E.J.W. 1979. Causes of worldwide changes in sea level. J. Geol. 136: 87-192.

(7) Fairbridge, R.W. 1961. Eustatic changes in sea level. In Physics and Chemistry of the Earth, eds. L.H. Ahrens et al., vol. 4, pp. 99-185. London: Pergamon Press.

(8) Fischer, A.G., and Arthur, M.A. 1977. Secular variations in the pelagic realm. In Deep-water Carbonate Environments, eds. H.E. Cook and P. Enos. Spec. Publ. Soc. Econ. Paleont. Min. 25: 19-50.

(9) Hager, B.H. 1980. Eustatic sea level and spreading rate are not simply related. E.O.S. 61: 374.

(10) Hallam, A. 1978. Eustatic cycles in the Jurassic. Paleogeog. Paleoclimat. Paleoecol. 23: 1-32.

(11) Hallam, A. 1981. A revised sea-level curve for the Jurassic. J. Geol. Soc. Lond. 138: 735-743.

(12) Hancock, J.M. 1975. The sequence of facies in the Upper Cretaceous of northern Europe compared with that in the Western Interior. In The Cretaceous System in the Western Interior of North America, ed. W.G.E. Caldwell. Spec. Paper Geol. Assoc. Can. 13: 83-118.

(13) Hancock, J.M., and Kauffman, E.G. 1979. The great transgression of the Late Cretaceous. J. Geol. Soc. Lond. 136: 175-186.

(14) Hardenbol, J.; Vail, P.R.; and Ferrer, R. 1981. Interpolating paleoenvironments, subsidence history and sea-level changes of passive margins from seismic and biostratigraphy. Proceedings of the 26th International Geological Congress, Geology of Continental Margins Symposium, Paris, July 7-17, 1980. Oceanol. Act.: 33-44.

(15) Harrison, C.G.A.; Brass, G.W.; Saltzman, E.; Sloan, II, J.; Southham, J.; and Whitman, J.M. 1981. Sea level variations, global sedimentation rates and the hypsographic curve. Earth Planet. Sci. Lett. 54: 1-16.

(16) Hays, J.D., and Pitman, III, W.C. 1973. Lithosphere plate motion, sea-level changes and climatic and ecological consequences. Nature 246: 18-22.

(17) Jeletsky, J.A. 1977. Causes of Cretaceous oscillations of sea level in Western and Artic Canada and some general geotectonic implications. Paleon. Soc. Japan Spec. Paper 21: 233-346.

(18) Kominz, M.A. 1983. Oceanic ridge volumes and sea level changes - an error analysis. In Interregional Unconformities and Hydrocarbon Accumulations, ed. J. Schlee, AAPG Memoir. Tulsa, OK: AAPG, in press.

(19) Kuenen, P.H. 1940. Causes of eustatic movements. 6th Pacific Science Congress Proceedings, vol. 2, pp. 833-837. Berkeley: University of California.

(20) Loutit, T.S., and Kennett, J.P. 1981. New Zealand and Australian Cenozoic sedimentary cycles and global sea-level changes. Am. Assoc. Petr. Geol. Bull. 65: 1586-1601.

(21) Mörner, N.-A. 1976. Eustasy and geoid changes. J. Geol. 84: 123-151.

(22) Mörner, N.-A. 1980. Relative sea level changes, tectono-eustasy, geoidal eustasy and geodynamics during the Cretaceous. Cretaceous Res. 1: 329 340.

(23) Mörner, N.-A. 1981. Revolution in Cretaceous sea-level analysis. Geology 9: 344-346.

(24) Olsson, R.K.; Miller, K.G.; and Ungrody, R.E. 1980. Late Oligocene transgression of middle Atlantic coastal plain. Geology 8: 549-554.

(25) Parsons, B. 1982. Causes and consequences of the relation between area and age of the ocean floor. J. Geophys. Res. 87: 289-302.

(26) Pitman, III, W.C. 1978. The relationship between eustacy and stratigraphic sequences of passive margins. Geol. Soc. Am. Bull. 89: 1389-1403.

(27) Rochow, K.A. 1981. Seismic stratigraphy of the North Sea "Palocene" deposits. In Petroleum Geology of the Continental Shelf of Northwest Europe, eds. L.V. Illing and G.D. Hobson, pp. 255-266. London: Heyden and Sons, Ltd.

(28) Schlanger, S.O.; Jenkyns, H.C.; and Premoli-Silva, I. 1981. Volcanism and vertical tectonics in the Pacific Basin related to global Cretaceous transgressions. Earth Planet. Sci. Lett. 52: 435-449.

(29) Sleep, N.H. 1976. Platform subsidence and eustatic sea level changes. Tectonophys. 36: 45-56.

(30) Sloss, L.L. 1963. Sequences in the cratonic interior of North America. Geol. Soc. Am. Bull. 74: 93-114.

(31) Sloss, L.L., and Speed, R.C. 1974. Relationships of cratonic and continental-margin tectonic episodes. In Tectonics and Sedimentation, eds. W.R. Dickenson. SEPM Spec. Publ. 22: 98-119.

(32) Steckler, M.S. 1981. The thermal and mechanical evolution of Atlantic-type continental margins. Ph.D. Thesis, Columbia University.

(33) Steckler, M.S. 1984. Two-dimensional considerations in the subsidence of continental margins. J. Geophys. Res., in press.

(34) Steckler, M.S., and Watts, A.B. 1982. Subsidence history and tectonic evolution of Atlantic-type continental margins. In Dynamics of Passive Margins, ed. R.A. Scrutton, vol. 6, pp. 184-196. Washington, D.C.: A.G.U.

(35) Suess, E. 1906. The Face of the Earth, vol. 2. Oxford: Clarendon Press.

(36) Terry, K.D., and Tucker, W.H. 1968. Biologic effects of supernovae. Science 159: 421-423.

(37) Turcotte, D.L., and Kenyon, P.M. 1983. Synthetic seismic stratigraphy. AAPG Bull., in press.

(38) Vail, P.R., and Hardenbol, J. 1979. Sea-level changes during the Tertiary. Oceanus 22: 71-79.

(39) Vail, P.R.; Hardenbol, J.; and Todd, R.G. 1982. Jurassic unconformities and global sea-level changes from seismic and biostratigraphy. Houston Geol. Soc. Bull. (Sept.): 3-4.

(40) Vail, P.R.; Mitchum, R.M.; Todd, R.G.; Widmier, J.M.; Thompson, III, S.; Sangree, J.B.; Bubb, J.N.; and Hatlelid, W.G. 1977. Seismic stratigraphy and global changes of sea-level. In Seismic-stratigraphy Applications to Hydrocarbon Exploration, ed. C.E. Payton, pp. 49-212, AAPG Memoir 26. Tulsa, OK: AAPG.

(41) Vail, P.R., and Todd, R.G. 1981. Northern North Sea Jurassic unconformities, chronostratigraphy and sea-level changes from seismic stratigraphy. In Petroleum Geology of the Continental Shelf of North-west Europe, pp. 216-235. London: Heyden and Son, Ltd.

(42) Van Hinte, J.E. 1978. Geohistory analysis-application of micropaleontology in exploration geology. Am. Assoc. Petr. Geol. 62: 201-222.

(43) Vogt, P.R. 1972. Evidence for global synchronism mantle plume convection and possible significance for geology. Nature 240: 338-342.

(44) Watts, A.B. 1982. Tectonic subsidence, flexure and global changes of sea level. Nature 197: 469-474.

(45) Watts, A.B., and Steckler, M.S. 1979. Subsidence and eustasy at the continental margin of eastern North America. In Deep Drilling Results in the Atlantic Ocean: Continental Margins and Paleoenvironments, eds. M. Talwani, W. Hay, and W.B.F. Ryan, pp. 218-234, Ewing Symposium Series 3. Washington, D.C.: AGU.

(46) Wood, R.J. 1981. The subsidence history of Conoco well 15/30-1, central North Sea. Earth Planet. Sci. Lett. 54: 306-312.

(47) Wood, R.J. 1982. Subsidence in the North Sea. Ph.D. Thesis, University of Cambridge.

(48) Wise, D.U. 1974. Freeboard and the volumes of continents and oceans through time. In The Geology of Continental Margins, eds. C.A. Burk and C.L. Drake, pp. 45-58. New York: Springer Verlag.

(49) Yanshin, A.L. 1973. About the so-called global transgressions and regressions. Byull. mosk. Obshch. Ispyt. Prir. (Otdel. geol.) 48: 9-45.

Patterns of Change in Earth Evolution, eds. H.D. Holland and A.F. Trendall, pp. 123-143.
Dahlem Konferenzen 1984. Berlin, Heidelberg, New York, Tokyo: Springer-Verlag.

Gradual and Abrupt Shifts in Ocean Chemistry During Phanerozoic Time

W.T. Holser
Dept. of Geology, University of Oregon
Eugene, OR 97403, and University of New Mexico
Albuquerque, NM 87131, USA

Abstract. The clearest records of changes in chemistry of the exogenic cycle are found in mineral inventories (NaCl, $CaSO_4$, C_{carb}, C_{org}, P), isotope ratios ($\delta^{34}S_{sft}$, $\delta^{13}C_{carb}$, $^{87}Sr/^{86}Sr_{carb}$, $^{87}Sr/^{86}Sr_{apt}$), and trace elements (Ce/La_{apt} and heavy metals in black shales) vs. age. While these variations can be simplistically modelled in the long-term to confine all variations to the larger sedimentary reservoirs, there are several reasons to assert that some of the variability is internal to the smaller oceanic (and atmospheric) reservoirs, especially for short-term events. These are controlled by complex feedback loops, perhaps ultimately forced by plate-tectonic activity cycles. Many links are only speculative.

INTRODUCTION

There is good reason to believe that the world ocean as a whole has not undergone radical long-term changes in most aspects of its chemical composition during Phanerozoic time (e.g., (24)). However, considerable changes certainly did take place in the geochemical cycles of certain elements, in certain areas, and at certain times. The variations range from secular changes on a time scale of 10^8 years to abrupt events that are virtually instantaneous in the geological record.

The chemistry of the oceans is intimately involved, on even the shortest of these time scales, with that of the biota, of the atmosphere, and of sediments (at least those newly deposited), and these should all be considered together as an interacting system, sometimes referred to as the "exogenic cycle" of the Earth. In particular instances, interactions with volcanism, especially that of newly forming mid-ocean ridges, may be important. In many cases evidence adduced for changes will refer directly only to the exogenic system as a whole, and only indirect evidence or modelling will be able to distinguish what part of the variations apply to the ocean specifically.

THE GEOCHEMICAL RECORD OF CHANGES IN OCEAN CHEMISTRY

Mineralogical Markers

Chemical changes may be monitored by qualitative but decisive differences in the mineralogy of sediments deposited during different geological ages. Some of these "mineralogical markers" are summarized in Table 1; this overview demonstrates that while the changes in the mineralogy of sediments are rather clear, the interpretation of their causes in terms of changes in oceanic or atmospheric chemistry is not.

Rock Inventories

In addition to the above qualitative shifts of mineralogy, one may also examine the variations in the mass of the several rock components with time (Fig. 1).

The mass of some evaporite bodies is so large and they can be deposited so quickly that their formation may be an important removal event for NaCl and $CaSO_4$ in the world ocean (Fig. 1a). Vast volumes of oceanic salts accumulated in a few hundred thousand years in giant, preexisting basins in the Upper Permian. As a result, mean oceanic salinity (mainly due to NaCl) dropped abruptly by 1 to 4‰ four or five times between the Permian and the Cretaceous (26, 43). The deposition of any one of these giant evaporites was a geological accident depending on the concatenation of unlikely paleogeographic, tectonic, and sea level conditions; the record clearly demonstrates that these removals were not periodic. Many authors (e.g., (32, 45)) have suggested that the drops in salinity during the Permian were the main cause of the dramatic biotic extinctions in the Late Permian; the data of Fig. 1a suggest, however, that even greater salinity drops occurred during the Mesozoic.

Transfers from the oceans to evaporites (Fig. 1a) was even more important for the sulfate content of seawater than for its chloride content; the shifts in the operation of the sulfur cycle that are indicated in the sulfur isotope age curve (see below) probably owe a good deal to variations in the rate of sulfate output to evaporites.

TABLE 1 - Some mineralogical markers of possible changes in ocean chemistry.

	Observation	Interpretation and Reference
1.	Mineralogy and its sequence invariable in marine evaporites of all ages.	Main elemental ratios in seawater have not varied by more than a factor of two or three (24, 25).
2.	High dolomite/calcite in Proterozoic, decreasing through Paleozoic, to zero in Cretaceous.	Higher Mg/Ca, or higher $p(CO_2)$, or higher shelf/deep sedimentation (48), or competition by high MOR for Mg (hence lower Mg/Ca), in Paleozoic seas (25).
3.	Calcite in pre-Carboniferous öoides, aragonite thereafter.*	Higher $p(CO_2)$ (at constant marine Mg/Ca) in earlier atmosphere (34).
4.	K-rich illite in pre-Carboniferous shales, Na-rich smectite thereafter.	Higher K/Na in earlier seas, or less leaching by plants, or high $p(CO_2)$ (50), or differing igneous provenance (24).

*D.K. Richter (unpublished data) has determined a much more complex distribution of aragonite and calcite acids with time.

Evaporite basins were also important in redistributing chemical components in the ancient oceans, by the effect of the reflux of saline brine on deep oceanic circulation. The brine sources had a substantial buoyancy flux (product of mass flux and density difference) that was able to drive the consequent water packet to oceanic depths. Outpourings from Permian evaporite basins had buoyancy fluxes estimated to range from that of the present outflow of the Red Sea to twice that of the Mediterranean Sea (26). In the absence of a strong latitudinal temperature contrast, warm salty bottom waters generated in evaporite basins (1, 37), or even on shallow subtropical shelves (1, 8, 51), replaced or inhibited cold polar waters and led to a more stable stratification of the deep ocean. All such bottom waters would have had a relatively low initial oxygen content.

Carbon (Fig. 1b and c) is a central element in the exogenic geochemical cycle (7), in particular with respect to the distribution of carbon through time between organic carbon (C_{org}: living, dead, or buried) and oxidized carbon (C_{carb}: atmospheric, dissolved in seawater, precipitated, or buried). The distribution of C_{carb} between atmosphere and ocean, at any one

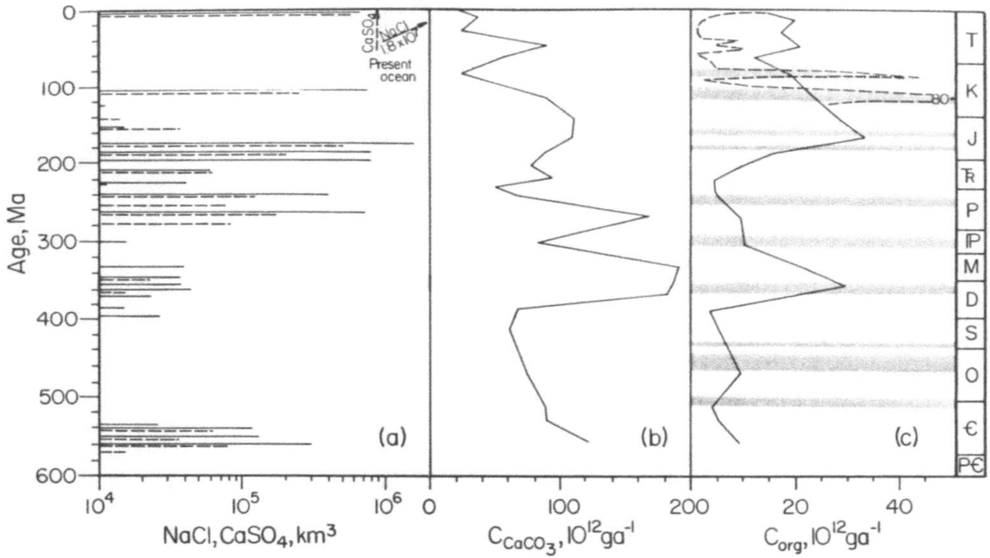

FIG. 1 - Historical inventories of some marine chemical sediments. (a) NaCl and $CaSO_4$ in marine evaporites, scaled logarithmically; content in the present ocean is also shown. Data from Zharkov (54) for Cambrian through Pennsylvanian time, and from Holser et al. (26) for the Permian through the Cenozoic. (b) Rate of deposition of C_{carb}. (c) Rate of deposition of C_{org}. Solid lines in (b) and (c) are carbon in sediments present today on the continents, aggregated to epochs (11); dashed line in (c) is based mainly on deep-sea cores and aggregated by stages or substages (2). Shaded areas in (c) are times of recognized anoxic events (1, 2, 6). Note that the scale in (c) is expanded relative to (b).

time, depends on the requirement that two HCO_3^- in seawater dissolve or precipitate one mineral $CO_3^=$, so that the seawater and atmospheric reservoirs of C_{carb} increase while the mineral reservoir decreases. In the long term the balance of C_{org} and C_{carb} is shifted by the deep burial of both kinds of carbon, and by the volcanic or metamorphic release of CO_2.

Dominant peaks in both C_{org} and C_{carb} (solid lines in Fig. 1b and c) are evident in Middle Jurassic and Late Devonian time. Such massive burial could not have been accomplished without substantial resupply of CO_2 and probable large variations of atmospheric CO_2 and O_2 (10), and these data may be questioned. The dashed line in Fig. 1c shows shorter term trends, from the Jurassic to the present, by more precise and finely-timed data from deep-sea cores. On an even finer scale, rates for C_{carb} vary by a factor of five and those of C_{org} by a factor of one hundred, in a few million years (43). Figure 1c also shows qualitatively the times of "anoxic events," for which abnormally large (but mostly unmeasured) amounts of black shales rich in C_{org} have been observed. Where these manifestations can be closely correlated with analyses for C_{org} in the Cretaceous (see Fig. 1c) it is clear that they represent very short but very high rates of carbon accumulation (1).

Phosphorite is another chemical sediment whose rate of deposition has varied significantly with time; economic accumulations have important maxima in Cambrian, Ordovician, Permian, Late Jurassic-Early Cretaceous, Paleocene-Eocene, and Miocene (14). The deposition rate of phosphorite is qualitatively correlated with the deposition of biogenic silica, but no clear relation to anoxic events (3).

Isotope Age Curves

"Age curves," displaying variations of isotope ratios with time, have been one of the most fruitful approaches to the study of seawater paleochemistry. Hoefs (23) reviewed these relations, and my Fig. 2 recapitulates them in terms of the most recent information. The basic data, of, e.g., the sulfur curve, are analyses of $\delta^{34}S$ (deviations of $^{34}S/^{32}S$ from a meteoritic standard, in parts per thousand) for sulfate minerals from marine evaporite rocks. The primary source of variation in sulfur isotopes of marine surface waters, as exemplified in the curve, is the varying degree of biologically mediated reduction of sulfate to sulfide, with a consequent fractionation of about -40‰. S is distributed among the participating reservoirs: sulfate (S_{sft}) in the world ocean and new

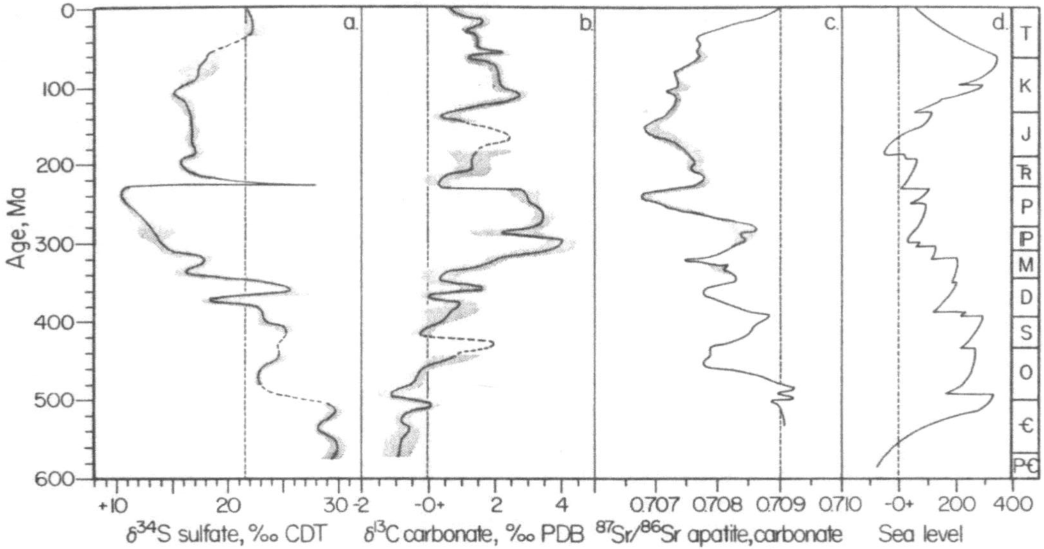

FIG. 2 - Age curves of (a) sulfur isotopes in evaporite sulfate; (b) carbon isotopes in carbonate; (c) strontium isotopes in carbonate and in fossil apatite; and (d) sea level. Shading gives range of uncertainty; dashed lines, lack of data. Sources: (a), (b), all published data aggregated in 10 Ma intervals (38), with finer details in (a) for the Devonian (12) and for the Triassic from Fig. 3; (c) for the Cambrian through the Permian from conodont and ichthyolith data (31) and for the Permian to present from mainly limestone data (11); and (d) second-order cycles of Vail and Mitchum (47).

evaporites on the one hand and sulfide (S_{sfd}) in newly deposited sulfides (eventually pyrite) on the other hand. Variations of $\delta^{34}S$ internal to an evaporite basin, due to local $SO_4^{=}$ reduction to $S^{=}$, can usually be spotted by comparing the $\delta^{34}S$ of sulfates from contemporaneous but widely separated evaporite basins.

The isotope age curve of C_{carb} (Fig. 2b) shows variations that are smaller, less well defined, and subject to variations of taxa (Berger et al., this volume), mineralogy, and diagenesis, but that are nonetheless real. The fundamental source of the variations in carbon isotopes is parallel to that of sulfur. Reduction of bicarbonate in seawater to organic carbon through photosynthetic pathways fractionates carbon about -25‰, and the proportion that is buried and preserved as C_{org} (actually net over erosional input) determines the magnitude of isotopic shift. Isotopic shifts in the carbon system may be complicated by diagenetic inputs from fermentative methane generation, which is usually subordinate but may be important locally.

Unlike S and C, whose biological fractionations could generate their isotope age curves internally to the exogenic cycle, Sr isotopes do not fractionate; Sr inputs to the oceans of differing isotopic composition are therefore required to explain the variations in the $^{87}Sr/^{86}Sr$ ratio of seawater. The main inputs are young Sr, largely from basaltic rocks, from mid-ocean ridges (MOR), with $^{87}Sr/^{86}Sr \simeq 0.704$, old Sr from granitic rocks on the world's cratons, with $^{87}Sr/^{86}Sr \simeq 0.720$ (e.g., (16)), and Sr from Phanerozoic limestones with an $^{87}Sr/^{86}Sr$ ratio of 0.706 - 0.709.

In these age curves the existence of long-term secular variations are shown by the sweep of the S and Sr curves to highs in the early Paleozoic and lows in the Permian-Triassic, and in the C_{carb} curve from a low in the early Paleozoic to a high in the Pennsylvanian-Permian. In accordance with the first-order model mentioned above, this implies that in the early Paleozoic, more S was removed from the oceans as S_{sfd}, leaving the measured S_{sft} isotopically heavy, while less C was transferred into the C_{org} reservoir, allowing the measured C_{carb} to be light. Toward the end of the Paleozoic, S shifted to S_{sft}, and C shifted to C_{org} (and Ca shifted from C_{carb} to S_{sft}). The marine Sr reservoir was influenced more strongly by old granitic sources in the early Paleozoic and by new basaltic sources in the Permian.

The best documented short-term excursion of the S-isotope age curve is also the most puzzling. At the end of Early Triassic time, $\delta^{34}S$ rose from its record low of $\delta^{34}S$ = +10.5‰ in the Late Permian, to about +28‰(Fig. 2a). As shown in Fig. 3, the worldwide extent of this anomaly has been confirmed abundantly. Nearly all of the rise in $\delta^{34}S$ occurred within the Spathian substage of the Scythian stage, and detailed correlations indicate that the rise took place within a fraction of Spathian time, a time interval probably no longer than 500 ka, and possibly much shorter. Other sharp rises in the sulfur isotope age curve occur during the Frasnian (Late Devonian) and Albian (Early Cretaceous) (12).

The best documented excursions of the carbon isotope age curve are provided by data from deep marine carbonates in DSDP cores and related onshore exposures. When allowance has been made for differences related to taxa and water depth, some of the important shifts of $\delta^{13}C$ are: (a) a rise of 2.4‰(Fig. 2b) or locally as much as twice that (Fig. 4a) during 30 Ma of the Early Cretaceous, accompanied by large-scale deposition and preservation of C_{org} (29); (b) a sharp downward but very brief deflection at the Cretaceous/Tertiary boundary (discussed by Padian et al., this volume); and (c) a rise of nearly 1‰ during about 3 Ma of the early Miocene and an equally sharp drop at the end of the Miocene (Fig. 4b; see also (39)). These and other events in the Cenozoic are more clearly identified by their $\delta^{18}O_{CaCO_3}$ signature, and by their relationship to oceanic circulation, to glaciation, and to other climatic factors that are discussed in detail elsewhere (4, 9). An extraordinarily large and fast rise of $\delta^{13}C$, near the end of Permian time, occurs in both the Zechstein Basin in northwestern Europe (36) and in the Delaware Basin of southwestern USA, where a rise of 8‰ can be demonstrated by varve chronology to have occurred in only 4.5 ka (35). Presumably this reflects a massive deposition of C_{org}.

One would expect the age curves of the oxidized forms of sulfur and carbon (Fig. 2a, b) to be paralleled at lower values by analogous isotope age curves for the reduced sulfide and organic carbon. Unfortunately, these curves are not well-defined. Claypool et al. (13), using data from black shales and oils, presented a preliminary curve of $\delta^{34}S_{sfd}$ that mimics many features of Fig. 2a. Only a very generalized curve has been compiled for $\delta^{13}C_{org}$ (40).

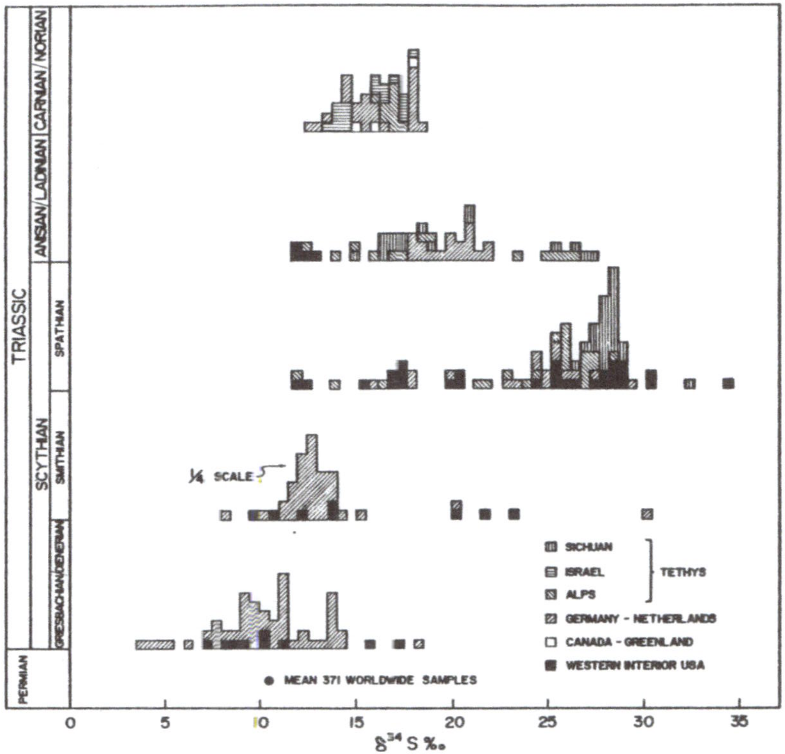

FIG. 3 - A short-term shift in $\delta^{34}S$: the Röt Event, a sharp rise of $\delta^{34}S$ at the end of the Early Triassic (12). Each square is one analysis, except for the Smithian, which is factored by four. Note greatly expanded time scale in the Scythian Stage (Early Triassic). The initial detection of the event in Germany and the Netherlands (references in (12)) is confirmed by abundant recent analyses from widely separated areas (Holser in (21)).

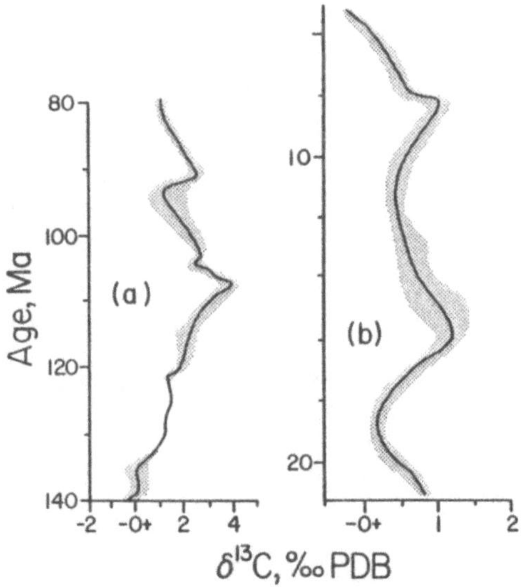

FIG. 4 - Examples of short-term shifts in $\delta^{13}C$: (a) Rise and subsequent decline of $\delta^{13}C$ in the Cretaceous, as measured by deep-sea sediments exposed at Peregrina Canyon, Sierra Madre Oriental, Mexico (42); (b) Variations of $\delta^{13}C$ in the Miocene, as measured in benthic foraminifera from DSDP Site 289, western equatorial Pacific ((53) - the paper also included data on $\delta^{18}O$). Note that both the time and the isotope scale are larger in diagram (b) than in (a).

Trace Elements

Trace elements in marine sediments, and particularly certain element ratios, may contain useful information about changes in ocean chemistry (25). Table 2 reviews some of the variations that seem most consistent.

Correlations

Garrels and Perry (19) recognized that the continuity of the fossil record puts limits on the possible variations of O_2 and CO_2 in the atmosphere, and that consequently major shifts of oxidation-reduction in the sulfur cycle must have been balanced by complementary reduction-oxidation shifts in the carbon cycle. A nominal overall equation is:

$$8(SO_4)^{2-} + 2Fe_2O_3 + 8H_2O + 15C_{org} = 4FeS_2 + 15CO_2 + 16(OH)^-. \quad (1)$$

TABLE 2 - Variations of trace element content of sedimentary rocks with time.

Elements	Variations	Interpretation and Reference
Sr/Ca in foraminiferal calcite	Gradual rise of 15% in Cenozoic; substantial dips in Eocene and Late Miocene.	Shifts in river chemistry, or rate of Ca input from MOR, or changes in deposition rate of Sr-rich aragonite (20).
Ce/La in marine fossil apatite	Negative Ce anomaly (or low Ce/La) absent in Ordovician through Mississippian, present in Late Paleozoic and Recent; Th/U similar.	Prevailing reduction in in Early Paleozoic seas releases insoluble Ce^{4+} as Ce^{3+} or precipitates soluble U^{6+} as U^{4+} (52).
Metals in black shales	Mo/C_{org}, U/C_{org}, and concentrations of metals in shales over seawater, all higher (up to 10x), especially in Paleozoic, than in present Black Sea.	Higher metal content of Early Paleozoic seas, or undetermined factors (reviewed in (25)).

They proposed several mechanisms that might couple the two systems and stabilize atmospheric $p(O_2)$, at least in the long term. Then, assuming a perfect balance of the two systems and well-mixed sedimentary reservoirs, they hypothesized that the S_{sft} isotope age curve should be mirrored, with a multiplication factor appropriate to the stoichiometry of Eq. 1, by an isotope age curve for C_{carb}. A substantial confirmation of this relation was found in a statistical treatment of all available sulfur and carbon isotope data (49), an updated version of which is shown in Fig. 5. When smoothed for 10 Ma intervals, and within the relatively large scatter of the carbon data, the two isotope age curves are negatively correlated, with a slope corresponding approximately to Eq. 1. This implies that the carbon and sulfur systems are nearly balanced.

However, the negative correlation of $\delta^{34}S$ and $\delta^{13}C$ is poor even during some intervals as long as 10 Ma. In Fig. 5 an age in Ma marks those correlation points that lie inexcusably far (more than $3\bar{\sigma}$) from the correlation line. A prominent example occurs in Late Devonian time, where the dip at 365 Ma and the peak at 355 Ma in $\delta^{34}S$ are matched

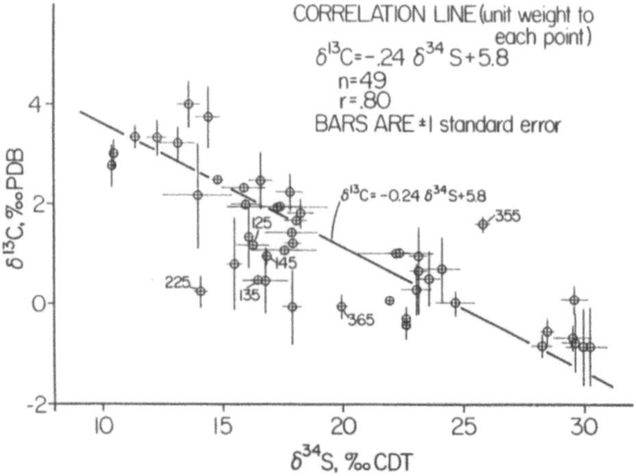

FIG. 5 - Correlation of $\delta^{13}C_{carb}$ and $\delta^{34}S_{sft}$ for published data aggregated in 10 Ma intervals (the fine structure in the Devonian and Triassic added to the $\delta^{34}S$ curve of Fig. 2a is not included here). Numbers indicate mean ages for intervals that plot more than three standard errors from the correlation lines (49).

by a like wobble in $\delta^{13}C$ (Fig. 5). During time intervals shorter than 10 Ma, variations in these two isotopic systems can go in any direction. In Late Permian time the very sharp rise in $\delta^{13}C$ (35) coincides with a period of substantially constant $\delta^{34}S$; in Early Triassic time a sharp rise in $\delta^{34}S$ (Fig. 3) is contemporaneous with a period of substantially constant $\delta^{13}C$; and in mid-Cretaceous time a rise in $\delta^{34}S$ (Fig. 2a) is matched by a rise in $\delta^{13}C$ (Fig. 4a). These examples dramatize the possible importance of short-term events as causes of shifts in atmospheric and/or oceanic chemistry.

The rises of $\delta^{13}C$ and $\delta^{34}S$ were probably caused by an excessive rate of deposition of the reduced species of these elements; the consequence of such excesses should therefore be present in the sedimentary record. Indeed, a positive correlation is evident between $\delta^{13}C$ (Fig. 2b) and mean C_{org} (Fig. 1c), although the contrary has been asserted (30). These are also at least qualitatively correlated with mean C_{carb} (Fig. 1b, and (43)) - what this probably means is that all are a result of total sedimentation rate. Deposition of S_{sft} (Fig. 1a) is evidently episodic and not correlated with any of the other measured variables, although correlations have been suggested (19, 30, 34). Correspondingly detailed data for S_{sfd} are not available.

The maxima in the Ce/La (and possibly the Th/U) ratio measured in fossil apatite, postulated as related to oceanic anoxia, are correlated approximately with $\delta^{34}S$ (52). Approximate correlations of the age curves of sulfur and strontium isotopes, and also of world sea level, were discovered independently by several investigators (e.g., (23)). Factor analysis (21) of an earlier, more time-smoothed version of Fig. 2 reiterates the inverse relation of $\delta^{34}S$ and $\delta^{13}C$. This analysis also concludes that external control by tectonic processes, as expressed by both sea level and the contribution of weathering to $^{87}Sr/^{86}Sr$, seems to be coupled through the carbon system and only indirectly to the sulfur oxidation-reduction system.

PROXIMATE CAUSES AND SIMPLE MODELS

To recapitulate the proximate causes mentioned in the previous sections: Marine $\delta^{34}S$ and $\delta^{13}C$ rise because of a net excess output of S_{sfd} and C_{org}, respectively; $^{87}Sr/^{86}Sr$ rises because of a net excess of input of old (granitic) compared to young (basaltic) material. A simple box model might contain a single-box mixed ocean with input/output (I/O) to reduced and oxidized sedimentary reservoirs. The I/O to the reduced reservoir (mainly shales) for both S and C is biologically mediated and is controlling for the isotope ratios of all reservoirs; the I/O with the oxidized reservoirs (evaporites and limestone, respectively) is essentially passive with respect to isotope ratios (no fractionation). Both I/O's are controlling with respect to the composition of the ocean (plus atmosphere). Historically only the changes of $\delta^{34}S$ were known; consequently, in order to model the transfers of S (and the history of reservoir masses) that accounted for the curve, it was necessary to assume some combination of the following: Fixed mass of oceanic sulfate, fixed flux and/or $\delta^{34}S$ of river input, or evaporite sulfate output proportional to mass of oceanic sulfate. Most models include the carbon cycle as well, with the additional assumptions that Eq. 1 has been exactly balanced, and that the O_2 and CO_2 content of the atmosphere has been constant.

As a recent example, Garrels and Lerman (18) started from a steady state model of the present S/C system and calculated backward through time the sulfur transfers between evaporite and sulfide mandated by the shifts of $\delta^{34}S$. Then, assuming Eq. 1 and a constant atmosphere, they calculated the corresponding shifts of carbon between C_{carb} and

C_{org}. Because of the correlation of Fig. 5, they therefore reproduced the carbon isotope age curve. The calculated evaporite reservoir varied from 125% of its size in Carboniferous time to 46% in Cambrian time; the fluxes of sulfate deposition required to attain these reservoir (and $\delta^{34}S$) levels varied from 150% to 15% of the putative present steady state. The other reservoirs and fluxes varied less than those, on a percentage basis, as the reservoirs involved were larger. Schidlowski and Junge (41) obtained similar results with a different assortment of assumptions and a more sophisticated model.

While these calculations are very interesting, they do not tell us much about possible variations in the composition of the ocean and atmosphere, because they assume both a uniform and constant composition of both reservoirs. The feedback mechanisms that have been proposed for maintaining such uniformity and constancy are operable at best only qualitatively in the long term. Furthermore, these models ignore what little data are available on the actual changes with time of these reservoirs (see Fig. 1). The models do not pretend to calculate the effects of short events such as illustrated in Figs. 3 and 4, or the departures from Eq. 1 that are evident even on intermediate time scales in Fig. 5. Real deviations from Eq. 1, as between 355-365 Ma (Fig. 5), indicate a pulse of O_2 (or CO_2) production (above the line) or consumption (below the line).

The models described above also assume a well mixed ocean. At least during anoxic events (Fig. 1c), stratification of the ocean may be important for some constituents, especially for short-term changes (e.g., (12, 46)). However, a theoretical model of the deep-sea cycles of C and O (44) indicates that stagnation (low circulation velocity) is not a requirement for oceanic anoxia, especially when it occurs high in the water column. These models will find application in the explanation of anoxic events and related short-range changes of chemistry when they have been expanded to include quantitative estimates of the flux of phosphate and of carbon and sulfur and their isotopes. So far, no models of this sort have been designed.

Additional I/O with MOR hydrothermal systems may be important for the cycles of some elements. The observational and experimental data are too voluminous to be covered here. In essence, water that has circulated through recent basalts and has emerged at hot springs is greatly depleted in Mg and SO_4 and enriched in Ca, K, Li, etc.; Sr is essentially invariant (15). If the rate of seawater circulation in all ridges is calculated from heat flow and anomalous He data, the changes measured in the hot springs add up to a substantial fraction of their river input to the oceans, and could be the mysterious missing sink for the river input of Mg. However, several facts indicate that the transfer into oceanic crust is not massive (27): a) The sulfur content of altered basalts is not more but a little less than that of fresh basalts; b) Retrograde solubility of anhydrite at even moderate temperatures should have reduced the SO_4 content of seawater to a fraction of its present value, and deposited large masses of anhydrite in sub-sea basalt; or c) If the circulating sulfate underwent reduction, that would have required the oxidation of all silicate FeO in several kilometers of basalt. Control or buffering of seawater composition by MOR activity is potentially substantial, but not yet demonstrated. If confirmed, then major shifts of the composition of seawater should be consequent to variations in the rate of MOR activity. These variations may be recorded by shifts of sea level (Fig. 2d) via the thermal expansion associated with MOR activity.

FEEDBACK, OSCILLATIONS, AND EXTERNAL FORCINGS

Berger (4) summarized feedback systems in the water and carbon systems and suggested ways in which such interactions could reinforce or amplify weak external forcings (such as orbital perturbations). Positive feedback not only increases the sensitivity of the system to external stimuli, but given appropriate time constants, the system may overshoot its new equilibrium level and may even go into a damped self-oscillation. Opposing feedback systems may be difficult to sort out logically. For example: Sea level rise might decrease erosion and nutrient supply to the sea, which lowers C_{org} production and leaves more atmospheric CO_2, leading to "greenhouse" warming and (if ice is present) reinforced sea level rise (positive feedback). However, alternatively, sea level rise might increase the area of shelf and marginal seas, in which C_{org} production and preservation can be prodigious, leading to extraction of atmospheric CO_2, cooling, and slowed sea level rise (negative feedback).

Climatic oscillations with quasi-regular periods of 100 ka, 43 ka, and 21 ka have been ascribed to orbital forcings of changing insolation (28), and both these and shorter climatic cycles (especially 2.5 ka) leave records that are most evident in fluctuations of $\delta^{18}O_{carb}$ but that are also detected in $\delta^{13}C_{carb}$ and in C_{org} through various causal chains (9). Sedimentation and the composition of the atmosphere and oceans undergo changes related to these shifts.

Over a longer time span, the ultimate cause of cycles such as those recorded in Figs. 1 and 2 is commonly attributed to time variations of the rate of mantle convection - and its surface expression in plate motions, MOR activity, and sea level changes. According to this type of hypothesis (10, 17, 30, 34), the 300 Ma-cycle that is so evident in Figs. 1 and 2, as well as in the incidence of glaciation and many other paleoclimatic factors, is due to an alternation between two states. Mid-Paleozoic and mid-Mesozoic times were dominated by high MOR activity, high sea level, fast plate motion and subduction, high CO_2-release from both MOR and continental border volcanism, high surface temperatures, low latitudinal contrast, no glaciation, high incidence of oceanic anoxia, and high C_{org}. In contrast, the late Proterozoic and the late Paleozoic (plus Triassic) were characterized by low MOR activity, low sea level, plates collected into a single, motionless continent, low CO_2 release, low surface temperatures with high contrast and glaciation, good polar circulation to the deep ocean, and low C_{org}. Minor variations on this theme may also be responsible for modulation of this first-order cycle by oscillations on a quasi-periodic interval of about 30 Ma (17, 47).

In a specific sense, plate motions also affect oceanic circulation by the opening and closing of seaways (22, 46).

Unfortunately, some of the most explicit records are inconsistent with this grand scheme. During high MOR activity, the strontium isotope age curve (Fig. 2d) should move toward low values commensurate with a $^{87}Sr/^{86}Sr \cong 0.703$ for MOR basalts. This would indicate a high MOR effect at the end of the Paleozoic rather than the low MOR, based on the sea level curve, in the above hypothesis. Also, it is difficult to understand why $\delta^{13}C$ and C_{org} are down when $\delta^{34}S$ and S_{sfd} are up. Rises in both C_{org} and S_{sfd} should result from extensive oceanic anoxia. A qualitative positive correlation of high $\delta^{34}S$ in the early Paleozoic with high Ce/La and with observed black shale sections suggests that a high $\delta^{34}S$, rather than in $\delta^{13}C$, is a measure of the long-term swing into anoxia. Leventhal (33) proposed that under "Black Sea conditions," anoxia in the water column would add S_{sfd} in excess of the weight ratio C_{org}/S_{sfd} = 2.5 found in normal marine sediments. Berner and Raiswell (5) added that a deficit of S_{sfd} relative to this norm would obtain in sediments deposited in fresh water, which is sulfate-poor. Variations in the importance of these two departures from the oxic norm may go a long way towards explaining the anti-correlation of $\delta^{34}S$ and $\delta^{13}C$, but the mechanism by which they are linked to an ultimate forcing function such as tectonic activity is still not clear.

Acknowledgements. This work was supported by U.S. National Science Foundation Grants EAR 872-1819 and EAR 811-5985 to the University of Oregon, and by the Caswell Silver Distinguished Professorship in Geology at the University of New Mexico. For unpublished information included in this survey, I am indebted to R.Y. Anderson, M.A. Arthur, R.A. Berner, W.B.N. Berry, G.W. Brass, G.C. Claypool, H.D. Holland, J. Kovach, J.S. Leventhal, T.B. Lindh, M. Magaritz, W.H. Peterson, R. Raiswell, E.S. Saltzman, R.S. Seymour, J.R. Southam, J. Veizer, P. Wilde, C.K. Wilgus, and J. Wright.

REFERENCES

(1) Arthur, M.A. 1979. Paleoceanographic events - recognition, resolution, and reconsideration. Rev. Geophys. Space Phys. 17: 1474-1494.

(2) Arthur, M.A. 1982. The carbon cycle: Controls on atmospheric CO_2 and climate in the geologic past. In Climate in Earth History, eds. W.H. Berger and J.C. Crowell, pp. 55-67. Washington: National Academy Press.

(3) Arthur, M.A., and Jenkyns, H.C. 1981. Phosphorites and paleoceanography. Ocean. Acta SP: 83-96.

(4) Berger, W.H. 1982. Deep-sea stratigraphy: Cenozoic climate steps and the search for chemo-climatic feedback. In Cyclic and Event Stratification, eds. G. Einsele and A. Seilacher, pp. 121-157. Berlin: Springer-Verlag.

(5) Berner, R.A., and Raiswell, R. 1983. Burial of organic carbon and pyrite sulfur in sediments over Phanerozoic time: A new theory. Geochim. Cosmochim. Acta 47: 855-862.

(6) Berry, W.B.N., and Wilde, P. 1978. Progressive ventilation of the oceans - an explanation for the distribution of the lower Paleozoic black shales. Am. J. Sci. 278: 257-275.

(7) Bolin, B.; Degens, E.T.; Kempe, S.; and Ketner, P., eds. 1979. The Global Carbon Cycle. New York: John Wiley and Sons.

(8) Brass, G.W.; Southam, J.R.; and Peterson, W.H. 1982. Warm saline bottom water in the ancient ocean. Nature 296: 620-623.

(9) Broecker, W.S. 1982. Ocean chemistry during glacial time. Geochim. Cosmochim. Acta 46: 1689-1705.

(10) Budyko, M.I., and Ronov, A.B. 1979. Chemical evolution of the atmosphere in the Phanerozoic. Geokhimiya 5: 643-653 (transl. Geochim. Internat. 16(3): 1-9).

(11) Burke, W.H.; Denison, R.E.; Heatherington, E.A.; Koepnick, R.B.; Nelson, H.F.; and Otto, J.B. 1982. Variation of seawater $^{87}Sr/^{86}Sr$ throughout Phanerozoic time. Geology 10: 516-519.

(12) Claypool, G.E.; Holser, W.T.; Kaplan, I.R.; Sakai, H.; and Zak, I. 1980. The age curves of sulfur and oxygen isotopes in marine sulfate and their mutual interpretation. Chem. Geol. 28: 199-260.

(13) Claypool, G.C.; Leventhal, J.S.; and Goldhaber, M.B. 1980. Geochemical effects of early diagenesis of organic matter and sulfur in Devonian black shales, Appalachian Basin. Abstract. Am. Ass. Petrol Geol. Bull. 64: 692.

(14) Cook, P.J., and McElhinny, M.W. 1979. A reevaluation of the spatial and temporal distribution of sedimentary phosphate deposits in the light of plate tectonics. Econ. Geol. 74: 315-330.

(15) Edmond, J.M.; Measures, C.; McDuff, R.E.; Chan, L.H.; Collier, R.; Grant, B.; Gordon, L.I.; and Corliss, J.B. 1979. Ridge crest hydrothermal activity and the balances of the major and minor elements in the ocean: The Galapagos Rift. Earth Planet. Sci. Lett. 46: 1-18.

(16) Faure, G. 1977. Isotope Geology. New York: John Wiley and Sons.

(17) Fischer, A.G. 1982. Longterm climatic oscillations recorded in stratigraphy. In Climate in Earth History, eds. W.H. Berger and J.C. Crowell, pp. 97-103. Washington: National Academy Press.

(18) Garrels, R.M., and Lerman, A. 1981. Phanerozoic cycles of sedimentary carbon and sulfur. Proc. Natl. Acad. Sci. USA 78: 4652-4656.

(19) Garrels, R.M., and Perry, E.A. 1974. Cycling of carbon, sulfur, and oxygen through geologic time. In The Sea, ed. E.D. Goldberg, vol. 5, pp. 303-336. New York: John Wiley and Sons.

(20) Graham, D.W.; Bender, M.L.; Williams, D.F.; and Keigwin, Jr., L.D. 1982. Strontium-calcium ratios in Cenozoic planktonic foraminifera. Geochim. Cosmochim. Acta 46: 1281-1292.

(21) Gregor, C.B., ed. 1984. Chemical Cycles in the Evolution of the Earth. New York: John Wiley and Sons, in press.

(22) Haq, B.U. 1981. Paleogene paleoceanography: Early Cenozoic oceans revisited. Oceanol. Acta SP: 71-82.

(23) Hoefs, J. 1981. Isotopic composition of the ocean-atmosphere system in the geologic past. Am. Geophys. Union Geodynam. Ser. 5: 110-118.

(24) Holland, H.D. 1974. Marine evaporites and the composition of seawater during the Phanerozoic. Soc. Econ. Paleontol. Mineral. Spec. Paper 20: 187-192.

(25) Holland, H.D. 1984. The Chemical Evolution of the Atmosphere and Oceans. Princeton: Princeton University Press, in press.

(26) Holser, W.T.; Hay, W.W.; Jory, D.E.; and O'Connell, W.J. 1980. A census of evaporites and its implications for oceanic geochemistry. Abstract. Geol. Soc. Am. Abstr. Progr. 12: 449.

(27) Holser, W.T.; Kaplan, I.R.; Sakai, H.; and Zak, I. 1979. Isotope geochemistry of oxygen in the sedimentary sulfate cycle. Chem. Geol. 25: 1-17.

(28) Imbrie, J., and Imbrie, J.Z. 1980. Modelling the climatic response to orbital variations. Science 207: 943-953.

(29) Kennett, J.P. 1982. Marine Geology. Englewood Cliffs: Prentice Hall.

(30) Keith, M.L. 1982. Violent volcanism, stagnant oceans and some inferences regarding petroleum, strata-bound ores and mass extinctions. Geochim. Cosmochim. Acta 46: 2621-2637.

(31) Kovach, J. 1980. Variations in the strontium isotope composition of seawater during Paleozoic time determined by analysis of conodonts. Abstract. Geol. Soc. Am. Abstr. Progr. 12: 465.

(32) Lantzy, R.J.; Dacey, M.F.; and Mackenzie, F.T. 1977. Catastrophe theory: Application to the Permian mass extinction. Geology 5: 724-728.

(33) Leventhal, J.S. 1983. An interpretation of carbon and sulfur relationships in Black Sea sediments as indicators of environments of deposition. Geochim. Cosmochim. Acta 47: 133-137.

(34) Mackenzie, F.T., and Pigott, J.D. 1981. Tectonic controls of Phanerozoic sedimentary rock cycling. J. Geol. Soc. Lond. 138: 183-196.

(35) Magaritz, M.; Anderson, R.Y.; Holser, W.T.; Saltzman, E.S.; and Garber, J. 1984. Isotope shifts in the Late Permian of the Delaware Basin, Texas, precisely timed by varved sediments. Earth Planet. Sci. Lett., in press.

(36) Magaritz, M., and Turner, P. 1982. Carbon cycle changes in the Zechstein Sea: isotopic transition zone in the Marl Slate. Nature 297: 389-390.

(37) Roth, P.H., and Bowdler, J.L. 1981. Middle Cretaceous calcareous nannoplankton biography and oceanography of the Atlantic Ocean. Soc. Expl. Paleontol. Mineral. Spec. Publ. 32: 517-546.

(38) Saltzman, E.S.; Lindh, T.B.; and Holser, W.T. 1982. $\delta^{13}C$ and $\delta^{34}S$, global sedimentation, pO_2 and pCO_2 during the Phanerozoic. Abstract. Geol. Soc. Am. Abstr. Progr. 14: 607.

(39) Savin, S.M., and Yeh, H.W. 1981. Stable isotopes in ocean sediments. In The Oceanic Lithosphere, ed. C. Emiliani, pp. 1521-1554. New York: John Wiley-Interscience.

(40) Schidlowski, M. 1982. Content and isotopic composition of reduced carbon in sediments. Phys. Chem. Sci. Rep. 1982 3: 103-122.

(41) Schidlowski, M., and Junge, C.E. 1981. Coupling among the terrestrial sulfur, carbon and oxygen cycles: Numerical modeling based on revised Phanerozoic carbon isotope record. Geochim. Cosmochim. Acta 45: 589-594.

(42) Scholle, P.A., and Arthur, M.A. 1980. Carbon isotope fluctuations in Cretaceous pelagic limestones: potential stratigraphic and petroleum exploration tool. Am. Ass. Petrol. Geol. Bull 64: 67-87.

(43) Southam, J.R., and Hay, W.W. 1981. Global sedimentary mass balance and sea level changes. In The Oceanic Lithosphere, ed. C. Emiliani, pp. 1616-1684. New York: Wiley-Interscience.

(44) Southam, J.R.; Peterson, W.H.; and Brass, G.W. 1982. Dynamics of anoxia. Paleogeo. P. 40: 183-198.

(45) Stevens, C.H. 1977. Was development of brackish oceans a factor in Permian extinctions? Geol. Soc. Am. Bull. 88: 133-138.

(46) Thierstein, H.R., and Berger, W.H. 1977. Injection events in ocean history. Nature 276: 461-466.

(47) Vail, P.R., and Mitchum, Jr., R.M. 1979. Global cycles of relative changes of sea level from seismic stratigraphy. Am. Ass. Petrol. Geol. Mem. 29: 469-472.

(48) Veizer, J. 1978. Secular variations in the composition of sedimentary carbonate rocks, II. Fe, Mn, Ca, Mg, Si and minor constituents. Precambrian Res. 6: 381-413.

(49) Veizer, J.; Holser, W.T.; and Wilgus, C.K. 1980. Correlation of $^{13}C/^{12}C$ and $^{34}S/^{32}S$ secular variations. Geochim. Cosmochim. Acta 44: 579-587.

(50) Weaver, C.E. 1967. Potassium, illite and the ocean. Geochim. Cosmochim. Acta 31: 2181-2196.

(51) Wilde, P., and Berry, W.B.N. 1982. Progressive ventilation of the oceans - potential for return to anoxic conditions in the post-Paleozoic. In Nature and Origin of Cretaceous Carbon-rich Facies, eds. S.O. Schlanger and M.G. Cita, pp. 209-224. New York: Academic Press.

(52) Wright, J.; Seymour, R.S.; and Shaw, H.F. 1984. REE and Nd isotopes in conodont apatite: variations with geological age and depositional environment. Geol. Soc. Am. Spec. Paper, in press.

(53) Woodruff, F.; Savin, S.M.; and Douglas, R.G. 1981. Miocene stable isotope record: A detailed deep Pacific Ocean study and its paleoclimatic implications. Science 212: 665-668.

(54) Zharkov, M.A. 1981. History of Paleozoic Salt Accumulation. Berlin: Springer-Verlag.

Patterns of Change in Earth Evolution, eds. H.D. Holland and A.F. Trendall, pp. 145-157.
Dahlem Konferenzen 1984. Berlin, Heidelberg, New York, Tokyo: Springer-Verlag.

Biological Innovations and the Sedimentary Record

A.G. Fischer
Dept. of Geological and Geophysical Sciences
Princeton University, Princeton, NJ 08544, USA

Abstract. Biological innovations which are likely to have influenced the evolution of the biosphere and to have been recorded in the sedimentary record include the invention of oxygen-liberating photosynthesis, those of iron oxidation and of the other chemolithotrophic activities of bacteria (nitrogen fixation, nitrate reduction, and sulfate reduction), the development of metazoans, the invention of burrowing, the evolution of calcareous and of siliceous skeletons, the colonization of the lands, the appearance of calcareous skeletons in the plankton, the appearance of siliceous skeletons in the phytoplankton, the invention of flowers, and the evolution of conceptual thought. Some of these are well recorded in sediments, others remain to be identified or seem not to have left the predicted effects. Inventions appeared in batches correlated with events in lithospheric evolution.

INTRODUCTION

In a Universe that follows the second law of thermodynamics, tending toward greater and greater dispersion and randomness of matter and energy, organisms are small eddies of high order and improbability, maintained against the current by deflecting energy from the general stream. From one initial eddy of this sort, in the dawn of Precambrian time, these eddies have multiplied in kind (there are now probably in excess of two million species), in strength (e.g., the complexity of Homo), and in numbers (estimated at 5×10^{22} individuals).

This has come about by continued experimental innovation which has resulted in a) a progressively more efficient utilization of available energy, b) a progressively wider use of available materials (nutrients), and c) a progressive spread of organisms to wider ranges of environment.

In sexual organisms, every individual is a minor experiment in innovation. But in this paper we are concerned with major innovations, with evolutionary inventions, using that term in its literal meaning of "coming into," without implying a conscious search. The inventions discussed here are those which may be expected to have left direct effects in the record, whether or not these effects have been observed.

The inventions of autotrophy and heterotrophy are lost in dimness of early history: we do not even know with certainty whether the earliest organisms were heterotrophs (as now generally believed) or autotrophs (as maintained by some heretics). Whatever the history, autotrophy has been the base of the food chain during most of life history, and heterotrophs play a vital role in completing the organic carbon cycle. But the qualitative history of these processes has not been smooth and has involved the innovative steps discussed below. They appear amongst the other inventions which are taken up in the order of historical appearance.

1. OXYGEN-LIBERATING PHOTOSYNTHESIS

The invention of a photosynthetic process by which water is split and oxygen is liberated has modified the Earth more profoundly than any other biological innovation, because it has provided the Earth with an aerobic atmosphere and hydrosphere. Not only is most life as we know it obligatorily aerobic, but weathering and sedimentation have changed with the advent of an oxidizing fluid envelope.

The first records of presumably photosynthetic oxygen-liberating organisms (stromatolite remains of presumed cyanobacteria) are from Archean rocks that are some 3,500 Ma old (2). Evidence of at least local aerobic conditions in the hydrosphere is seen in the Archean ferric banded iron formations (BIF) of 3400-2900 Ma ago (14). However, the very vastness of those iron deposits in later Archean time (about 2500 to 1800 Ma ago) may be evidence of a larger reservoir of unoxidized hydrosphere, which was able to store dissolved ferrous iron. Occurrence of redbeds - detrital, largely alluvial sediments - from 1900 Ma ago on (9) suggests that the atmosphere had become oxidizing by that date.

Parts of the hydrosphere have remained in the reduced state up to the present. This is attributable largely to the activity of biotic reduction processes: heterotrophy by bacteria, fungi, and animals as well as to bacterial sulfate reduction. Whether hydrothermal activity has been a significant factor remains to be seen. The distribution of the hydrosphere between reduced and oxidized portions has varied through time (8, 10, 15, 16). This may in part be recorded in the sulfur isotope curve (Fig. 1) (10, 16). Waves of iron deposition, such as the three waves of BIF deposition in the Precambrian (14), and waves of chamosite-glauconite deposition in Phanerozoic time (9) may mark times when large volumes of reduced water became oxidized in response to inorganic factors such as changes in the vigor of oceanic circulation (6).

2. IRON OXIDATION AND OTHER CHEMOLITHOTROPHIC ACTIVITIES OF BACTERIA

The invention of chemolithotrophy by bacteria has led to a biotic modification of many geochemical cycles. They include, above all, the precipitation of ferric iron in sediments (18). Less well understood is the history of the bacteria which mobilize iron in the zone of weathering by oxidizing pyrite. Manganese bacteria go back at least to 2,000 Ma (3). The activities of various sulfur bacteria (22, 23) also fit under this heading.

3. NITROGEN FIXATION AND NITRATE REDUCTION

Nitrogen fixation by various bacterial groups is essential for the maintenance of fertility and must have been invented early in the history of life. Reduction of such nitrates, in anaerobic waters and sediments, is less remarkable for what it did than for what it prevented: a) the conversion of the oceans into a nitrate brine, from which saltpeter would have precipitated in evaporite settings; b) depletion of the atmosphere in nitrogen, resulting in a greatly diminished atmospheric volume and pressure. The Earth would have evolved very differently under these conditions.

4. SULFATE REDUCTION

Sulfate reduction (22, 23) by bacteria saved the world from accumulating a skin of gypsum. Unlike nitrate, the reduction of sulfate results indirectly in the precipitation of a mineral, pyrite, which selectively removes the lighter isotope, as recorded in the famous sulfur isotope curve (Fig. 1). Isotopic values suggest that sulfate reduction may have been initiated about 2,750 Ma ago. Sulfate reduction through history preserved parts

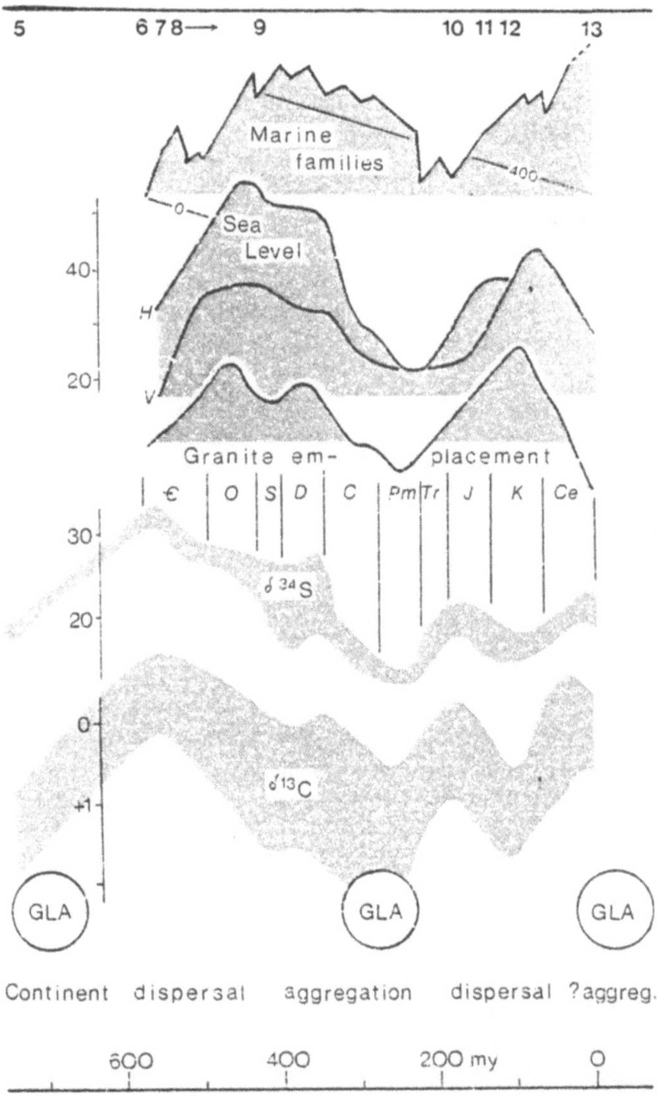

FIG. 1 - Patterns in Earth history. From top: inventions, keyed to text. Numbers of marine metazoan families, plotted on slant to remove trend (21). Sea level curves in % of continents flooded (5, 6); H: Hallam curve (12); V: Vail curve (24). Granite emplacement in North America (4). Sulfur isotope curve from marine sulfates, Carbon isotope curve from marine limestones (14). GLA: major glaciations.

of the hydrosphere in a highly reduced, hydrogen sulfide-rich state (sulfuret). In the Phanerozoic, this is recorded as the finely laminated carbon-rich "pontic" or "euxinic" facies most commonly exemplified by the pyritic "black shales." The proportion of the hydrosphere in the sulfuret state has varied markedly through time (10), partly as a result of physical changes such as the area and the circulation rate of the marine hydrosphere, partly due to biotic inventions such as photosynthetic oxygen release and bioturbation - factors which induced reciprocal fluctuations in the size of the sulfate and sulfide reservoirs and therefore in the isotopic composition of dissolved sulfur at sea.

5. INVENTION OF METAZOANS - GENERAL

About 650-600 Ma ago, a major series of biological inventions produced macroscopic, mobile, heterotrophic organisms - the metazoans. These first organisms as preserved in the Ediacarian faunas were soft-bodied, fed on the existing masses of planktonic and presumably also benthonic protists and bacteria, and must have greatly accelerated the organic carbon cycle by shortcutting the route from autotrophs and tiny protistan predators to decomposers. This was speeded even more when metazoans evolved their own predator-prey cycles. From Ediacarian time on, rapid turnover of marine producer organisms has been the rule. The invention of grazing may well have caused the decline in stromatolite diversity which has been noted in the latest Precambrian. The concentration of phosphorus in the Late Precambrian-Phanerozoic phosphorite deposits is probably a direct outcome of this recycling of organic matter through animals whose tissues are preferentially buried in poorly aerated bottoms associated with belts of high productivity (upwelling).

6. INVENTION OF BIOTURBATION

While the invention of metazoans speeded the return of organic carbon to the inorganic state in the open water masses, organic matter buried in the sediment was safe from them through Ediacarian time. Intense bioturbation begins at the base of the Cambrian; since that time aerobic sediments the world over have been churned up. The mechanical inventions which permitted them to invade sediments were chiefly of two kinds: the development of exoskeletons, such as those possessed by arthropods, which allowed them to dig into sediments by means of their appendages; and the evolution of coelomes that could be used a "hydroskeletons," allowing worm-shaped organisms to penetrate sediments in the manner of the earthworm. Metazoans penetrated the sediment for two purposes: in search of food, and in search of protection from predators. The effect in both instances was a churning of sediments and the loss of the finer sedimentary structures. An indirect effect was the return of more buried organic matter into the carbon cycle: leading to a decrease in the amount of organic matter buried, and a corresponding decrease in net generation of oxygen per unit of organic matter photosynthesized. The effectiveness of sediment-scavenging is demonstrated by the wide spread of red fossiliferous limestones in the Early Cambrian (Tommotian and Adtabanian of Siberia; similar red limestones in the British Isles and in Newfoundland).

7. EVOLUTION OF CALCAREOUS SKELETONS

Through much of Precambrian time, the degree of supersaturation of the oceans must have resembled that of modern carbonate lakes (e.g., Pyramid Lake, Nevada, or Green Lake, New York). They were so highly supersaturated with calcium carbonate that benthic photosynthesizers became encrusted with tufa to form stromatolites, while planktonic organisms precipitated carbonate mud. This state was terminated by the evolution of calcareous skeletons in many groups of organisms in the course of Cambrian and Ordovician time. Skeletal limestones became the predominant marine carbonate sediments in the middle Ordovician, coincident with a great decline of stromatolites in marine sediments. Only occasionally, such as in the Jurassic and in the Pleistocene, have chemically precipitated carbonates (in this case, oolite) reached notable proportions.

8. INVENTION OF SILICEOUS SKELETONS

The influx of silica to the oceans must have been balanced in Precambrian time by the precipitation of silica from a supersaturated system. Such is presumably the origin of the jaspers associated with the BIF, and of the cherts; delicate microorganisms have been preserved in exquisite detail in the cherts of the Gunflint and Bitter Springs formations.

Beginning about 550 Ma ago in early Cambrian time, sponges and (?) radiolarians began to withdraw silica for the construction of siliceous skeletons. Sponge-bearing cherts are very abundant throughout Paleozoic platform limestones, and sponge spiculites are common in Paleozoic basin deposits. Radiolarites are found in such deep-sea deposits as have been preserved from Ordovician time on. The youngest marine chert on record that preserves soft-bodied organisms (filamentous algae, delicate internal structures of graptolites) in the manner of the famous Precambrian cherts is the lower Ordovician chert of the Holy Cross Mountains in Poland.

9. COLONIZATION OF THE LANDS

The colonization of lands by higher organisms took place in Silurian-Devonian time. It involved the development of large plants with massive supporting structures (wood), low in phosphorus content and other nutrients (17). Productivity and oxygen-output of the biosphere must have doubled fairly rapidly. With the presence of a food source on land, animals soon followed. One would expect the development of a vegetative cover to have had a significant effect on sediments:

1) One might expect to find a new kind of sediment composed largely or entirely of vegetable matter. This did indeed appear in the form of peat and coal.

2) A turn from less to more clayey and micaceous sediments in normal shelf settings, for the following reason: in unvegetated lands, dessicated soils and sediments are at mercy of winds. Continents without vegetation may have exported much clay to the oceans as atmospheric dust and may have transported large quantities of sediment as dunes and loess, grinding up such soft minerals as the micas in the process. To my knowledge, no such change has been observed (or sought).

3) A change from less to more chemically mature sediments, in response to a) increased biogenic generation of carbon dioxide in soils, increasing rates of chemical decomposition; and b) longer weathering of soils protected from erosion by plant cover. This effect has also not been observed. Indeed, observations support the opposite: mature quartzites are more conspicuous in the Proterozoic (e.g., the Huronian quartzites of the Lake Superior country) than in the Paleozoic. This paradox may possibly be reconciled in the following ways: a) atmospheric carbon dioxide pressure may well have been many times that of today's in Proterozoic times, in which case the lack of organic generation of carbon dioxide in soils may have been of no consequence; b) if eolian processes played a large part in moving sediments in pre-vegetation time, and if rivers tended more to braiding than to meandering, the transportation time of sediments from source to final deposition and the exposure of sediments in transit to weathering in the vadose zone may well have been longer, providing more time for alteration of unstable minerals. A thin algal land cover may also have affected the mechanical properties of soils.

4) An increase (doubling?) of the global rate of carbon fixation and oxygen release due to the presence of higher plants and of their communities on land. How might one expect this to be recorded in the rocks? Direct calculations of the rate of carbon precipitation per time interval - in the form of carbonate or organic matter (Holser, this volume, Fig. 1) - are difficult and unreliable, especially for times for which we have no deep-sea record. The isotope curves in Figure 1 show a fairly steady increase in the carbon 13/12 ratio of marine limestones: this suggests that photosynthetically-fixed (isotopically-light) "organic" carbon was withdrawn into the lithosphere at almost steady rates from Cambrian to Permian times, with the exception of an anomalous drop in the Devonian. It would seem that in general the increase in carbon fixation that resulted from the colonization of the lands was counterbalanced by an increase in the rate at which fixed carbon was returned to the atmosphere as carbon dioxide - as a result of increased atmospheric oxygen content which is self-limiting by way of forest fires. Indeed, one may speculate on the possibility that the reversals of the isotope trend in Devonian and Late Permian times occurred because high atmospheric oxygen levels suddenly found access to large and easily oxidized (inflammable) parts of the lithospheric carbon reservoir. Perhaps gigantic fires left a recognizable sedimentary clue - widespread charcoal. The alternative - some strange event that wiped out most of the vegetation for a part of Devonian time and resulted in markedly decreased burial of organic matter - should have left a crisis in the paleobotanic record. Neither has been recognized.

The story of the carbon isotopes is borne out by the sulfur isotope curve which mirrors the oxidation state. Accumulation of organic carbon in the lithosphere and of oxygen in the atmosphere-hydrosphere system resulted in a progressive decrease in the sulfide (pyrite) reservoir, and a consequent lightening of the sulfate reservoir, from Cambrian to Permian times. In this case, too, there is a reversal in the Devonian, which suggests a reduction in atmospheric oxygen and consequent temporary re-expansion of the sulfuret - an event dramatically illustrated by the great incidence of Middle and Late Devonian marine black shales.

10. INVENTION OF CALCAREOUS MICROPLANKTON

The plankton of the Mesozoic evolved a number of groups of organisms provided with calcareous armor: the coccolithophyceans (Triassic?), the calpionellids (Jurassic, after a precocious appearance in the Devonian), and the globigerinacean foraminifera (Late Jurassic onward). From late Jurassic time onward (11), they diverted an appreciable part of the world's carbonate deposition from the shelves to the deep ocean. This led to a marked reduction of carbonate shelves and increased the transfer of carbonates to the lower crust and mantle, via crustal subduction. To the extent to which the batholiths of the mountain belts are derived from or contaminated by oceanic slabs undergoing subduction, they may be expected to have become more calcium-rich since Jurassic time. This remains to be tested.

11. INVENTION OF SILICEOUS TESTS BY PHYTOPLANKTON

At some time in the Jurassic or early Cretaceous, diatoms appeared and began to compete with radiolarians for silica. More efficient at utilizing silica at low levels of concentration, they are largely responsible for the great present-day silica depletion in the upper (photic) ocean waters. Their particular success since mid-Tertiary time in cold, nutrient-rich waters has shifted the bulk of marine silica deposition to the high latitudes and the belts of upwelling waters.

12. INVENTION OF FLOWERS

The development of angiosperms in Cretaceous time is responsible for the great diversity and the productivity of the present lands. One of the main factors in this evolutionary explosion of sedentary organisms was the invention of the flower as a means of enticing mobile organisms - animals - into the role of sexual messengers. The long-distance fertilization which became possible through the action of insect and other animal vectors allowed plant species to reproduce sexually in much more rarified populations. This allowed a greater variety of plants to coexist in a given area, and to utilize resources more efficiently. It also provided greater potential for migration and colonization. While no direct geological changes have been attributed to the rise of angiosperms, the very economical use of phosphorus by these plants should in theory have added to their productive potential, and should have caused a second measurable drop in the mean C/P ratios of buried organic matter: the first of these steps must have accompanied the development of forests in Devonian time. These effects remain to be documented.

13. INVENTION OF CONCEPTUAL THOUGHT

One of the most revolutionary biological innovations has been the development of the central nervous system. In the Hominidae this resulted in the power of conceptual thought. This invention allowed Man to expand his use of available energy in ways that have given him ever more power over the biosphere, and it is modifying the natural geophysical-geochemical cycles. Landscape and seafloor are being changed at an ever increasing rate; limnic systems have been severely affected over much of the world; many elements are cycling in new and different pathways; new chemical compounds are being created in vast numbers. Radioactive chain reactions have been achieved; controlled fusion may not be far away. Climatic changes appear likely. More and more of the biosphere's energy and resources are being funneled into the maintenance of this one species. Inquiring into the nature of the Universe, of the Earth, and of Life, the species is beginning to realize the danger of destroying its own environment by three new riders of the apocalypse: Unchecked population growth, industrial chemical pollution, and chemical-biological-atomic warfare. Whatever may happen, for better or worse, the future of the Earth will always bear the imprint of this species.

OTHER ASPECTS

In this discussion we have only looked at biological innovations as originators of changes in the sedimentary record. However, the biosphere is a product of co-evolution between the physicochemical environments on the one side and the organisms on the other. In many cases cause and effect are not separable. The appearance of families in time (Fig. 1) has been episodic (21) and appears linked to eustasy, incidence of vulcanism, the state of continental dispersion or aggregation, and the nature of global climates and atmospheric chemistry (5-7, 19). It also bears resemblance to the isotope curves of sulfur and of carbon. The bimodal structure of the Phanerozoic suggests the possibility of a cycle with a period of around 400 Ma. Some have sought the origin of such cycles in galactic rotation, others in mantle convection. Of the Phanerozoic inventions, **6**, **7**, **8**, and **9** occurred during the great evolutionary outburst of the early Paleozoic, **10**, **11**, and **12** during that of the later Mesozoic.

A cycle in the 30 Ma range seems to be reflected in global diversity of planktonic species as well as in the complexity of individual pelagic communities, as expressed by changes in the size and occupance of the top predator niche (8). At a lower level, instabilities in the marine environment are abundantly demonstrated by certain kinds of bedding, such as limestone-shale alternations. Such phenomena may be enhanced by diagenetic processes. Some of them are regular (1, 20), show a hierarchy of superimposed cycles, and match the periodicities of present-day orbital variations (eccentricity at 413,000 and ca. 100,000 years; obliquity at 41,000 years; axial precession relative to perihelion at 19,000 and 23,000 years). The climatic influence of these variations has been demonstrated, as has their role in forcing the advances and retreats of the Pleistocene ice sheets (13). In pelagic sediments, cycles of this sort take the form of alternations in clay content, in degree of oxidation, in the relative productivity of different planktonic organisms, and in seafloor dissolution of carbonate.

Environmental stresses of yet shorter period include those due to tidal, diurnal, lunar, and annual cycles. Beside these more or less periodic changes in the physical environment, there are stochastic changes - some induced by random causes such as the impact events discussed by Shoemaker (this volume), others engendered by the multiple intersections of periodic causes having different periods.

These environmental stresses on organisms have time and again changed the direction in which natural selection has operated; they continue to empty old niches and to provide new ones, making the biosphere receptive to biological innovation. We may thus conclude that much biological innovation is a consequence of the spasmodic nature of Earth history and has in turn contributed thereto.

REFERENCES

(1) Anderson, R.Y. 1982. A long geoclimatic record of the Permian. J. Geophys. Res. 87 C-9: 7285-7294.

(2) Awramik, S.M. 1982. The pre-Phanerozoic fossil record. In Mineral Deposits and the Evolution of the Biosphere, eds. H.D. Holland and M. Schidlowski, pp. 67-82. Dahlem Konferenzen. Berlin-Heidelberg-New York: Springer-Verlag.

(3) Crerar, D.A.; Fischer, A.G.; and Plaza, C.L. 1980. Metallogenium and the biospheric deposition of manganese from Precambrian to Recent time. In Geology and Geochemistry of Manganese, ed. I.M. Varentsov, vol. 3, pp. 285-303. Budapest: Hungarian Academy of Science.

(4) Engel, A.E.J., and Engel, C.G. 1964. Continental accretion and the evolution of North America. In Advancing Frontiers in Geology and Geophysics, eds. A.P. Subramaniam and S. Balakrishna, pp. 17-37e. Hyderabad, India: India Geophysical Union.

(5) Fischer, A.G. 1981. Climatic oscillations in the biosphere. In Biotic Crises in Ecological and Evolutionary Time, ed. M. Nitecki, pp. 103-131. New York: Academic Press.

(6) Fischer, A.G. 1982. Long-term climatic oscillations recorded in stratigraphy. In Climate in Earth History, eds. W.H. Berger and J.C. Crowell, pp. 97-104. Washington, D.C.: National Academy of Science Press.

(7) Fischer, A.G. 1983. The two Phanerozoic supercycles. In Catastrophies in Earth History: The New Uniformitarianism, eds. W. Berggren and J. Van Couvering. Princeton: Princeton University Press, in press.

(8) Fischer, A.G., and Arthur, M. 1977. Secular variations in the Pelagic realm. In Deep Water Carbonate Environments, eds. H.E. Cook and P. Enos. Soc. Econ. Paleontol. Min. Spec. Publ. 25: 19-50.

(9) Folinsbee, R.E. 1982. Variations in the distribution of mineral deposits with time. In Mineral Deposits and the Evolution of the Biosphere, eds. H.D. Holland and M. Schidlowski, pp. 219-236. Dahlem Konferenzen. Berlin-Heidelberg-New York: Springer-Verlag.

(10) Garrels, R.E., and Lerman, A. 1981. Phanerozoic cycles of sedimentary carbon and sulfur. Proc. Natl. Acad. Sci. USA 78: 4652-4656.

(11) Garrison, R.E., and Fischer, A.G. 1969. Deep-water limestones and radiolarites of the alpine Jurassic. In Depositional Environments in Carbonate Rocks, ed. G. Friedman. Soc. Econ. Paleontol. Min. Spec. Publ. 14: 20-55.

(12) Hallam, A. 1977. Secular changes in marine inundation of USSR and North America through the Phanerozoic. Nature 194: 1121-1132.

(13) Hays, J.D.; Imbrie, J.; and Shackleton, N.J. 1976. Variations in the Earth's orbit: pacemakers of the ice ages. Science 194: 1121-1132.

(14) James, H.L., and Trendall, A.F. 1982. Banded iron formation: distribution in time and paleoenvironmental significance. In Mineral Deposits and the Evolution of the Biosphere, eds. H.D. Holland and M. Schidlowski, pp. 199-217. Dahlem Konferenzen. Berlin-Heidelberg-New York: Springer-Verlag.

(15) Jenkyns, H.C. 1980. Cretaceous anoxic events from continents to oceans. J. Geol. Soc. Lond. 137: 491.

(16) Keith, M.L. 1982. Violent vulcanism, stagnant oceans and some inferences regarding petroleum, strata-bound ores and mass extinctions. Geochim. Cosmochim. Acta 46: 2621-2637.

(17) Lerman, A. 1982. Sedimentary balance through geological time. In Mineral Deposits and the Evolution of the Biosphere, eds. H.D. Holland and M. Schidlowski, pp. 237-256. Dahlem Konferenzen. Berlin-Heidelberg-New York: Springer-Verlag.

(18) Nealson, K.H. 1982. Microbiological oxidation and reduction in iron. In Mineral Deposits and the Evolution of the Biosphere, eds. H.D. Holland and M. Schidlowski, pp. 51-65. Dahlem Konferenzen. Berlin-Heidelberg-New York: Springer-Verlag.

(19) Mackenzie, F.T., and Piggot, J.D. 1981. Tectonic controls of sedimentary rock recycling. J. Geol. Soc. Lond. 138: 183-196.

(20) Schwarzacher, W., and Fischer, A.G. 1982. Limestone-shale bedding and perturbations of the Earth's orbit. In Cyclic and Event Stratification, eds. G. Einsele and A. Seilacher, pp. 72-95. Berlin-Heidelberg-New York: Springer-Verlag.

(21) Sepkoski, J.J., Jr. 1981. A factor analytic description of the Phanerozoic marine fossil record. Paleobiology 7: 36-53.

(22) Trudinger, P.I., and Williams, N. 1982. Stratified sulfide deposition in modern and ancient environments. In Mineral Deposits and the Evolution of the Biosphere, eds. H.D. Holland and M. Schidlowski, pp. 199-236. Dahlem Konferenzen. Berlin-Heidelberg-New York: Springer-Verlag.

(23) Trueper, N.G. 1982. Microbial processes in the sulfur cycle through time. In Mineral Deposits and the Evolution of the Biosphere, eds. H.D. Holland and M. Schidlowski, pp. 5-38. Dahlem Konferenzen. Berlin-Heidelberg-New York: Springer-Verlag.

(24) Vail, P.R.; Mitchum, R.M.; and Thompson, S. 1977. Seismic stratigraphy and global changes in sea level, Part 4. In Seismic Stratigraphy, ed. C.E. Peyton. Amer. Assoc. Petrol. Geol. Mem. 26: 83-97.

Patterns of Change in Earth Evolution, eds. H.D. Holland and A.F. Trendall, pp. 159-168.
Dahlem Konferenzen 1984. Berlin, Heidelberg, New York, Tokyo: Springer-Verlag.

Late Precambrian and Early Cambrian Metazoa: Preservational or Real Extinctions?

A. Seilacher
Geologisches Institut der Universität Tübingen
7400 Tübingen, F.R. Germany

Abstract. The interpretation of Ediacara-type body fossils in terms of modern soft-bodied metazoans must be questioned. Their morphology rather suggests foliate, non-locomotory quasi-autotrophs. Their mode of preservation, which has no counterpart in comparable post-Vendian rocks, remains problematical, since associated trace fossils attest to oxic conditions and the presence of worm-like heterotrophic burrowers in the same environment. It seems that Vendian biota mark not simply a non-skeletal start of metazoan evolution, but a distinct episode in the history of life that was followed by a major extinction.

INTRODUCTION

Paleontologic research during the last decades has changed our views of early metazoan evolution tremendously. We recognize not only a sudden appearance of a diversified Cambrian megafauna, but also a pre-trilobite (Tommotian) shelly fauna, which was preceded by a phase (Vendian-Ediacarian), in which larger, presumably metazoan organisms were present that had not yet developed mineralized skeletons. We can only speculate about a still earlier phase of metazoan history. One could, for instance, interpret the presence of similar planktonic larvae in different phyla of modern marine organisms, not simply as a convergently evolved means of dispersal, but as an archaic heritage from a hypothetical planktonic and larviform stage of metazoan evolution. Metazoans of this stage would not have left a fossil record because of their microscopic size and non-skeletal nature. The Vendian radiation might have expressed the metazoan conquest of the benthic realm, in which size was no longer limiting. The resulting diversified extension of ontogenetic histories could therefore follow diverse pathways of adaptation to the new habitat: Wormlike infaunal organisms left a trace fossil record and epifaunal organisms left a record as body fossils. Cambrian radiations followed the introduction of mineralized skeletons that opened new possibilities of constructional design.

In this paper we do not want to emphasize the appearance, but rather the disappearance of these early metazoan faunas, and to discuss whether they are preservational in nature and hence reflect major changes in the necrolytic and diagenetic regimes, or whether they are due to major extinctions.

THE PROBLEM OF THE EDIACARAN FOSSILS

Several decades ago the discovery of impressions of distinctive soft-bodied organisms in sandy deposits of late Precambrian age in South Australia created a sensation. The Cambrian appearance of trilobites and other "shelly" organisms with mineralized skeletons no longer marked the beginning of metazoan radiation but could be regarded as a second major step in the evolution of multicellular animals. Since the discovery of the Ediacara fauna, similar or even identical impressions have been found in similar stratigraphic positions in more than a dozen localities - particularly in Australia, South Africa, the Soviet Union, England, and Newfoundland (1, 2, 7). Without a doubt we are dealing not with a local preservational "bonanza" such as the Burgess Shales, but with the distinctive fauna of an Ediacarian period that preceded the Cambrian radiation.

This development has raised a problem as vexing as the fauna itself: that of its preservation. Most authors agree that Ediacara-type organisms were soft-bodied animals, similar to the jellyfishes that are so common in most occurrences. Such organisms have existed ever since. Nevertheless, no remains comparable to the Ediacara soft-bodied fossils, at least like their more characteristic members, have been found in younger rocks. Individual "Fossil-Lagerstätten" with an extraordinary level of preservation, including that of soft-bodied organisms, have formed repeatedly since late Precambrian time. These deposits usually consist of very fine-grained sediments deposited in continuously or intermittently stagnant basins, or of coarser deposits in which whole organisms became smothered. In contrast, Ediacara fossils are found in sandstones and sand/shale sequences whose sedimentary structures and trace fossil content are characteristic of normal and well aerated shallow marine environments (6). A question that is particularly relevant in the context of this workshop is this: "Why did the Ediacaran mode of preservation become 'extinct' with the beginning of the Cambrian?" The recent discovery of non-skeletal metazoans in the Precambrian of the Oleniok Uplift (Fedonkin and Rozanov, Moscow, personal communication) will be of great interest in this connection, since they are preserved in thinly bedded dolomites, possibly of the lithographic type, i.e., in a facies with different preservational properties.

We must first test the reality of the "extinction" by answering the following questions:

1. Could the extinction be an artefact in the sense that we are dealing with pseudo-fossils of inorganic origin? This is certainly true of many so-called "Precambrian fossils." The supposed jellyfishes from the Hakatai Shales of the Grand Canyon (3, 14) are surely nothing more than compactional haloes around sand salt crystals. Another example of pseudo-fossils are the sinuous, wormlike ridges of "Manchuriophycus" (3, 13), which are only shrinkage patterns of thin mud lenses deposited in ripple troughs. The complex forms of Ediacaran impressions clearly exclude the possibility of a mechanical origin.

2. Do similar impressions occur in post-Vendian deposits, and have they so far escaped attention? Most Ediacaran fossils are faint but distinctive impressions on sole faces of sandstone beds; these are also optimal sites for the preservation of burrows. Students of trace fossils have probably searched square miles of such sole faces in rocks of all ages without ever coming across a truly Ediacaran type of preservation. It is therefore most unlikely that any similar impressions occur in post-Vendian deposits.

A full reevaluation of Ediacaran fossils is clearly needed; the following section of this paper is limited, however, to comments on published material which are immediately relevant to the final discussion and suggestions.

Ediacaran Medusoids

The most common elements of the Ediacara fauna are round impressions with radial and concentric structures that have been classified as medusoids. Most published pictures are small and lighted in a variety of different directions so that is is difficult to judge critical details. We therefore owe a great deal to M. Wade (18) for a critical analysis of preservational details. The mode of preservation of medusoids is consistently different from that of other groups of Ediacaran fossils. Unfortunately, Wade's reconstructions of preservational processes are based on the unproven assumption that these were really jellyfishes.

Stratinomy. Fossil-bearing sole faces record the change of pelitic to sandy sedimentation during an episodic rise in turbulence (commonly a storm event) in which coarser material was deposited in a graded fashion following an initial phase of mud erosion. Since medusoids are mainly found on these sole faces, they must have reached the bottom together with benthic types of Ediacaran organisms. This is not what we might expect from such light bodies. In the lithographic limestones (Upper Jurassic) of Pfalzpaint near Solnhofen - one of the few localities where undoubted jellyfishes did become fossilized - they are invariably found within the event-generated flinz beds and not at their base, where the heavier bodies of arthropods, echinoderms, squids, and fishes are located. If this is the case in a coccolithic mud, how much more should it apply to sand!

Postmortem deformation. Solnhofen specimens also show the characteristic alteration that occurs when a jellyfish shrinks by dehydration within sediments. Radial and concentric wrinkles develop in the peripheral zone of the umbrella, where the radial and circular muscle fibers must be located in medusoid swimmers, while the central part of the impression remains smooth or carries a regular number of radial furrows corresponding to the mesentheria within the stomach cavity. Ediacaran medusoids, in contrast, show strong concentric folds in the center and radial structures at the periphery.

Radial feeding burrows? Radial backfill burrows that reflect the probing of wormlike sediment feeders around a vertical burrow are among the structures that are commonly mistaken for fossil jellyfishes. Medusina, Palaeosemaeostoma, Kirklandia, and Gyrophyllites (8) are familiar examples from the Phanerozoic record. Brooksella from the late Precambrian of the Grand Canyon may be another example. Among the Ediacara medusoids, Mawsonites spriggi (see (5), Figs. 1-2 and (1), cover picture) is clearly such a burrow system. Like the Cretaceous Kirklandia, it shows several tiers of sharply separated radial lobes that extend from the central shaft with a downward inclination. Seleniform backfill lamallae can also be seen in the lobes as well as in the vertical shaft of published Mawsonites specimens.

Actinian (?) burrows? Circular burrows, possibly made by actinians that lived in sandy sediments, are another source of confusion. Paleozoic examples (Bergaueria) are usually preserved as smooth and dome-shaped casts, sometimes with a central depression; Mesozoic representations (Solicyclus) have regular radial grooves around the smooth central field. Without careful examination of the original material, it is difficult to assign Ediacaran medusoids to this type of trace fossils; however, the separation of specimens of Medusinites, Edicaria, and Cyclomedusa from the surrounding bedding surface by a sharp groove and the eccentrically arranged annular furrows in some specimens (18) may indicate that they belong to this kind of trace fossils.

Sandy skeletons of actinians? A last possibility is related to a new interpretation of Protolyella, a three-dimensional fossil of Cambrian to Ordovician age with a globular base and radial plus concentric ridges on the truncated upper surface. In the past, it has been interpreted as the sandy filling of a medusoid stomach cavity. New material from the Ordovician of Jordan (10), however, shows that the concentric grooves of the upper surface continue inside the body as concentric hemispherical laminae that cannot be explained by sedimentation within a cavity. Rather they reflect the laminar growth, possibly within the gastral cavity of an actinian, of a heavy sandy skeleton that stabilized the animal on the soft substrate. The occurrence of a central depression at the base of many Protolyella casts suggests that the organisms were similar to those responsible for Bergaueria burrows.

A three-dimensional character is also suggested by some Ediacara "medusoids." In one specimen ((18), Fig. 2) they are aligned in the troughs of oscillation ripples in the same manner as Protolyella in Cambrian and Ordovician sandstones. The animals may have accumulated on and burrowed in the sand during the event and were then smothered by the muddy tail of the same tempestite (see (15), Fig. 5b for a similar behavior in ophiuroids). The round basal disk of the Pennatula-like Charniodiscus ((1), Fig. 2A) could be a similar sandy weight belt.

Conclusion: None of the numerous "medusoids" in the Ediacaran fauna can be interpreted with certainty as a jellyfish. Rather we seem to be dealing with a heterogeneous group of trace fossils and remains of unidentified benthic organisms.

Non-medusoid Body Fossils
Other enigmatic fossils of the Ediacaran fauna have been variously assigned to pennatulid coelenterates, annelids, and arthropods, i.e., to phyla still living today (5). Only one author (11, 12) referred them to an extinct phylum Petalonamae. Affiliation with modern creatures, based on crude morphologic similarities, should, however, be considered tentative. Sea pens, for instance, are colonial organisms in which the branches must be separated so that the polyps can function properly. Nevertheless, the branches are never spread apart in Ediacaran "sea pens," not even in specimens that are preserved within the sandstone in a three-dimensionally twisted fashion. It is also strange that among the "annelids" only very unusual, large and foliate forms (Dickinsonia; Spriggina) should be preserved, while normal cylindrical worms, whose presence is well documented by associated trace fossils, have never been found as body fossils in Ediacara-type deposits. The only vaguely worm-like body fossil (4) is a three-dimensional internal cast and not the flattened impression that one would expect.

Nature of the cuticle. In our present discussion, questions of taxonomic affiliation are of secondary importance. It suffices to state that the non-"medusoid" Ediacaran impressions are neither pseudo-fossils nor trace fossils, but true body fossil impressions of highly organized organisms. Their mode of preservation is characteristically different from that of the medusoids (18): instead of convex casts they form concave external molds on sandstone sole surface. This is not what we would expect, because a cavity resulting from burrowing, or from the decay of a soft body, at a sand/mud interface should be filled with the noncohesive sand that casts the mold left in the more cohesive mud. The situation is reversed only after the sand has become somewhat immobilized by cementation. Since non-medusoid impressions are also preserved within the sandstone, we conclude that they reflect a cuticular skeleton resistant enough to survive until the sand had become diagenetically altered.

On the other hand, the way in which these fossils have become deformed in particular situations (Dickinsonia draping over Arborea, (5), plate 103, Fig. 1; twisted Pteridinium within the sandstone, (5), plate 101, Figs. 1-3) suggests that the cuticle was flexible but rigid enough not to become wrinkled. It must have consisted of a highly elastic material. In sandy marine sediments of later ages such biomaterials have become digested by microorganisms - except under anoxic conditions. The presence of burrowing organisms in the Ediacara sediments rules out this possibility.

Functional morphology. A feature that Ediacaran "pennatulids," "chondrophores," and Dickinsonia have in common is their quilt-like repetitive structure which has been interpreted as expressing metameric segmentation. It could, however, be simply a means to stiffen a hydraulic skeleton biomechanically, somewhat in the manner of an air mattress; it would also have provided an internal compartmentalization that facilitated metabolic processes (17). The air mattress model does not apply to Spriggina, because some of the specimens are laterally bent, nor to Tribrachidium, Parvancorina, and Praecambridium, in all of which the boundaries between radial sectors stick out as sharp ridges instead of forming reentrant grooves. Nevertheless, these forms were also quilted and frond-like.

The frond-like shape of most Ediacaran fossils could be explained in terms of metabolism, since it maximizes the external surface. This may have been necessary for respiration at a time when the atmosphere and seawater contained less oxygen than today. One might even speculate that dissolved food could have been absorbed directly through the enlarged body surface, since orifices and intestines have not been observed in Ediacaran fossils. Another possibility is an association with photosymbiotic algae, which would benefit from a frond-like shape and could have provided food and oxygen directly to the host.

A mode of nutrition unfamilar in modern Metazoa is also indicated by the non-locomotory character of most Ediacara-type organisms. Morphological distinction between front and rear end - a universal attribute of mobile benthic organisms except regular echinoids - is found only in Spriggina and Parvancorina. The rest, including the benthic "medusoids," were probably more or less sessile. We do not, however, have indications of the presence of lophophores or other filtering devices that are the base for sedentary life in modern marine organisms.

All these considerations are, of course, based on the assumption that we are really dealing with the Metazoa in the modern sense, and this is not at all certain.

Conclusion: Non-medusoid body fossils of the Ediacaran fauna seem to be the remains of benthic organisms that are not referable to extant phyla. They had a flexible cuticle whose survival suggests that it was composed of biomaterial indigestible for the contemporary microorganisms.

EVOLUTIONARY SCENARIO

The problem that appeared to be mainly a matter of preservation at the outset of this review has thus turned out to be of basic significance for the theme of this Dahlem workshop. If one views the Ediacaran fauna as the initial stage of metazoan evolution, it could be incorporated into the sigmoidal part of the Cambrian diversity explosion that was unique in the sense that it took place in an ecological vacuum. In terms of diversity, this curve is still valid, because the number of species and the degree of provincial differentiation was very low in the Vendian biota. But if the Ediacara-type fossils turn out to be basically different from Metazoa in the modern sense and if their disappearance is not only an artefact of preservation, then the Cambrian radiation was preceded by a major extinction event in the benthic realm.

One could go a step further and ask whether a similar relationship might not exist between the shelly faunas of the Tommotian and the trilobite-dominated stages of the Cambrian. After all, animals with a taxonomy based on that of modern organisms could have suppressed the basic constructional and ecological differences of this earliest shelly fauna. The problem of the Tommotian biota also has preservational overtones, because the shelly fossils of this stage are all very small and phosphatized. But this time-specific mode of preservation continues throughout the remainder of the Cambrian. It became "extinct," however, in later periods, when suitable micromolluscs were still present but are significantly absent in the phosphatic residues studied by conodont workers (O. Walliser, personal communication).

An unusually high phosphorous content in Cambrian seas was probably responsible for the easy phosphatization of mineralized as well as non-mineralized skeletons (9, 10), as well as for the high percentage of originally phosphatic shells in Tommotian faunas.

We conclude that metazoan evolution was no smoother in its initial than in its later phases. Extinctions caused by changes in the physical environment preceded radiations that increased the overall diversity in a step-like mode. Since the levels of diversity were generally low, these early extinctions had little in the way of quantitative effects. Nevertheless, their consequences played a major role in the course of metazoan evolution.

SUGGESTIONS FOR FUTURE RESEARCH

This review of the literature has been admittedly cursory. We have come to the conclusion that the nature of Ediacara-type body fossils of Vendian age may have been misinterpreted. Not only their affiliation with extant phyla, but also their truly metazoan nature should remain open to question. More definite answers regarding their affiliations can be given only after a very critical reexamination of the fossil material. Such a study should include associated trace fossils, which seem to be less different from post-Vendian forms and that may represent the stocks from which marine epibenthic phyla radiated at a later time. The study of trace fossils and associated sedimentary structures will also provide better information about the depositional environments and the distribution of food in sediments during those early times.

REFERENCES

(1) Cloud, P., and Glaessner, M.F. 1982. The Ediacarian period and system: Metazoa inherit the earth. Science 217: 783-792.

(2) Fedonkin, M.A. 1981. White sea biota of the Vendian. Precambrian non-skeletal fauna of the northern Russian platform. Trans. Akad. Nauk. SSR. 342: 1-100.

(3) Glaessner, M.F. 1969. Trace fossils from the Precambrian and basal Cambrian. Lethaia 2: 369-393.

(4) Glaessner, M.F. 1979. An echiurid worm from the Late Precambrian. Lethaia 12: 121-124.

(5) Glaessner, M.F., and Wade, M. 1966. The late Precambrian fossils from Ediacara, South Australia. Paleontology 9: 599-628.

(6) Goldring, R., and Curnow, C.N. 1967. The stratigraphy and facies of the late Precambrian at Ediacara, South Australia. J. Geol. Soc. Austral. 14: 195-214.

(7) Jenkins, R.J.F. 1981. The concept of an "Ediacaran Period" and its stratigraphic significance in Australia. Tran. Roy. Soc. S. Austral. 105: 179-194.

(8) Moore, R.C. 1956. Treatise on invertebrate paleontology. Part F. Coelenterata. Lawrence, KS: University of Kansas Press.

(9) Müller, K.J. 1979. Phosphatocopine ostracodes with preserved appendages from the Upper Cambrian of Sweden. Lethaia 12: 1-

(10) Müller, K.J. 1982. Weichteile von Fossilien aus dem Erdaltertum. Naturwissenschaften 69: 249-254.

(11) Pflug, H.D. 1972. Systematik der jung-präkambrischen Petalonamae. Paläont. Z. 46: 56-67.

(12) Pflug, H.D. 1974. Feinstruktur und Ontogenie der jung-präkambrischen Petalo-Organismen. Paläonet. Z. 48: 77-109.

(13) Schindewolf, O.H. 1956. Über präkambrische Fossilien. In Geotektonishes Symposium zu Ehren von Hans Stille, ed. F. Lotze, pp. 455-480. Stuttgart: F. Enke.

(14) Seilacher, A. 1956. Der Beginn des Kambriums als biologische Wende. N. Jb. Geol. Paläont. 103: 155-180.

(15) Seilacher, A. 1982. Distinctive features of sandy tempestites. In Cyclic and event stratification, eds. G. Einsele and A. Seilacher. Heidelberg: Springer-Verlag.

(16) Seilacher, A. 1982. Paleozoic sandstones in southern Jordan: trace fossils, depositional environments and biogeography. Contribution, 1st Jordanian Geological Congress, Amman.

(17) Sernetz, M.; Rufeger, H.; and Kindt, R. 1982. Interpretation of the reduction law of metabolism. Exp. Biol. Med. 7: 24.

(18) Wade, M. 1968. Preservation of soft-bodied animals in Precambrian sandstones at Ediacara, South Australia. Lethaia 1: 238-267.

Standing, left to right:
Dick Holland, Holger Kulke, Michael Sarnthein, Hans Füchtbauer, Tony Lasaga, Otto Walliser, Gerold Wefer.

Seated, left to right:
Bill Jenkins, Ida Valeton, Dolf Seilacher, Bill Holser, Wolf Berger.

Patterns of Change in Earth Evolution, eds. H.D. Holland and A.F. Trendall, pp. 171-205. Dahlem Konferenzen 1984. Berlin, Heidelberg, New York, Tokyo: Springer-Verlag.

Short-term Changes Affecting Atmosphere, Oceans, and Sediments During the Phanerozoic
Group Report

W.H. Berger, Rapporteur
H. Füchtbauer
H.D. Holland
W.T. Holser
W.J. Jenkins
H.G. Kulke
A.C. Lasaga
M. Sarnthein
A. Seilacher
I. Valeton
O.H. Walliser
G. Wefer

INTRODUCTION

The view of Earth history which emphasized gradualism and a narrow form of uniformitarianism has lost appeal. This view was encouraged by Charles Lyell and Charles Darwin over a century ago, in an effort to combat a catastrophism that derived its strength from mythology rather than from observation. Lyellian uniformitarianism has served its purpose well and for a long time (some think too long). It is now being superceded by concepts emphasizing rapid change in historical development. While this approach is less timid, less dogmatic, and perhaps more realistic than strict gradualism, it is not exactly new. There have always been those who were impressed by the evidence for sudden change in the geologic record (see (83)). However, their views were not generally in the mainstream of geologic thought.

The emerging consensus on the nature of Earth history recognizes an alternation, on many different time scales, of periods of stable or slowly changing conditions with periods of rapid change. Some of these episodes or punctuations in history may be mere accelerations of normal trends, due to nonlinearities in the system, or they may be extreme conditions produced by an unusual constellation of normal environmental factors. Others may call for special explanations such as asteroid impact, intense vulcanism, or other large short-term energy input to the system.

The new emphasis on rapid change in the geologic record owes much to the triumph of plate tectonics, which recently vindicated Alfred Wegener's concepts regarding continental drift. When two continents part, or unite, a new situation can arise rather suddenly, with potentially profound influences on climate and the evolution of life. Wegener was keenly interested in these interactions, and modern discussions of the subject are, to a large extent, variations on themes in his work.

The recent dramatic increase in our knowledge of deep-sea sediments through ocean-drilling also contributed greatly to the new focus on periods of rapid change. Rather than confirming gradual change, the deep-sea record yielded excellent evidence for pronounced steps in the evolution of late Cretaceous and Cenozoic climate and ocean chemistry. The most spectacular of these steps, of course, is the Cretaceous-Tertiary boundary, whose origin is the most hotly debated topic in stratigraphy today.

Another important factor in drawing the attention of geologists to periods of rapid change is the example of our own time. Mankind is now altering the global environment on a scale unprecedented in human history, and, quite possibly, in the history of the entire Cenozoic. Geochemists have begun to model the interactions of the ocean, the atmosphere, the biosphere, and of soil formation and sedimentation in an effort to describe and extrapolate the changes wrought by the development of agriculture and industry. Geologists are in a position of providing background information on fast transitions in the geologic past. As yet, the Earth is its own best model to provide tests for our computer simulations.

What is a "short-term change"? What kind of phenomenon qualifies for study under this heading? What is the minimum rate and amplitude of change necessary for us to take note? How are rates to be measured? We are not ready to provide answers to such questions, however reasonable they may be. The study of "events" is at a beginning, the time for classifications has not yet come. As a working definition in the present context, we use "short-term change" in the sense of "establishment of substantially different conditions on the surface of the Phanerozoic Earth during a time span of about ten million years or less." We can do no more here than to draw attention to the fact that short-term changes have occurred, and that their study holds great promise for advancing our understanding of the exogenic system. During periods when the ocean-atmosphere-biosphere-pedosphere system changes quickly and suddenly, there is a chance that we can isolate a few important driving mechanisms from the normal background noise.

Our task is to scrutinize the record for evidence of short-term change. When confronted with such evidence, we should ask: a) Is it a global phenomenon? b) What exactly happened to the geologic record? c) What mechanisms were at work, and how were they coupled? d) What was the effect of the event on subsequent Earth history, and especially on evolution? The types of events of interest range from the Permian and Triassic excursions in the global carbon and sulfur isotope curves (Holser, this volume) to the stratigraphically well documented Eocene-Oligocene transitions that are seen in deep-sea sediments. The resolution of the former signal (which is based on compilations) is between one to several million years. The resolution of the latter is about 0.1 million years.

THE EVIDENCE FOR SHORT-TERM CHANGE

Field Geology

Field geologists are familiar with the fact that sedimentary bodies in any one basin are commonly well defined by distinct boundaries which can be mapped without ambiguity over long distances. Even rather thin layers have been followed for hundreds of miles in ancient deposits and on the modern seafloor; turbidite layers in the Alps and certain Eocene chert layers in the North Atlantic are examples of such layers. The sharp lower and upper limits of such sediment bodies demonstrate that changes occurred rapidly within the basin in question. The contemporaneity of such changes can often be demonstrated. While this demonstration is by no means trivial, either as a task or as a result, we are here concerned with the demonstration of rapid facies change on a global scale.

There are many instances of field geologists discovering exactly the same stratigraphic sequence in coeval sedimentary rocks on different continents. Certain ones of such parallelisms have found adequate explanation. In the case of Paleozoic rocks in South Africa and South America, for example, the original contiguous sedimentary realm has been torn apart by continental drift. However, other instances remain puzzling and intriguing: parallel sequences of black shale sedimentation in the Ordovician of several continents, synchronous widespread coal formation in the Pennsylvanian, salt bodies in the Permian, redbeds in the Triassic, the iron ores and cherts of the Jurassic, the black shales and glauconite-rich sandstones of the middle Cretaceous, the chalks of the late Cretaceous, the bauxites in the Eocene, the diatomites of the Miocene.

Few geologists familiar with a particular portion of the stratigraphic record in one part of the world seem to be able to escape a feeling of déjà-vu when looking at rocks of the same age in different parts of the world. There are, then, some indications that the overall patterns of sedimentation - the geochemical cycles of weathering, erosion, fractionation, and final deposition - vary on a global scale in such a fashion as to imprint the record with signatures which are characteristic for certain periods. The terms "privileged facies" and "time-specific facies" have been used to describe the phenomenon. Climate-related facies seem to be particularly predestined to carry a strong global overprint. Lateritic soil formations and associated sedimentation patterns in terrestrial and marine basins provide one outstanding example (99).

Correlation and Resolution

It is very difficult, of course, to go beyond the recognition of such long-term fluctuations in the geochemical setting, and to demonstrate that the lower and upper boundaries of a privileged facies are globally contemporaneous. However, this is absolutely necessary if a fast transition is to be documented.

Difficulties arise from several sources. The problem of hiatuses, that is, uneven rock preservation, is perhaps the most familiar one (30, 94). Differences in sedimentation-rate fluctuations, between otherwise matched sequences, is a more subtle hindrance to correlation. Whatever the problems, a close match in the successions both of lithologic facies and fossil content commonly is very suggestive of the contemporaneity of sedimentary sequences. However, exact contemporaneity frequently remains open to doubt, because fossils are facies-dependent and do not necessarily provide a truly independent clock. Radiometric dating on the scale here considered is too imprecise to be of use in improving the biological resolution.

The problem of high resolution on a global basis has been solved on a scale of $n \cdot 10^3$ years for the Pleistocene, and on a scale of 10^4 to 10^5

years for well preserved, continuously cored deep-sea sediments of Pliocene to late Cretaceous age (13, 46, 71). Similarly, high resolution (2×10^5 years) may be possible in well preserved pelagic strata of the entire Phanerozoic, where fossils are abundant. Such resolution refers to correlation, not to absolute dating. Absolute ages, for the last 100 million years, are uncertain by perhaps 5 to 10%. Differences between absolute ages (which are of interest here) are usually quite unreliable. Sedimentation rates over short time spans, therefore, are extremely difficult to establish in pre-Pleistocene sediments. Rates of change recorded in sediments with unknown rates of sedimentation cannot be measured.

Clocks

There are two kinds of clocks which have received much attention in working on Cenozoic sedimentation rates and which hold promise for use even as far back as the Paleozoic.

The first clock has a quasi-random beat. It is the stratigraphy of magnetic reversals. To it we owe the mapping of seafloor ages, the rates of spreading and of continental drift, the exact duration of Pleistocene climatic cycles, and the confirmation of the synchroneity of stable isotope excursions in the Cenozoic and late Cretaceous. The resolution of this clock depends on the frequency of reversals and on the confidence with which particular reversals can be identified. Many a dispute has arisen over correlations which matched reversal sequences that were offset by one "cycle."

The second timing device is a true clock, with at least two, perhaps five, frequencies of oscillation. For Pleistocene sediments this clock, the Milankovitch mechanism, has been used to measure sedimentation rates to a hitherto unknown precision (see Fig. 1). This is done by matching a climate-related signal within the record to the signal expected from the orbital frequencies of the Earth.

The promise of such precise measurement, and associated lead-lag analysis (70), is that we shall be able to reconstruct the ocean's response to climatic change on a scale of about 1000 years on a global basis. Thus, for the first time, we shall be able to study rates of change within short-term transitions from one climatic state to another.

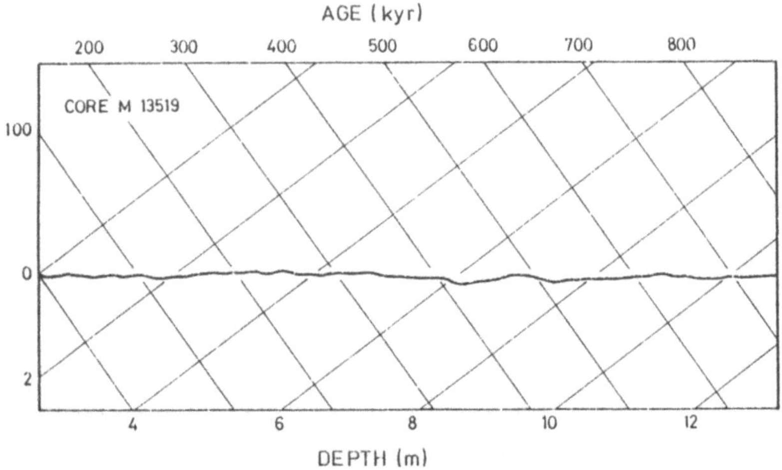

FIG. 1 - Fine-tuning of sedimentation record based on the Milankovitch hypothesis (Herterich and Sarnthein, unpublished manuscript).

There is no reason, in principle, why Milankovitch tuning could not be used in pre-Pleistocene sediments. The search for Milankovitch cycles has begun for the Tertiary and earlier epochs (see Schwarzacher and Fischer in (30)). For Mesozoic and older rocks the oscillations of the tuning fork must be established from the record itself: the various cyclicities in the spin and orbit of the Earth cannot be assumed to have been identical in the distant past. However, recent evidence does suggest that a precessional cycle with a period of about 20,000 years may be present in Permian salt deposits (2).

Stable Isotope Record

The best evidence for short-term changes on a global scale derives from the stable isotope record of deep-sea sediments of the last 100 million years or so ((4, 15, 60, 82, 87), and see Table 1).

With few exceptions, the evidence for these events is based on oxygen and carbon isotopes in planktonic and benthic foraminifera, following the method pioneered by Emiliani (31). Changes in the signals are taken to be synchronous on a global scale whenever they correlate well between ocean basins (based on biostratigraphy and magnetostratigraphy) (47). They are then assumed to reflect changes in ocean chemistry and in

TABLE 1 - Compilation of short-term changes in deep-sea sediments (Berger in (30)). For recent discussions of the deep-sea record see Emiliani (32).

1) Cretaceous-Tertiary Boundary Event (~ 65 m.y.). Large-scale extinction of oceanic plankton and other organism groups. Large climatic fluctuations probable.

2) Paleocene-Eocene Boundary Event. Short-lived peak warming. Expansion of tropical faunas and floras.

3) Early-to-Mid-Eocene cooling steps. Shift of climate zones toward equator. Cooling of deep water. Chert formation.

4) Eocene Termination Event (~38 m.y.). Cooling in high and low latitudes, expansion of polar highs. Significant changes in deep-sea benthic fauna. Rapid drop of the carbonate compensation depth.

5) Mid-to-Late Oligocene Oxygen Shift (?) (~30 m.y.). First occurrence of rather heavy oxygen isotope values in deep-sea benthic foraminifera, presumably due to polar bottom water formation.

6) Mid-Miocene "Oxygen Shift" (~15 m.y.). Oxygen isotope ratios shift to heavier values, presumably due to Antarctic ice buildup.

7) 6-million-year "Carbon Shift." Isotope ratios of the ocean's carbon shift to lighter values, presumably due to organic carbon input from regression and erosion. pCO_2-change.

8) Messinian "Salinity Crisis" (~5.5 m.y.). Isolation of Mediterranean through regression, strong cooling.

9) 3-million-year Event. Onset of Pleistocene-type climatic fluctuations.

10) 1-million-year Event. Onset of large climatic fluctuations after period of quiescence.

11) "Terminations" of the Late Pleistocene (last 0.8 m.y.). Rapid deglaciations, warming of mid-latitudes. pCO_2-change.

climate (temperature, salinity), rather than purely regional effects. The fact that stable isotope stratigraphies from different continents can be correlated in great detail supports the inference (85).

Nevertheless, it is always necessary to take into account local and regional influences on stable isotope signals, as is true for any other type of stratigraphic record.

The advantages of using the stable isotopes of common marine elements (O, C, S, Sr) to characterize the geochemical state in the Earth's exogenic system have been detailed by Holser and by Fischer (both this volume). The "global curves" presented there are, in essence, lithologic compilations similar to, say, limestone/shale ratios through time. However, the curves of variations in isotopic ratios have the advantage that a global signal is likely to be strongly represented in all sedimentary basins, no matter what their setting, as long as they are open to the ocean (and as long as diagenetic overprints are small).

To what degree compilations based on bulk analyses of scattered rock samples can establish a convincing case for short-term change is still questionable. For example, the differences in the stable isotopic composition of carbon in different components of the same carbonate sediment can be substantial (Fig. 2). Thus, a sudden change in isotopic composition in a carbonate rock could simply reflect a change in the proportion of, say, molluscs and algae. These difficulties can be overcome, in principle, by using well-defined sediment fractions (or by correcting for changes in particle ratios), and by using only reliably dated rock samples. Even with relatively low resolution ($n \cdot 10^6$ years), the long-term intercorrelations between the different isotope curves, and comparisons with other geochemical indicators, with sea level, and with diversity curves contain useful guidelines for modeling the system, as

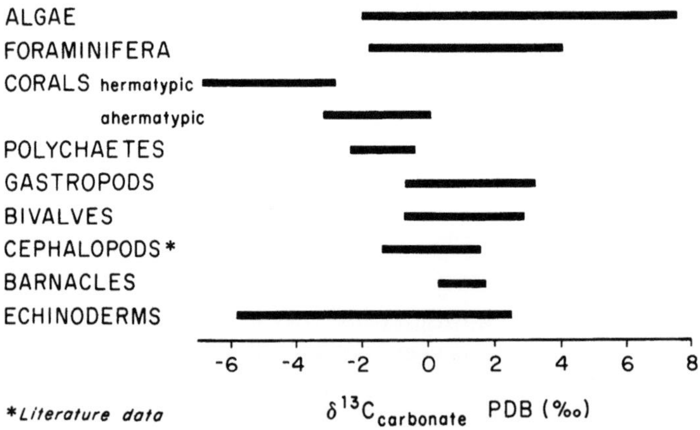

FIG. 2 - Differences in C isotopes, contemporaneous sediments (Wefer, unpublished data).

discussed by Holser and Fischer (both this volume). The anticorrelation between ^{13}C and ^{34}S, seen in the Phanerozoic record, holds only over long time periods. On a short time scale in the marine realm, the deposition of reduced carbon and sulfur are positively correlated. A combination of arguments based on steady-state (39) and on major shifts in geochemical setting through geologic time (18) may be necessary to resolve this paradox (and others like it).

THE ROLE OF SEA LEVEL CHANGE

Instability and Sea Level

The search for the causes of short-term changes in the Earth's environment is not usually hampered by a scarcity but by an overabundance of possible explanations. In principle, there are three types of causes for change: agents arriving from beyond the atmosphere (extraterrestrial, or ET), those related to processes in the mantle (including isostatic effects), and those inherent in the exogenic system itself. Our main focus here is on the last of these. The instabilities within the exogenic system (ocean-cryo-atmosphere-biosphere-surface deposits) derive from two sources: a) positive feedback, especially from changes in albedo and free CO_2, and b) transient reservoir effects (Berger, in (12)). Transient reservoir effects are those that involve surficial accumulations of materials (polar ice caps, salt bodies, deposits of organic carbon, isolated water masses) whose quick release to the ocean-atmosphere system affects climate in major ways (52, 92). Release can effect changes in sea level, in ocean stratification and circulation, in the hydrologic cycle, and in the concentration of CO_2 in the ocean and atmosphere. The result of large, sudden release is a rapid change in climate, with implications for geochemistry (redistribution of surficial sediments) and evolution (change of habitat).

One promising strategy is to explore the role of changes in sea level in generating instability of the system. Sea level change, from the beginnings of geologic mapping, has been recognized as a central theme of the Phanerozoic record. It is rare that a major change in sedimentary facies will not be associated with a change in water depth by the observer. Likewise, global ("eustatic") sea level changes have long been proposed as chief agents of change in climate and evolution (12, 35, 89). In principle, a change in sea level can produce short-term climatic changes in two ways: a) Sea level can rise or fall quickly and establish a new steady state, corresponding to the new level, in a short time. b) Sea level can rise or fall slowly, but a substantial short-term effect on the system is derived from amplification by positive feedback, or from perturbation of transient reservoirs, or both.

Why Sea Level Changes

There are several ways to produce rapid changes in global sea level. Best known of these is the buildup and decay of polar ice. Under present (Pleistocene) conditions the potential amplitude of the change is 200m (±10%); we are right now 70m below the maximum possible rise. (This fact is of some interest because the recent rise in pCO_2 could cause extensive melting of polar ice.) Throughout the late Pleistocene, sea level varied within closely defined limits, by about 120m to 140m (19, 88). A mechanism is (and has been) at work which prevents total deglaciation as well as continued ice buildup after a certain maximum is reached.

Sea level can also change rapidly due to the filling or emptying of large isolated basins. This mechanism has become a very attractive possibility since it became established that the Mediterranean dried out in the Late Miocene (79). The presence of salt deposits around the margins of the Atlantic Ocean make it likely that isolated basins were present throughout the early breakup of Pangaea. The amplitudes of expected sea level changes due to desiccation and filling of large basins are between 10m and 50m (14).

Changes in sea level on a scale of $n \cdot 10^6$ years can be related to orogenesis. The pushing-up of the Himalayas, for example, produces a sea level drop of about 50m (14). The late Tertiary mountain-building activity, which produced the Alpine chains and today's Cordilleras, is generally accepted to be responsible for the overall regression since the Miocene (45).

Long term changes in sea level have been tied to seafloor spreading and continental drift (51, 61, 77). Changes in the distribution of hotspots relative to continents and ocean basins have been proposed also as a means of changing sea level (1). The development of large oceanic plateaus in the western Pacific (Ontong-Java, Manihiki, Shatsky, etc.) and the associated volcanism have been linked to the late Cretaceous transgression (84). (Of course, it is not the volume of the volcanic masses which is important, but the evidence for increased heating of the lithosphere. For substantial change of sea level, areas of the seafloor much greater than plateaus have to change elevation.)

The manner in which changes in ocean capacity are translated into transgressions and regressions on land is by no means clear (3). Curray (28) and Pitman (74) have stressed the interaction between eustatic rise and fall, regional subsidence (or uplift), and rate of sedimentation (Fig. 3). In particular, Pitman has argued that the sequence of transgressions and regressions which is seen in the Tertiary reflects an overall drop of sea level which proceeds at varying rates. Changes in the hypsographic curve through geologic time must also be taken into account in relating eustatic sea level fluctuation to transgression and regression (42).

Sea Level Curves

Independently of a better understanding of the causes of sea level changes, the need is for a reliable reconstruction of global variations of sea level during the entire Phanerozoic. This is a most difficult task, especially for a time resolution of better than 10 million years. One problem is epeirogeny, the vertical movement of large crustal areas (67). Others are changes in the geoid (68) and migration of continents through the equatorial (bulge) zone (33). The traditinal method has been to compile the areal extent of marine rocks, integrated for certain time periods. The weaknesses of this method are well-known. Differences between

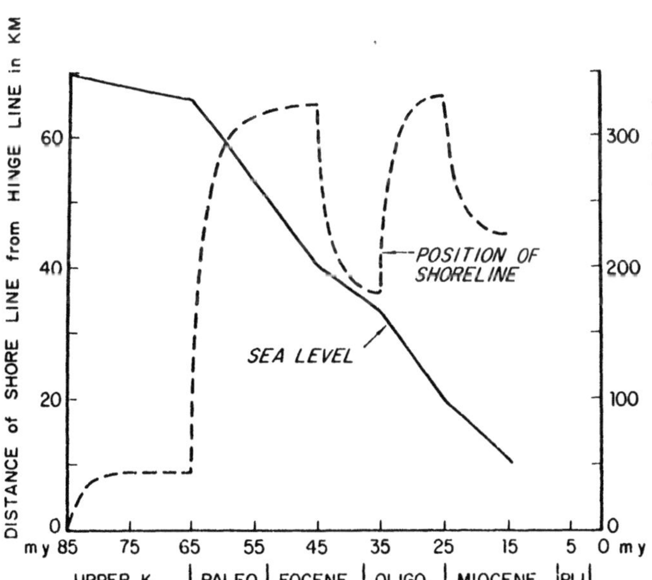

FIG. 3 - Transgressions during regression as proposed by Pitman (74).

periods are smoothed because of lack of resolution in dating, later erosion removes the evidence for previous transgressions, subsurface rocks are not usually well studied (or when studied, the information is not necessarily available), and disconnected outliers of marine rocks have undue influence on the reconstruction of shelf flooding. The most severe limitation is the one owing to dating. Thus, reconstructions of Phanerozoic sea level variation tend to be smooth and generalized curves. The curves resemble a double-arch or a capital M ("M-curve"). Two versions of the M-curve are reproduced by Fischer (this volume). The first arch of the M is centered on the Ordovician, the second on the Cretaceous. The lowest stands are in the Permian and at the present.

Generalized M-curves are not particularly relevant to the question of short-term change. Of more interest in this context are the second-order fluctuations of sea level. The ones produced by Vail and Mitchum (figured by Holser, this volume) have received much attention, especially with regard to the instantaneous regressions (for discussions see (74) and (104)). The reliability of this Phanerozoic sea level curve is difficult to assess, to say the least. One reason is that, unlike the classical M-curve, the "Vail Curve" shows the lowest point not in the Permian but well after it, within a low stand of sea level spanning the entire period from the middle Permian to the middle Jurassic.

Effects of Sea Level Change

The usefulness of the various versions of the Vail sea level curves lies in the details they offer for comparison with stratigraphic signals describing rapid change. A recent example of such use is given in Fig. 4.

There is a correspondence between hiatuses in deep-sea carbonate deposition of the Cenozoic and regression events in the Vail Curve. It could be argued that a change in sea level leads to a rearrangement of areas accessible for chemical erosion and deposition, and that such rearrangements must result in hiatuses in some places and increased sedimentation rates in others. The task, of course, is to reconstruct the mechanisms which lead to the redistribution (including changes in bottom water circulation). In this case, redistribution involves carbonate sediments, that is, the geochemistry of carbon. It seems likely, therefore, that atmospheric pCO_2 and hence climate and evolution were also affected during these periods of redistribution.

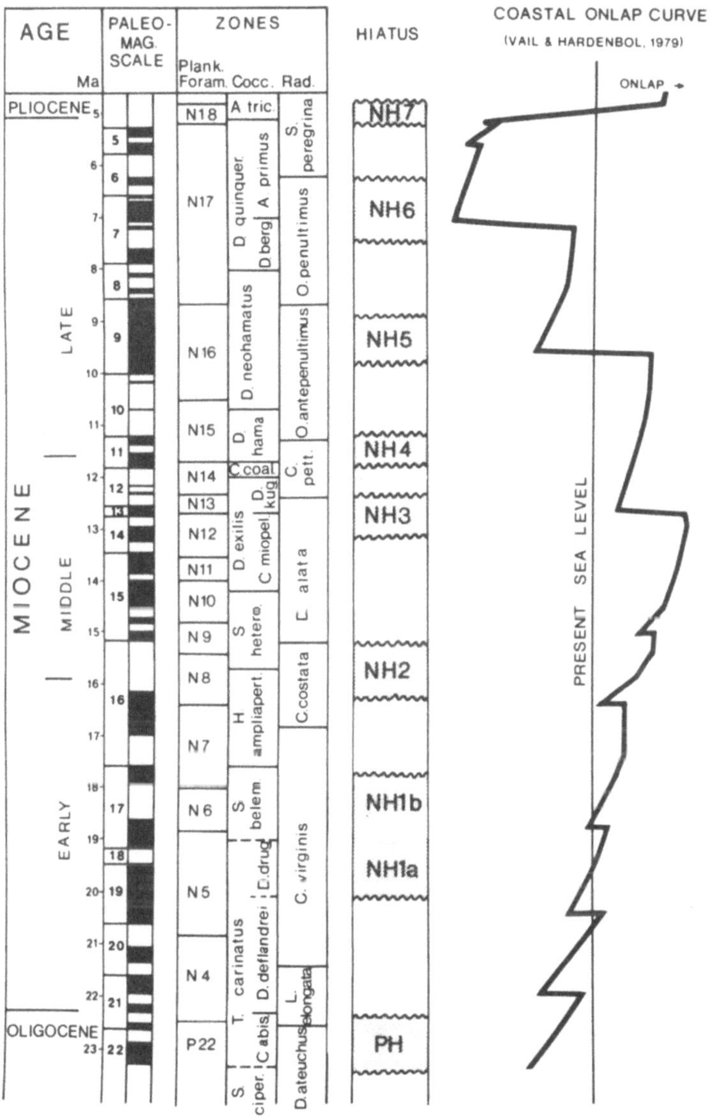

FIG. 4 - Cenozoic hiatuses in deep-sea sediments and Vail Curve (8).

Sea level change can produce rearrangement of materials (and therefore geochemical change) simply by providing new shelf surface for deposition or for erosion of shallow water sediments. Such changes can interfere substantially in the carbon cycle of the ocean, and hence in the ocean's fertility. Holland proposes the following scenario for transgression during conditions favorable for black shale development on shelves (strong oxygen minimum):

The amount of organic carbon deposited per unit time on Earth may be taken as

$$dM_c/dt = f \cdot P \ , \tag{1}$$

where P is the photosynthetic productivity of the biosphere in moles of carbon per year, and f is the fraction of organic carbon buried within sediments. f depends greatly on the oxygen content of seawater. It is approximately 0.003 for the modern ocean as a whole, but near 0.05 for anaerobic basins. During transgressions of an ocean with a strong and shallow oxygen minimum, the area newly covered by low-oxygen water is greatly increased. On this area (the deepened shelf) f is now greatly increased. Hence, if the rate of burial of organic carbon is to remain approximately constant (for geochemical balance), P must be decreased correspondingly. Transgression may therefore greatly reduce the fertility of the ocean. (In contrast, Fischer and Arthur (35) have suggested increased fertility during high stands of sea level.)

A mechanism for linking transgression and reduction of fertility has been given by Broecker (23). He proposes that phosphorus is extracted by sedimentation on the newly flooded shelf, especially within estuaries, and becomes unavailable for recycling within the ocean (until the next regression). Phosphorus is an essential nutrient. Another possibility is that nitrate is more easily destroyed under conditions of widespread shelf anoxia (11). This would lower the fertility of the ocean due to lack of nitrate and hence allow increased phosphate concentrations in the ocean, without affecting the productivity of the ocean. This phosphate could then be used for regional concentration by phosphatization of limestones and other deposits.

The significance of the above geochemical models involving carbon deposition and ocean fertility lies in the fact that even a slow change in sea level could produce results rather quickly, once a critical shelf area is attained. Lowered ocean fertility can be expected to have a multitude of effects on the chemistry of the ocean, on sedimentation patterns, on climate, and on life.

The relationships between anaerobism and fertility are by no means clear. Indeed, the entire complex of questions surrounding the occurrence of organic-rich deep-sea sediments (in the Cretaceous) is the subject of intense discussion and modeling ((6, 10, 56, 78, 90, 91), and Brass et al. in (12)).

A more direct physical means of amplifying and translating sea level changes into global changes of the environment is through albedo. The ocean surface is much darker than land. On the average, it reflects only between 4 and 10% of the light incident on it from the sun overhead. The land reflects much more: 10% from dark forests, 30% from deserts, 50-80% from snow and ice. A transgression makes the Earth darker and therefore warmer. Transgression also provides an increased surface for the evaporation of water. More heat is trapped by the greenhouse effect through an increase in water vapor. Heat storage and transport are facilitated both in the ocean and the atmosphere: the planetary climate becomes more equable. Climatic warming feeds back into geochemistry: bottom waters of the ocean become warmer (and slightly saltier) and hence hold less oxygen. Chemical erosion increases relative to physical erosion (although not necessarily in absolute terms). Other mechanisms linking sea level and climate also exist, although they probably have less impact than albedo (see Berger in (30)).

Sea Level and Diversity

Discussions about the correlation of sea level fluctuations with the diversity (i.e., number of taxa) of marine fossils have a long tradition. Recent examples are by Fischer and Arthur (35) and Fischer (this volume). In the present context, we are especially interested in whether the correlation holds true on short time scales. There is some evidence that this is so (see Fig. 5).

Before the physical and biological causes of such a correlation can be discussed intelligently, it has to be realized that a) there is a sampling problem, and b) diversity is a complex phenomenon which can be changed in several ways (36, 43).

The sampling problem, stated in crude simplicity, is this: few marine sediments are deposited during regression, hence few marine fossils are found, hence the observed diversity is low. The obvious answer is to

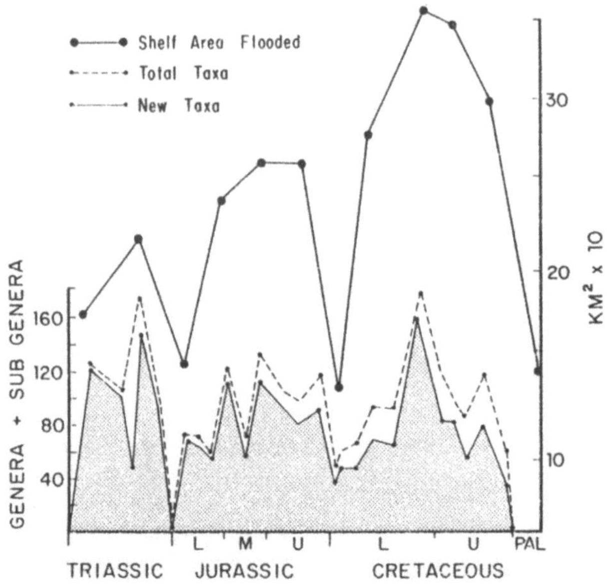

FIG. 5 - Diversity of Ammonites compared with sea level changes (57).

normalize the sampling space through some kind of rarefaction technique, as described in ecology textbooks. A more subtle problem could, however, remain: since shelf seas are shallower during regression, the type of environment (not just the area) preserved on the shelves during transgression differs from that preserved during regression. The argument boils down to one concerning bias in preservation. It leads back to the well-known imperfection of the record which plagues the terrestrial plant record perhaps even more than the record for marine fossils (see, e.g., Niklas et al. (72)).

Sampling problems and the imperfection of the record cannot be the whole story, however. There is a positive correlation between plankton diversity and sea level change in the Tertiary (16, 25, 102). It is quite clear that this correlation is real. It is also clear that whatever controls diversity changes in the benthic or bentho-pelagic organisms on the shelf is not necessarily an important factor in producing diversity changes in the plankton organisms of the open ocean.

On the shelves, species-area relationships and provinciality presumably are important (see (96) for discussions). In the open ocean, stability of stratification and fertility patterns are probably crucial (63). Both regions are surely sensitive to global temperature gradients, and especially to fluctuations of such temperature gradients on ecological time scales (life spans to succession spans). It is not necessarily true, incidentally, that higher fertility engenders higher diversity as has been suggested. In the present open ocean, food chains are longest (and diversity greatest) in the oligotrophic regions (81).

The compound nature of diversity fluctuation derives from the fact that diversity is the result of a dynamic balance between rates of speciation and rates of extinction. As discussed by Padian et al. (this volume), changes in either factor (or different combinations in both) can produce a given change in diversity, whereas compensating changes in both factors produce no obvious result in the diversity index. Therefore, while the overall correlation between diversity and sea level draws our attention to interesting phenomena, there is no guarantee whatever that there is one cause or set of causes which is responsible for correlations seen in any one time interval. Cause-and-effect relationships are most readily studied where the signal is strong, that is, where diversity changes significantly. This is the case at periods of extinction and of radiation.

Sea Level and Mass Extinction

A survey of mass extinctions in Phanerozoic oceans has recently been provided by Sepkoski ((86); see also (76)). He suggests that as many as fifteen mass extinctions are present in the fossil record, of which the Late Permian one was by far the most severe. More than one half of all families of marine taxa became extinct at the end of the Paleozoic. Sepkoski classifies four mass extinctions as intermediate: the Ashgillian event (terminal Ordovician), the Frasnian event (Middle to Late Devonian), the Norian event (Late Triassic), and the Maestrichtian event (K/T boundary). Each of these "events" saw a decrease of between 15 to 22% in global familial diversity. Less severe extinction events are seen in the Cambrian (five events), in the Toarcian (Early Jurassic), in the Cenomanian (mid-Cretaceous), in the late Eocene (Eocene-Oligocene boundary), and in the Pliocene. The earliest mass extinction has been claimed for the late Precambrian (650 Ma) by Vidal and Knoll (101), involving unicellular algae. The Ediacaran radiation of metazoans (Cloud and Glaessner (26)) may have to be seen in the context of a preceding mass extinction, just like to many other radiations.

A comparison of the timing of the major mass extinctions with the appropriate "sea level" curves by Vail et al. (95) suggests that there is no clear relationship between extinction events and regression of the sea. While it is true that the most prominent extinction, which ended the Permian, took place during a low stand of sea level, the major drops in sea level were before and after the Permian termination. During the Ashgillian and Frasnian events, sea level was high and regressions (if any) were minor in the Vail scheme. The Norian event preceded a strong drop in sea level shown for the earliest Jurassic. The K-T event is, indeed, marked by a distinct regression, but the Eocene-Oligocene event, again, preceded a major regression in the middle Oligocene.

An attempt to relate diversity patterns to sea level patterns, then, is unlikely to produce much insight into the driving forces of accelerated extinction. McLaren (66) has suggested that each period of mass extinction may have its own unique set of causal factors. He finds that the Frasnian-Famennian boundary may well be one of the best documented mass extinction events of the Paleozoic (65, 66). The disappearance of shallow-water reefs and the wholesale extinction of brachiopods are especially noteworthy. The tropical and subtropical zones appear most affected. Interestingly, a decrease of diversity precedes the mass extinction event, suggesting that the event may have taken place within a setting of environmental stress. A brief period of glaciation appears to have been centered on the event (Crowell in (12)), which may account for stressful conditions (climate, ocean circulation, level of oxidation; see Legget et al. (62)).

Changes of sea level must be seen against the climatic and geochemical background of the period in which they occur. In a Frasnian ocean, with possibly widespread euxinic conditions in shallow waters (24), the effects of a change in sea level should have been different from those of a change during the Maestrichtian, when shallow seas were well oxidized. The example of "anaerobic transgression" discussed earlier illustrates this point. A low pO_2 in the early Paleozoic would have provided a set of conditions for climatic fluctuations which was entirely different from that of the Cenozoic. It is noteworthy, in this context, that the substantial and rapid sea level fluctuations of the Pleistocene did not result in the widespread extinction of marine shelf organisms. Climatic fluctuations did, however, affect organisms on land (e.g., (41)). In contemplating the course of evolution, transgression (or regression) must be seen in the context of paleogeographic and oceanographic background conditions.

Conversely, low or high sea level positions may precondition the system so that it reacts differently to a given set of disturbances. In viewing the overall association of low sea level stand and decreased diversity, for example, one is tempted to dismiss a call for special extinction agents and invoke instead unusual reactivity of the system to otherwise "normal" disturbances, including sporadic bombardment from outer space. In the search for "the cause" of rapid change in the geologic record, the influence of synergetic effects between tectonics, geochemistry, and astronomy may have been underestimated.

THE ROLE OF PLATE TECTONICS

General Importance

Plate tectonics has become the reigning paradigm of geological research. It provides the basis for reconstructions of the geography of continents and ocean basins, their configurations, and their positions with respect to each other and the equator. It provides the first plausible explanation for the onset of the ice ages (29), it helps to explain the continuing diversification of higher organisms following the Permian (97), and it accounts for the salt deposits at the margins of the Atlantic (49). In the context of short-term changes on the surface of the Earth, plate motions are important in changing the sensitivity of the exogenic system to amplifying mechanisms (albedo, pCO_2), for example, through position and elevation of continents. Plate tectonics provide or destroy opportunities for transient reservoirs and for interbasin exchange, and affect sea level and the intensity of volcanism.

The effects of continental drift and orogenesis on climate and climate sensitivity were modeled by Donn and Shaw (29). They showed semi-quantitatively how the polar position of Antarctica and the northward drift of the Eurasian and North American land masses could account for the overall cooling trend observed on Earth since the Mesozoic. Orogenesis, by lowering sea level, increases the land area, which in turn increases the surface albedo. As the land masses move poleward, their snow-cover increases and less of the sun's radiation is retained. Ultimately, conditions for ice buildup are reached, first on Antarctica and subsequently in the northern hemisphere. At this point the system becomes highly sensitive to disturbance (transient reservoir ice) and to relatively small variations in the distribution of seasonal irradiation (Milankovitch mechanism). More recently, Barron et al. (9) and Thompson and Barron (93) have addressed the questions of the effects of land-sea distributions and continental drift on Earth albedo and climate.

Evolution and Tectonics

Marked changes in the evolution of complex organisms may be closely tied to plate tectonics. Diversity on the family level (see Fig. 1 of Fischer, this volume) is characterized by a rise during the Cambrian from near zero to a Paleozoic plateau, followed by the well-known late Permian crash, a low stand in the Triassic, and since then a steady increase that was interrupted briefly at the end of the Cretaceous.

The timing of the breakup of Pangaea in the Triassic invites speculations as to cause-and-effect relationships between its antecedents and the Permo-Triassic crisis. Whatever the reasons for the late Permian extinctions, ET or otherwise, the system was in an unusual state and sensitive to disturbance. The late Paleozoic glaciations may have been part of such conditioning (27). The buildup of large salt deposits at the end of the Paleozoic has been invoked as a cause for extinction, through effects on overall salinity or on density stratification (34). Funnell (38) has related the deposition of $CaSO_4$ to increases in pCO_2. In essence, he argues that the availability of Ca governs $CaCO_3$ deposition and hence the ease with which carbon can escape the ocean-atmosphere system.

The chief significance of the evaporite deposits, in the present context, may lie in the evidence they give for the presence of isolated basins which provide for transient reservoirs of "bad water" (fresh or salty). Pulsing brine sources, for example, could introduce great fluctuations into the fertility of the ocean. Quasi-stagnant high salinity pools on the deep seafloor would quickly trap all nutrients, through removal of settling organic matter, preventing recycling. Conversely, the intermittent erosion of such pools could generate short-term eutrophication of surface waters and hence produce stress for oligotrophic organisms.

Conceivably, large freshwater invasions from the Arctic might adversely affect marine life in upper waters through changes in salinity and climate, as proposed by Gartner and Keany (40). Whatever the mechanisms, semi-isolated basins throughout the earliest Mesozoic must have increased the sensitivity of the system to disturbance. The basic reason why the Triassic fauna remained "grubby" (96), and diversity failed to increase, may be this instability of ocean fertility and climate which is expected from the breakup situation. The quality of the fossil record is such that the entire Permo-Triassic diversity crisis can be argued to be a result of decreased origination rates, presumably due to long-lasting environmental stress (54).

The post-Triassic diversification was explained by Valentine and Moores (98) as a result of increased shelf area and provinciality. Alternatively, or in addition, it may be seen as a relaxation phenomenon. Continued drift and continued generation of young seafloor eliminated many of the semi-isolated basins with their potential for collecting chemical deposits (extraction of salt, nutrients, carbon) and may have diminished their role as intermittent sources of instability.

Paleogeography and Circulation

The exchange of water masses between different ocean basins is an important factor in determining sediment distribution, fertility patterns, the heat budget, and biotic provinciality. Continental drift and seafloor spreading constantly alter the exchange configurations, opening or closing straits and raising or lowering sills. Overall, the post-Permian history of ocean configuration can be described as a replacement of latitudinal circulation (Tethys) with a meridional circulation, and the placement of bottom-water sources of the poleward extremes of the Atlantic (17). The opening and closing of ocean gateways can generate a short-term change in the system. That certainly occurred six million years ago, when access to the Mediterranean was restricted, with substantial feedback into the climate (103). However, the importance of gateways as a global factor for geochemical and climatic change has not been demonstrated unequivocally. A lack of precise dating combined with a lack of physical modeling are the main limiting factors in such an attempt.

Volcanism

Volcanism has been another forcing factor related to plate tectonics. Today, volcanism is concentrated around subduction zones. Not so evident are the effects of the Mid-Ocean Ridge, where acid is pumped into the sea (releasing CO_2). High rates of seafloor spreading will tend to produce a CO_2-rich atmosphere. Volcanism has two effects: it tends to increase the pCO_2 (greenhouse effect) and it tends to increase the dust content of the atmosphere. If one could show that the rate of volcanic emissions has been particularly high during certain periods of Earth history, one might find possible causes for short-term change. So far, the best evidence for the effects of volcanism on climate has come from studies of Late Pleistocene records (20, 21). Perhaps volcanism helps to produce ice ages. However, there is the intriguing possibility that ice ages produce volcanism: do the growth and decay of ice sheets produce sufficient crustal stress to trigger volcanism? (75)

The history of volcanism has been extracted largely from the ash-derived sediments, although it should also be possible to date the construction of volcanic edifices and estimate their volume. This has been done, for example, for the Columbia River Basalt (53). The cessation of these flows, incidentally, coincides with the growth of the Antarctic ice sheets as seen in the oxygen isotope stratigraphy of deep sediment (82, 87). Flood basalt provinces typically contain a volume of basalt between 10^5 and 10^6 km^3 (69). The associated emission of acids quite possibly had important effects on climate during flood basalt buildups through the increase of atmospheric CO_2.

A typical volcanic ash curve for Cenozoic sediments is shown in Fig. 6. The intensity of volcanism appears to have been high during the Eocene and the Miocene and increased in the latest Tertiary. The meaning of the curve in terms of global activity is not entirely clear (73); even if it were, the conclusions to be drawn in terms of climate evolution would still be vague. This should not detract from the value of studying the history of volcanism. If intensities did change substantially through time (as argued in (58)), the effects should have been noticeable (7).

THE FEEDBACK FROM EVOLUTION

Gaia

There is a concept which asserts that the evolution of organisms on the planet regulates the geochemistry of the exogenic system in favor of survival of life on Earth. This is the "Gaia Hypothesis" (64). Disproof of this "hypothesis" is hard to imagine as long as le Châtelier's Principle is at work and as long as life is able to adapt to changing conditions on the surface of the planet. The philosophical underpinnings of the "hypothesis" appear to be teleological and lie outside the scope of our discussion. However, the "hypothesis" illustrates an important point: the geochemistry of the ocean, the air, and the surficial deposits are dominated by life processes. This concept is by no means new: the works of Vernadsky (100), Hutchinson (55), and Harvey (48) come to mind. It has perhaps not always received the attention it deserves.

Atmosphere and Evolution

The fact that the atmosphere consists largely of nitrogen attests to the efficiency of denitrifying bacteria which produce nitrogen gas from nitrogen oxides. Conversely, nitrogen is fixed into organic matter by other bacteria. This cycle is of prime importance for the fertility of the ocean because fixed nitrogen is a limiting nutrient. Any change

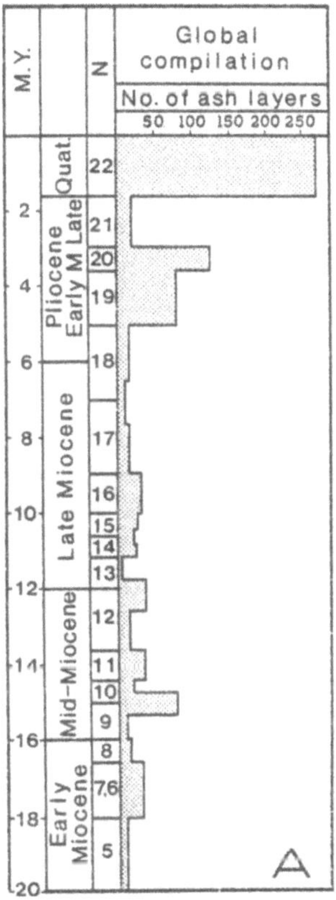

FIG. 6 - Late Cenozoic volcanism as seen in deep-sea ash layers (59).

of conditions which change the rates in this cycle would have far-reaching effects. The development of symbioses between land plants and nitrogen-fixing bacteria and that between certain diatoms and blue-greens are examples. Unfortunately, the geologic record is unlikely to contain an explicit signal of such evolutionary changes.

The history of the oxygen content of the atmosphere is closely tied to the evolution of photosynthetic organisms, as is well-known (see Fischer, this volume). The evolution of land plants in the late Paleozoic, which soon provided resistant and bulky carbon for burial, presumably could have helped induce the decreasing importance of anaerobic marine deposition in subsequent epochs. It must be pointed out, however, that there is no significant ^{13}C shift seen in the geologic record at the time when land plants became important. Thus, gradual change rather than rapid change may be involved. Life systems and sedimentation are highly sensitive to threshold levels of oxygen concentration. Thus, gradual changes have the potential to produce short-term steps when a critical point is crossed. For example, many marine organisms do quite well in seawater with an oxygen content greater than 0.2 ml. per liter. Below this level, however, the environment becomes effectively "anaerobic," at least for metazoans. Also, small differences at low oxygen levels determine whether the sediment will be stirred through bioturbation. Cessation of bioturbation over a large area can remove large amounts of carbon (and associated nutrients) from recycling and hence can decrease fertility considerably. The change needed for this effect near critical levels of oxygen may be as small as 1% of the range of oxygen content in seawater. It is highly probable, against this background, that the early evolution of land plants (and later of macroalgae) continuously affected the level of instability in the system, via oxygen content in the ocean.

The CO_2 content of the atmosphere has been recognized since the last century as a major factor in climate evolution. Because of the current rapid increase in CO_2 level (through feedback from primate evolution), much attention has been given to the factors controlling it. Through the linkage of pCO_2 to the erosion and sedimentation cycles of organic carbon, carbonate, sulfide, sulfate, and iron minerals, the long-term carbon cycle is highly complex. It is well appreciated that the various cycles are biologically mediated. Bacterial decomposition within bioturbated layers, calcareous shell formation, carbonate dissolution through CO_2-enriched groundwater, bacterial sulfate reduction, and bacterial iron oxidation are important topics in this context.

For short-term change in pCO_2, two factors are especially relevant. The first is a buildup of a biosphere whose carbon mass (with associated soil carbon) exceeds that of the atmosphere. The average residence time of carbon in the biosphere (mainly woody plants) is less than 100 years. That is, this reservoir is potentially highly volatile and a source of instability. Growth and decay of vegetative cover, of course, also affect surface albedo in major ways: another source of instability. The second factor is the partitioning of inorganic carbon between the ocean and the atmosphere. At present, the marine reservoir of carbon is about 60 times that of the atmosphere. To become a system that is in thermodynamic equilibrium (a "Sillén-Ocean"), the ocean would have to give up about 1.5% of its CO_2 and deliver it to the atmosphere. The pCO_2 would then be double. Life processes in the ocean tend to keep CO_2 in the ocean by producing respirative CO_2 below the surface waters. One manifestation of this process is the oxygen minimum. If, in the course of geologic history, plankton organisms became more (or less) resistant to decay, or if bacteria changed their strategy of decomposition, then pCO_2 would have changed accordingly.

Carbonate, Silica, Phosphate

Possible effects of biotic evolution on the cycles of carbonate and silica are mentioned by Fischer (this volume). Seilacher (this volume) discusses the importance of fossil preservation as a monitor of, and a factor in, the phosphate cycle. In the present ocean, the "forced" precipitation of carbonate by corals, molluscs, forams, coccoliths, etc., depresses the level of dissolved carbonate. The deep ocean is therefore much more undersaturated than it would be in the absence of carbonate-secreting organisms. This fact profoundly influences the distribution of carbonate sediments on the face of the planet. Major changes in carbonate precipitation, from evolution, apparently occurred twice: in the late Cambrian and Ordovician, when corals, molluscs, and many other metazoans acquired calcareous skeletons, and in the late Jurassic, when foraminifera and coccoliths started to invade the plankton. The efficiency with which coccolithophores removed carbonate from the shelves to the deep sea in the Cretaceous and Cenozoic may have resulted in a calcium shortage through subduction of deep-sea carbonate (50).

The silica cycle in the present ocean is dominated by diatoms. They are the reason why surface waters are nearly free of dissolved silica, and why glass sponges are rare in shallow waters, except in upwelling regions.

The predominance of planktonic diatoms is a relatively recent phenomenon. Their evolution may have greatly influenced that of the radiolarians which were obliged to keep building lighter and lighter skeletons throughout the Cenozoic. The evolution of grasses in the early Cenozoic, with their ability to dissolve and incorporate silica into their blades, may have been important in the leaching of soils. Eocene deep-sea cherts, in this fashion, might be linked to evolution on land. The separation of silica-free shallow water from silica-rich deep water introduces an instability which can produce short-term changes and a rapid alteration of facies in the record through small fluctuations in water mass characteristics or water mass boundaries.

The phosphorus cycle is crucial for ocean fertility because P may be the ultimately limiting nutrient. In this view, nitrogen fixation (and reduction) follows P abundance (22). The evolution of organisms precipitating phosphatic skeletons could have influenced the availability of P for photosynthesis. Although it is true that phosphate skeletons at present seem to represent but a small portion of the sink for P (37), the addition of such a sink to existing ones could have reduced P levels nevertheless (5). The preservation of phosphatic fossils may be taken as a clue to the availability of dissolved phosphate (Seilacher, this volume).

The general idea which emerges from this discussion is a sense that organisms, by precipitating solids from air and water, introduce disequilibrium into the system. If reservoirs created by organisms are large and active, instabilities are created. Disequilibrium and instability are not the whole story, however. Much of biotic evolution provides a negative feedback on the trends in geochemical evolution. For example, when oxygen became abundant, metazoans arose which produced bioturbation and kept organic matter recycling within the system; this created an additional oxygen demand. Disequilibrium provides energy gradients which can be colonized through evolution. This stabilizing influence follows from Le Châtelier's principle. This principle also implies that changes wrought by one organism (or a group of organisms) will eventually help limit that organism's activity. The diatom story is an obvious example. Other examples may be found as mankind finds its own limitations which are imposed by negative feedback in a changing exogenic system.

THE FUTURE

What are the most promising avenues for research leading to a better understanding of short-term change? Obviously, such research must first of all focus on periods of short-term change, that is, it must identify such periods through detailed stratigraphic work. This is by no means a trivial task, because of the necessity to demonstrate sufficient completeness of the record. Beyond this basic need for high-resolution stratigraphy, it is unlikely that geologists will agree on any one best way to study short-term change. Field geologists will insist on more field evidence, paleontologists on better analysis of larger collections, geochemists on more research into chemical signals and diagenesis, and modelers will stress that quantitative hypotheses (though wrong) are better than a narrative of historial events (albeit correct).

Perhaps the most encouraging trend in stratigraphy is that so many scientific disciplines are tackling geologic problems together, the best example being the full-scale assault on the K/T event, that is, the end of the Mesozoic (Padian et al., this volume). Similar interdisciplinary cooperation (and competition) should prove fruitful in pursuing the classic problems of Phanerozoic history touched on in this report. A first step might be to identify particular sections of the record worth studying in great detail. Times of worldwide facies changes and times of mass extinctions offer themselves as the most obvious points of attack. The necessary inspiration for the charge will have to come from the geologists involved rather than from a committee.

REFERENCES

(1) Anderson, D.L. 1982. Hotspots, polar wander, Mesozoic convection and the geoid. Nature 297: 391-393.

(2) Anderson, R.Y. 1982. A long geoclimatic record from the Permian. J. Geophys. R. 87: 7285-7294.

(3) Armstrong, R.L. 1969. Control of sea level relative to the continents. Nature 221: 1043.

(4) Arthur, M.A. 1979. Paleoceanographic events - recognition, resolution, and reconsideration. Rev. Geophys. 17: 1474-1494.

(5) Arthur, M.A., and Jenkyns, H.C. 1981. Phosphorites and paleoceanography. Oceanol. Act. (Suppl.) 4: 83-96.

(6) Arthur, M.A., and Natland, J.H. 1979. Carbonaceous sediments in the North and South Atlantic: the role of salinity in stable stratification of Early Cretaceous basins. In Deep Drilling Results in the Atlantic Ocean: Continental Margins and Paleoenvironment, Maurice Ewing Series, eds. M. Talwani, W. Hay, and W.B.F. Ryan, vol. 3, pp. 375-401. Washington, D.C.: American Geophysical Union.

(7) Axelrod, D.I. 1983. Role of volcanism in climate and evolution. Geol. Soc. Am. Spec. Paper 185, in press.

(8) Barron, J.A., and Keller, G. 1982. Widespread Miocene deep-sea hiatuses: coincidence with periods of global cooling. Geology 10: 577-581.

(9) Barron, E.J.; Sloan, J.L.; and Harrison, C.G.A. 1980. Potential significance of land-sea distribution and surface albedo variations as a climatic forcing factor: 180 m.y. to the present. Palaeogeo. P. 20: 17-40.

(10) Berger, W.H. 1979. Impact of deep-sea drilling on paleoceanography. In Deep Drilling Results in the Atlantic Ocean: Continental Margins and Paleoenvironment, eds. M. Talwani, W. Hay, and W.B.F. Ryan, vol. 3, pp. 297-314. Maurice Ewing Series. Washington, D.C.: American Geophysical Union.

(11) Berger, W.H. 1982. Deglacial CO_2 buildup: constraints on the coral-reef model. Palaeogeo. P. 40: 235-253.

(12) Berger, W.H., and Crowell, J.C., eds. 1982. Climate in Earth History. Studies in Geophysics. Washington, D.C.: National Academy Press.

(13) Berger, W.H., and Vincent, E. 1981. Chemostratigraphy and biostratigraphic correlation: exercises in systemic stratigraphy. Oceanol. Act. (Suppl.) 4: 115-127.

(14) Berger, W.H., and Winterer, E.L. 1974. Plate stratigraphy and the fluctuating carbonate line. In Pelagic Sediments on Land and Under the Sea, eds. K.J. Hsü and H. Jenkyns. Spec. Publ. Int. Ass. Sedimentol. 1: 11-48.

(15) Berger, W.H.; Vincent, E.; and Thierstein, H.R. 1981. The deep-sea record: major steps in Cenozoic ocean evolution. Soc. Econ. Pa. Spec. Publ. 32: 489-504.

(16) Berggren, W.A. 1969. Rates of evolution in some Cenozoic planktonic foraminifera. Micropaleontology 15: 351-365.

(17) Berggren, W.A., and Hollister, C.D. 1977. Plate tectonics and paleocirculation - commotion in the ocean. Tectonophysics 38: 11-48.

(18) Berner, R.A., and Raiswell, R. 1983. Burial of organic carbon and pyrite sulfur in sediments over Phanerozoic time: a new theory. Geochim. Cos. 47: 855-862.

(19) Bloom, A.L. 1971. Glacial-eustatic and isostatic controls of sea level since the last glaciation. In Late Cenozoic Glacial Ages, ed. K.K. Turekian, pp. 355-376. New Haven, CT: Yale University Press.

(20) Bray, J.R. 1974. Volcanism and glaciation during the past 40 millenia. Nature 252: 679-680.

(21) Bryson, R.A., and Goodman, B.M. 1980. Volcanic activity and climatic change. Science 207: 1041-1044.

(22) Broeker, W.S. 1971. A kinetic model for the chemical composition of sea water. Quatern. Res. 1: 188-207.

(23) Broecker, W.S. 1982. Ocean chemistry during glacial time. Geochim. Cos. 46: 1689-1705.

(24) Buggisch, W. 1972. Zur Geologie and Geochemie der Kellwasserkalke and ihrer begleitenden Sedimente (Unteres Oberdevon). Abh. hess. L.-amt Bodenforsch. 62: 1-68.

(25) Cifelli, R. 1969. Radiation of Cenozoic planktonic foraminifera. Syst. Zool. 18: 154-168.

(26) Cloud, P., and Glaessner, M.F. 1982. The Ediacarian period and system: metazoa inherit the Earth. Science 217: 783-792.

(27) Crowell, J.C. 1978. Gondwanan glaciation, cyclothems, continental positioning, and climatic change. Am. J. Sci. 278: 1345-1372.

(28) Curray, J.R. 1964. Transgressions and Regressions. Papers in Marine Geology, Shepard Commemorative Volume, pp. 175-203. New York: Macmillan.

(29) Donn, W.L., and Shaw, D.M. 1977. Model of climate evolution based on continental drift and polar wandering. Geol. Soc. Am. Bull. 88: 390-396.

(30) Einsele, G., and Seilacher, A., eds. 1982. Cyclic and Event Stratification. Berlin: Springer-Verlag.

(31) Emiliani, C. 1955. Pleistocene temperatures. J. Geol. 63: 538-578.

(32) Emiliani, C., ed. 1981. The Ocean Crust. The Sea, vol. 7. New York: Wiley-Interscience.

(33) Fairbridge, R.W. 1961. Eustatic changes in sea level. In Physics and Chemistry of the Earth, vol. 4, pp. 99-185. London: Pergamon Press.

(34) Fischer, A.G. 1964. Brackish oceans as the cause of the Permo-Triassic marine faunal crisis. In Problems in Paleoclimatology, ed. A.E.M. Nairn, pp. 566-577. New York: Interscience.

(35) Fischer, A.G., and Arthur, M.A. 1977. Secular variations in the pelagic realm. In Deep Water Carbonate Environments, eds. H.E. Cook and P. Enos. Soc. Econ. Pa. Spec. Publ. 25: 19-50.

(36) Flessa, K.W., and Imbrie, J. 1973. Evolutionary pulsations: evidence from Phanerozoic diversity patterns. In Implications of Continental Drift to the Earth Sciences, eds. D.H. Tarling and S.K. Runcorn, vol. 1, pp. 247-285. London: Academic Press.

(37) Froelich, P.N.; Bender, M.L.; Luedtke, N.A.; Heath, G.R.; and DeVries, T. 1982. The marine phosphorus cycle. Am. J. Sci. 282: 474-511.

(38) Funnell, B.M. 1981. Mechanisms of autocorrelation. J. Geol. Soc. (London) 138: 177-181.

(39) Garrels, R.M., and Lerman, A. 1981. Phanerozoic cycles of sedimentary carbon and sulfur. Proc. Natl. Acad. Sci. 78: 4652-4656.

(40) Gartner, S., and Keany, J. 1978. The Terminal Cretaceous event. A geologic problem with an oceanographic solution. Geology 6: 708-712.

(41) Grayson, D.K. 1977. Pleistocene avifaunas and the overkill hypothesis. Science 195: 691-693.

(42) Hallam, A. 1963. Major epeirogenic and eustatic changes since the Cretaceous, and their possible relationship to crustal structure. Am. J. Sci. 261: 397-423.

(43) Hallam, A. 1973. Diversity, provinciality and extinction of Mesozoic invertebrates in relation to plate movement. In Implications of Continental Drift to the Earth Sciences, eds. D.H. Tarling and S.K. Runcorn, vol. 1, pp. 287-294. London: Academic Press.

(44) Hallam, A. 1977. Secular changes in marine inundation of USSR and North America through the Phanerozoic. Nature 269: 769-772.

(45) Hamilton, W. 1968. Cenozoic climatic change and its cause. In Meteorological Monographs, vol. 8, no. 30, pp. 128-133. Boston: American Meteorological Society.

(46) Haq, B.U. 1981. Paleogene paleoceanography: Early Cenozoic oceans revisited. Oceanol. Act. (Suppl.) 4: 71-82.

(47) Haq, B.U.; Worsley, T.R.; Burckle, L.H.; Douglas, R.G.; Keigwin, L.D.; Opdyke, N.D.; Savin, S.V.; Sommer, M.A.; Vincent, E.; and Woodruff, F. 1980. Late Miocene marine carbon-isotopic shift and synchroneity of some phytoplanktonic biostratigraphic events. Geology 8: 427-431.

(48) Harvey, H.W. 1963. The Chemistry and Fertility of Sea Waters. Cambridge: Cambridge University Press.

(49) Hay, W.W. 1981. Sedimentological trends resulting from the breakup of Pangaea. Oceanol. Act. (Suppl.) 4: 135-147.

(50) Hay, W.W.; Barron, E.J.; Sloan, J.L.; and Southam, J.R. 1981. Continental drift and the global pattern of sedimentation. Geol. Rdsch. 70: 302-315.

(51) Hays, J.D., and Pitman, W.C. 1973. Lithospheric plate motion, sea level changes and climatic and ecologic consequences. Nature 246: 18-22.

(52) Holser, W.T. 1977. Catastrophic chemical events in the history of the ocean. Nature 267: 403-408.

(53) Hooper, P.R. 1982. The Columbia River basalts. Science 215: 1463-1468.

(54) Hüssner, H. 1983. Die Faunenwende Perm/Trias. Geol. Rdsch. 72(1): 1-22.

(55) Hutchinson, G.E. 1954. The biogeochemistry of the terrestrial atmosphere. In The Earth As a Planet, ed. G.P. Kuiper, pp. 371-433. Chicago: University of Chicago Press.

(56) Jenkyns, H.C. 1980. Cretaceous anoxic events: from continents to oceans. J. Geol. Soc. (London) 137: 171-188.

(57) Kennedy, W.J. Ammonite evolution. In Patterns of Evolution, ed. A. Hallam, pp. 251-304. Amsterdam: Elsevier.

(58) Kennett, J.P. 1981. Marine tephrochronology. In The Sea, ed. C. Emiliani, vol. 7, pp. 1373-1436. New York: John Wiley and Sons.

(59) Kennett, J.P., and Thunell, R.C. 1977. On explosive Cenozoic volcanism and climatic implications. Science 196: 1231-1234.

(60) Lancelot, Y. 1980. Environements sédimentaires océaniques: développement de la paléo-océanographie. Mem. Ser. Soc. Geol. France 10: 351-362.

(61) Larson, R.L., and Pitman, W.C. 1972. World-wide correlation of Mesozoic magnetic anomalies and its implications. Geol. Soc. Am. Bull. 83: 3645-3662.

(62) Leggett, J.K.; McKerrow, W.S.; Cocks, L.R.M.; and Rickards, R.B. 1981. Periodicity in the early Palaeozoic marine realm. J. Geol. Soc. (London) 138: 167-176.

(63) Lipps, J.H. 1970. Plankton evolution. Evolution 24: 1-22.

(64) Lovelock, J.E., and Margulis, L. 1974. Atmospheric homeostasis by and for the biosphere: the Gaia hypothesis. Tellus 26: 1-10.

(65) McLaren, D.J. 1982. Frasnian-Famennian extinctions. Geol. Soc. Am. Spec. Paper 190: 477-484.

(66) McLaren, D.J. 1983. Bolides and biostratigraphy. Geol. Soc. Am. Bull. 94: 313-324.

(67) Menard, H.W. 1973. Epeirogeny and plate tectonics. EOS 54: 1244-1255.

(68) Mörner, N.-A. 1981. Revolution in Cretaceous sea-level analysis. Geology 9: 344-346.

(69) Mohr, P. 1983. Ethiopian flood basalt province. Nature 303: 577-584.

(70) Moore, T.C.; Pisias, N.G.; and Heath, G.R. 1977. Climate changes and lags in Pacific carbonate preservation, sea surface temperature and global ice volume. In The Fate of Fossil Fuel CO_2 in the Oceans, eds. N.R. Andersen and A. Malahoff, pp. 145-165. New York: Plenum Press.

(71) Moore, T.C., and Romine, K. 1981. In search of biostratigraphic resolution. In The Deep Sea Drilling Project: A Decade of Progress, eds. J.E. Warme, R.G. Douglas, and E.L. Winterer. Soc. Econ. Pa. Spec. Pub. 32: 317-334.

(72) Niklas, K.J.; Tiffney, B.H.; and Knoll, A.H. 1980. Apparent changes in the diversity of fossil plants. In Evolutionary Biology, eds. M.K. Hecht et al., vol. 12, pp. 1-89. New York: Plenum.

(73) Ninkovich, D., and Donn, W.L. 1975. Explosive Cenozoic volcanism and climatic interpretations. Science 194: 899-906.

(74) Pitman, W.C. III. 1978. Relationship between eustacy and stratigraphic sequences of passive margins. Geol. Soc. Am. Bull. 89: 1389-1403.

(75) Rampino, M.R.; Self, S.; and Fairbridge, R.W. 1979. Can rapid climatic change cause volcanic eruptions? Science 206: 826-829.

(76) Raup, D.M., and Sepkoski, Jr., J.J. 1982. Mass extinctions in the marine fossil record. Science 215: 1501-1503.

(77) Russell, K. 1968. Oceanic ridges and eustatic changes in sea level. Nature 218: 861-862.

(78) Ryan, W.B.F., and Cita, M.B. 1977. Ignorance concerning episodes of ocean-wide stagnation. Marine Geol. 23: 197-215.

(79) Ryan, W.B.F.; Hsü, K.J.; Cita, M.B.; Dumitrica, P.; Lort, J.M.; Maync, W.; Nesteroff, W.D.; Pautot, G.; Stradner, H.; and Wezel, F.C. 1973. Initial Reports of the Deep Sea Drilling Project, vol. 13. Washington, D.C.: U.S. Government Printing Office.

(80) Ryan, W.F.; Cita, M.B.; Dreyfus Rawson, M.; Burckle, L.H.; and Saito, T. 1974. A paleomagnetic assignment of Neogene stage boundaries and the development of isochronous datum planes between the Mediterranean, the Pacific and Indian Oceans in order to investigate the response of the world ocean to the Mediterranean "salinity crisis". Riv. Ital. Pa. 80: 631-688.

(81) Ryther, J.H. 1969. Photosynthesis and fish production in the sea. Science 166: 72-76.

(82) Savin, S.M.; Douglas, R.M.; and Stehli, F.G. 1975. Tertiary marine paleotemperatures. Geol. Soc. Am. Bull. 86: 1499-1510.

(83) Schindewolf, O.H. 1950. Der Zeitfaktor in Geologie and Paläontologie. Stuttgart: Schweizerbart.

(84) Schlanger, S.O.; Jenkyns, H.-C.; and Premoli-Silva, I. 1981. Volcanism and vertical tectonics in the Pacific Basin related to global Cretaceous transgressions. Earth Plan. 52: 435-449.

(85) Scholle, P.A., and Arthur, M.A. 1980. Carbon isotopic fluctuations in pelagic limestones: potential stratigraphic and petroleum exploration tool. Am. Assoc. Pet. Geol. Bull. 64: 67-87.

(86) Sepkoski, J.J. 1982. Mass extinctions in the Phanerozoic oceans: a review. Geol. Soc. Am. Spec. Paper 190: 283-289.

(87) Shackleton, N.J., and Kennett, J.P. 1975. Paleotemperature history of the Cenozoic and the initiation of Antarctic glaciation: oxygen and carbon isotope analyses in DSDP Sites 277, 279, and 281. Initial Reports of the Deep Sea Drilling Project, vol. 29, pp. 743-755. Washington, D.C.: U.S. Government Printing Office.

(88) Shackleton, N.J., and Opdyke, N.D. 1973. Oxygen isotope and palaeomagnetic stratigraphy of equatorial Pacific core V28-238: Oxygen isotope temperatures and ice volumes on a 10^5 year and 10^6 year scale. Quat. Res. 3(1): 39-55.

(89) Tappan, H. 1968. Primary production, isotopes, extinctions and the atmosphere. Paleo. P. 4: 187-210.

(90) Thiede, J., and van Andel, T.H. 1977. The paleonenvironment of anaerobic sediments in the late Mesozoic South Atlantic Ocean. Earth Plan. 33: 301-309.

(91) Thierstein, H.R. 1979. Paleoceanographic implications of organic carbon and carbonate distribution in Mesozoic deep-sea sediments. In Deep Drilling Results in the Atlantic Ocean: Continental Margins and Paleoenvironment, eds. M. Talwani, W. Hay, and W.B.F. Ryan, vol. 3, pp. 249-274. Maurice Ewing Series. Washington, D.C.: American Geophysical Union.

(92) Thierstein, H.R., and Berger, W.H. 1978. Injection events in ocean history. Nature 276: 461-466.

(93) Thompson, S.L., and Barron, E.J. 1981. Comparison of Cretaceous and present Earth albedos: implications for the causes of paleoclimates. J. Geol. 89: 143-167.

(94) Tipper, J.C. 1983. Rates of sedimentation, and stratigraphical completeness. Nature 302: 696-698.

(95) Vail, P.R.; Mitchum, R.M.; and Thompson, S. 1977. Seismic stratigraphy and global changes of sea level. Part 4: global cycles of relative changes of sea level. Am. Assoc. Pet. Geol. Mem. 26: 83-97.

(96) Valentine, J.W. 1973. Evolutionary Paleoecology of the Marine Biosphere. Englewood Cliffs, NJ: Prentice-Hall.

(97) Valentine, J.W., and Moores, E.M. 1970. Plate-tectonic regulation of faunal diversity and sea-level: a model. Nature 228: 657-659.

(98) Valentine, J.W., and Moores, E.M. 1972. Global tectonics and the fossil record. J. Geol. 80: 167-184.

(99) Valeton, I. 1983. Klimaperioden lateritischer Verwitterung und ihr Abbild in den synchronen Sedimentationsräumen. Z. Deutsch. Geol. Ges. 134, in press.

(100) Vernadsky, V.I. 1924. La Géochimie. Paris: Félix Alcan.

(101) Vidal, G., and Knoll, A.H. 1982. Radiations and extinctions of plankton in the late Proterozoic and early Cambrian. Nature 297: 57-60.

(102) Vincent, E., and Berger, W.H. 1981. Planktonic foraminifera and their use in paleoceanography. In The Sea, ed. C. Emiliani, vol. 7, pp. 1025-1119. New York: Wiley-Interscience.

(103) Vincent, E.; Killingley, J.S.; and Berger, W.H. 1980. The Magnetic Epoch-6 Carbon Shift, a change in the oceanic $^{13}C/^{12}C$ ratio 6.2 million years ago. Marine Micropaleo. 5: 185-203.

(104) Watts, A.B. 1982. Tectonic subsidence, flexure and global changes of sea level. Nature 297: 469-474.

Patterns of Change in Earth Evolution, eds. H.D. Holland and A.F. Trendall, pp. 207-219.
Dahlem Konferenzen 1984. Berlin, Heidelberg, New York, Tokyo: Springer-Verlag.

Patterns and Geological Significance of Age Determinations in Continental Blocks

S. Moorbath
Dept. of Geology and Mineralogy, Oxford University
Oxford OX1 3PR, England

Abstract. Continental blocks usually consist of juxtaposed tectonic provinces of very different ages. Age and isotope evidence suggest that the continental crust has grown spasmodically at a time-integrated nonlinear rate from about 3700 Ma to the present. However, there is no compelling evidence for global synchronicity of all crust-forming events.

INTRODUCTION

Hypotheses concerning the smooth or spasmodic occurrence of geological phenomena throughout Earth history often depend upon the time-constant and the resolving capacity of the methods used to date those phenomena. It is likely that each individual geological phenomenon is episodic in that relatively brief spasms of intense activity are separated by relatively long, quiescent intervals. However, time integration of such spasms, whether related to erosion, volcanic activity, earthquakes, plutonism, metamorphism, ocean-floor spreading, subduction, mountain-building (orogeny), crust formation, etc., can simulate a smooth evolutionary pattern when the applied dating method cannot resolve discrete events in a discontinuous sequence. In contrast, the ultimate source of energy of all geological activity, namely, radiogenic heat production in the Earth's interior, is continuous, except at the atomic level. The decline in terrestrial radiogenic heat production through geological time is quantitatively well understood from data on the concentration and distribution of radioactive elements in the mantle and in the crust. Spasmodic geological activity on many different time-scales is ultimately a consequence of the interplay between a continuous heat supply and the thermo-mechanical properties of those materials in the mantle and crust in which heat can be converted to motion. A vast literature exists on the geological history and the thermal history of the Earth, but attempts to model the complex relationship between them and to define the timing of major geological events are still unsatisfactory, in part because they are highly dependent on understanding the convection regime of the mantle through Earth history. The distribution of isotopic age patterns in continental blocks through geological time offers a complementary observational approach to these models. The age patterns allow us to see whether the Earth has been active continuously, or whether resolvable spasms were separated by relatively lengthy periods of tectonic quiescence. Asking the question in this form is almost certainly too simplistic, and we shall see that an approximation to the truth, as with many other initially opposing scientific hypotheses, probably lies somewhere in between.

GENERAL DISCUSSION

In any discussion of orogeny (often defined as "profound deformations of rock bodies along restricted zones and within a limited time interval," normally accompanied by uplift, folding, metamorphism, and igneous intrusion), there has long been a conflict between those who consider orogenesis to be periodic or cyclical, and those who deny any regularity to orogenesis. Stille, writing in the 1920s, considered that all orogenic phases were synchronous and worldwide in nature. Gilluly, in the 1940s, considered that there was no periodicity to orogenesis, that orogenies have been continuous throughout geological time, and that there have been no major peaks of orogenic activity.

These earlier discussions are mainly of historical interest because they preceded modern developments in global tectonics and in geochronology. More recently, several authors have discussed the periodicity of orogeny in terms of seafloor spreading, plate tectonics, and modern orogenic theory. For example, Brookfield (3) discusses possible synchronous relationships between mid-ocean rises, continental edge orogenies (i.e., Cordilleran, Andean, and Island-arc types), and continental collision orogenies (i.e., Alpine-Himalayan type) from first principles. All this has been broadly tested on the complex, well studied succession of worldwide Phanerozoic orogenies (<570 Ma) which reflect the occurrence of modern-style plate tectonics and the attendant continental drift, subduction, ocean opening and closing (Wilson Cycle), and continental collision, etc.

The earliest serious attempt to study the distribution of isotopic dates in time and space was made by Gastil (13). From a sample of about 400 K-Ar, Rb-Sr, and U-Pb dates on various types of minerals from Africa, North and South America, Australia, Southeast Asia, Europe, and Siberia, he postulated that "crustal adjustments" which set or reset mineral dates are periodic and roughly cyclic. Intervals for which abundant mineral dates have been preserved are about 175 to 250 Ma in length, with cycles of about 350 to 500 Ma. Intervals with an abundance of dates fall in the ranges: 2710 - 2490 Ma, 2220 - 2060 Ma, 1860 - 1650 Ma, 1480 - 1300 Ma, 1100 - 930 Ma, 620 - 280 Ma, and 120 Ma to the present. Gastil concludes that "the Earth's history is marked by episodes during which the spatial distribution of intense mineralogenic activity shifts, resulting in the selective preservation of 'withdrawal' dates for the area becoming stable." Writing before the widespread acceptance of continental drift, Gastil also states that "the spatial distribution of mineral dates in North America defines a sequence of age provinces which are younger nearer the continental margin but does not support the hypothesis of continental accretion. An hypothesis envisioning the gradual outward solidification of an originally mobile continent is more satisfactory but is not supported by the spatial distribution of mineral dates on other continents and does not explain repeated, widespread rejuvenation of stable areas." Gastil's investigation was soon repeated by Dearnley (7) on a sample of about 3400 mineral age determinations. A cumulative curve derived from his date frequency histogram showed three particularly well-defined changes of slope at ca. 1950, 1075, and 180 Ma, and two other less abrupt changes at 2750 and 650 Ma, which were all regarded as signifying the onset of worldwide tectonic regimes each with a duration of several hundred million years. Global reconstructions of continental masses based on morphological fits on continental shelves showed that dated orogenic belts are very well aligned. Dearnley regarded this as added support for continental drift (although he was at that time a strong advocate of an expanding Earth to explain drift).

These, and similar, compilations received much publicity and were made the basis of various episodic Earth models. Runcorn (29) and Bott (2) were early exponents of correlating mineral date peaks with changes in mantle convection patterns, which were regarded as responsible for continental drift and associated tectonic activity. These earlier models of mantle convection, however, bore little resemblance to those currently in vogue. In order to account for the supposedly related phenomena of changes in the distribution of mountain chains over continents, changes in the number and shape of continental masses, and changes in the nature of the convection system believed to exist in the mantle, Sutton (30) postulated the existence of "chelogenic cycles," consisting of a sequence of events leading to the displacement and disruption of the continents as they existed at the start of the cycle and later to the regrouping of these disrupted masses of continental crust. Sutton used global mineral date patterns to suggest that the main structural units on all continents began their tectonic development in one of the following periods: 2900 - 2700 Ma, 1900 - 1700 Ma, or 1200 - 1000 Ma. Sutton regarded each period as marking the start of a chain of connected events continuing for between 750 to 1250 million years; the youngest cycle is still continuing, and there was no break at the beginning of the Cambrian. I consider this hypothesis to be too broad and simplistic to accord with modern geochronological data and current ideas of Earth behavior.

It was soon realized (20) that the analysis of histograms of isotopic dates could be subject to serious misinterpretation unless the reported dates refer to the true age of a geological event, and unless the geological nature and significance of that event is fully understood. Published histograms included dates from a great variety of rock types, representing totally different and quite unrelated orogenic and anorogenic environments, and they included many dates that were not directly related to the time of igneous or metamorphic activity (for example, "cooling" dates), as well as dates without any geological significance at all (partially "overprinted" dates). Most of the histogram points were originally based on K-Ar and Rb-Sr measurements on micas, in which diffusion of radiogenic Ar and Sr continues to temperatures ("blocking temperatures") well below, and significantly postdating, igneous or metamorphic crystallization. Hence extreme caution is necessary when one compares isotopic dates obtained by different methods on different minerals with different blocking temperatures, and on different rock types. Ideally the histogram approach of identifying orogenic periodicity should only be used to compare dates obtained by one method on one mineral or rock type.

K-Ar and Rb-Sr mineral dates are still very widely used nowadays for general geochronological reconnaissance in different types of terrains, for studying the timing and rates of uplift and cooling in orogenic belts, for dating unmetamorphosed igneous rocks and mineral deposits as well as authigenic and detrital sedimentary minerals, etc.

Increasingly during the past 10 to 15 years, however, geochronologists have concentrated on the application of dating methods which provide close approximations to true ages of crystallization and rock formation; these dates are particularly useful in regionally metamorphosed Precambrian terrains with igneous protoliths (orthogneisses). Rb-Sr, Sm-Nd, and Pb/Pb whole rock isochron methods, as well as the U-Pb zircon method, have been very widely applied to dating a large variety of Precambrian and Phanerozoic rocks. Application of these methods (for general reviews, see (11, 24, 25)) can provide reliable constraints on the primary age and temporal evolution of rock units, even where parent-daughter systems have suffered disturbance in metamorphic and/or metasomatic events significantly postdating primary rock formation. A knowledge of the initial isotopic composition of Sr, Nd, and Pb is essential for the correct interpretation of isochron age data, to constrain the crustal residence time of the protoliths of the dated rock unit, and to assess the relative contribution of mantle and crust during magma genesis. These newer methods are being widely used to study the timing and petrogenesis of continental crust formation in a way that was not previously possible. There is no doubt, for example, that the igneous protoliths of many calc-alkaline orthogneisses, which make up such a high proportion of major Precambrian shield areas, were derived from a source region with mantle-type Sr, Nd, and Pb isotopic ratios. Furthermore, it is clearly demonstrable that extraction of the igneous precursors from the mantle, their petrological-geochemical-metamorphic differentiation, and their final stabilization as new, thick continental crust, occurred within a time span that rarely exceeded 100 to 200 million years, and sometimes much less (14, 19, 22, 23). Such immense, complex, multiphase, but relatively brief, continental crust-forming events ("accretion superevents") appear to have occurred at different times in different places during the past 3700 Ma. Globally the late Archean, some 3000 - 2500 Ma ago, was the most productive period of sialic crust formation.

There are large terrains of granitoid gneisses, greenstone belt supracrustals, and varied post-tectonic intrusive rocks that were formed in most major shield areas within this broad age range. They are, however, by no means globally contemporaneous within the errors of age measurement. Preliminary age and isotope studies of several individual accretion superevents furthermore demonstrate an overall igneous emplacement sequence from predominantly mantle-derived, calc-alkaline magmas to predominantly crust-derived, granitic magmas, as "ripening" of juvenile continental crust proceeded quasi-episodically during periods of tens or a few hundreds of millions of years.

A much more recent geological environment in which calc-alkaline magmas derived from the mantle and/or subducted oceanic crust have contributed substantially to continental growth is the circum-Pacific belt, particularly in the Andes and Cordilleras (continental edge orogenies). A clear geological and isotopic distinction can almost always be made in the Phanerozoic between tectonic environments dominated by juvenile crustal accretion leading to continental growth, and those dominated by the reworking of older crustal rocks, as in continental collision orogenies. As we go back in time, particularly into the Archean, recognition of "modern" tectonic environments becomes problematic and highly controversial. In addition, the time difference between closely spaced orogenies such as those which characterize the Phanerozoic (for example, Caledonian/Appalachian, Hercynian, Alpine, etc.) become increasingly difficult to resolve. Nonetheless, a clear distinction can often be made by means of modern age and isotope methods between primary crustal accretion and secondary crustal reworking of any age.

Let us now return to the problem of periodicity, with special reference to crustal accretion and continental growth. We note that Rb-Sr, Sm-Nd, and Pb/Pb whole rock isochron methods, as well as U-Pb techniques applied to zircon, are analytically more complex and time-consuming than most mineral dating techniques and that even at the time of this writing the geochronological and isotopic coverage is extremely patchy. Restricted parts of several continental regions (for example, in North America, Greenland, western Europe, Africa, and Australia) are becoming quite well known, at least in outline, whilst others (for example, Asia, South America, Antarctica) are an almost complete isotopic blank. With modern methods, detailed "age and isotope" mapping of continental areas is perfectly feasible, but at the present rate of progress this will take many more years, particularly in those territories where modern analytical and interpretative techniques are not yet being applied.

From a survey of age determinations relating to rock-forming events, particularly Rb-Sr whole rock isochron ages and initial $^{87}Sr/^{86}Sr$ ratios, it was suggested (21) that "relatively short periods of accelerated crustal growth ... may have occurred episodically rather than randomly or semicontinuously during the Earth's history" and that "they may well correspond with the most commonly observed groupings of radiometric dates relevant to rock-forming events, which are very approximately 3800 - 3500; 2800 - 2500; 1900 - 1600; 1200 - 900; 500 - 0 Ma ago." Caution was urged regarding the likelihood of inadequate sampling, but these time intervals have been very widely quoted in the scientific literature. For example, the author of a recent model (15) for the origin of "hot-spot" volcanism linked to mantle plumes states that "... one may correlate periods of intense continental magmatism at 3.6, 2.7, 1.8, 1.0 Ga ago with times of plume-triggered mantle convection and the intervening quiet periods with times of mostly stagnant conditions in the mantle, when the newly accumulated subducted crust was gradually heating up but had not reached the point of instability." In this model intermittent mantle convection alternates with stagnant conditions. Another model (28) postulates that these same episodes of rapid continental growth can be explained in terms of the interaction between major convective overturns in the lower mantle and "megaliths" of subducted oceanic lithosphere, suspended near the 650 km discontinuity.

Condie (5) surveyed global dates briefly and concluded that "existing data indicate that world-wide Precambrian orogenic periods are episodic averaging 200 to 400 million years in length and occurring about every 500 to 600 million years. Major orogenic periods occur at 3.0-3.8, 2.5-2.7, 1.5-2.0, 0.9-1.2, and 0.5-0.7 billion years. Major North American Phanerozoic orogenics average about 50 million years in length and occur at irregular spacings."

CONCLUDING DISCUSSION

The time is clearly not yet ripe for an authoritative answer to the main problem under discussion, but any future attempts must take into account the following observations:

1. The date ranges given by Condie (5) and Moorbath (21) are almost certainly too generalized and oversimplified, since they are based on data that are too limited. These, as well as previous, compilations have been widely quoted in support of the proposition that major periods of global tectonism were episodic and synchronous, and that they were separated by intervening periods of global tectonic "recession."

2. The occurrence of episodic crustal "accretion superevents" is established beyond reasonable doubt, but their global synchronicity throughout geological time is in doubt. In one of the best-studied regions, West Greenland, accretion superevents in adjacent terrains can be clearly demonstrated at ca. 3700 - 3600, 3000 - 2700, and 1900 - 1700 Ma ago (23, 32); the overlap is minor (31). A broadly similar range of dates has been found on the adjacent North American craton, although the oldest

group of ages in Greenland has not yet been positively identified in the remainder of the North Atlantic craton or on the Baltic Shield. In the Zimbabwean craton of southern Africa, accretion superevents occurred ca. 3500 - 3400 and 2900 - 2700 Ma ago. Preliminary evidence is accumulating ((1), and Oxford, unpublished work) that the earliest accretion superevent on the Indian subcontinent occurred about 3400 - 3200 Ma ago; this event is not synchronous with any others that have been recognized elsewhere. Several other major crust-forming events which do not fit previously proposed intervals occurred in the Arabian Shield and in southern Norway. Age and isotope studies have demonstrated that in the Arabian Shield (ca. 600,000 km^2) crustal accretion of mantle-derived granitoids was intense between ca. 850 and 550 Ma ago. (10); this event was partly contemporaneous with the major Pan-African orogeny in other parts of Africa, which was dominated by the reworking of older continental crust. In southern Norway, major crust-forming events commenced at ca. 1600 - 1500 Ma ago and were completed several hundred million years later (12).

3. The temporal and regional distribution of episodic accretion superevents throughout Earth history must be viewed in conjunction with our increasingly sophisticated understanding of the Earth's changing thermal regime (6, 17, 18, 33). Much recent geochemical and isotopic evidence from continental and oceanic rocks is compatible with the hypothesis that the evolution of the lithosphere is dominated by irreversible chemical differentiation of the mantle (and particularly of the upper mantle) and by the permanence of the greater part of the continental crust once it has accreted, differentiated, and stabilized. There have been several notable recent efforts, based mainly on isotopic data, to model continental growth by irreversible differentiation of the mantle through geologic time (8, 16, 26, 27). In the light of global thermal and isotopic models, I consider that the production of continental crust was strongly inhibited prior to about 3800 Ma ago, and that before this time the Earth had an essentially globe-encircling, impermanent, mainly mafic or ultramafic protocrust most of which was recycled through the mantle by disruption and foundering. This crust may have been akin to the skin of solids on a lava lake or slag furnace. Heat production was undoubtedly much more rapid and mantle convection more intense more than 3800 Ma ago. Relatively small amounts of acid (granitic) differentiates may already have existed at this early stage. Such rocks are certainly known from the ca. 3800-Ma-old Isua supracrustal succession of West Greenland (22). The creation of the first true, thick, calc-alkaline continental crust at

about 3700 - 3600 Ma ago probably resulted from protoplate tectonic processes, distributed over the whole globe, once the mafic protoplates had become sufficiently rigid, extensive, and thick to undergo plate interaction involving some form of subduction and further crustal thickening, analogous to modern intraoceanic island arcs. During the next few hundred million years the efficiency of the process which inhibited the production of true continental crust decreased. There is now general agreement that by about 2800 - 2500 Ma at least 50% and possibly as much as 80% of the present-day continental crustal mass, with average modern crustal thickness, was already in existence (9, 22, 27). Since the end of the Archean, there has been a decline in average continental growth rate to the present, in broad parallelism with the decay in radiogenic heat production of the mantle. There is a good deal of isotopic and other evidence to suggest that juvenile continental crust has been produced from the upper mantle during the past few hundred million years above subduction zones. In the Cordilleras and the Andes of the western Americas, juvenile continental crust in the form of calc-alkaline plutons is accreted onto, and intruded into, older continental crust characterized in different regions by isotopic ages ranging from Archean, through Proterozoic to Phanerozoic. Future geochronological investigations of individual sectors of this immense, complex accretionary terrain could yield grossly misleading generalizations regarding the global distribution of accretion superevents in space and time, particularly if such a study is done after continental masses have once more become redistributed.

4. The occurrence of regionally episodic accretionary and orogenic events is not incompatible with the overall smooth time-integrated parallelism between rates of crustal growth and heat production. Given our present understanding of global tectonics and the limited amount of available isotopic age data, there is no compelling reason to postulate that the Earth as a whole has suffered alternating periods of tectonic (sensu lato) activity and quiescence. It is more likely that at any given time some tectonic zones were highly active (for example, at sites of continental edge and continental collision orogenies), whilst other zones were passive (for example, within cratonic interiors). The regional focus of tectonic activity has changed throughout time, and the temporal and spatial juxtaposition of accreted and of reworked continental terrains is a consequence of the perpetual redistribution and rearrangement of continental crust over the surface of the globe by the convecting mantle through the medium of plate tectonics. Whether convection itself occurred

continuously or episodically is still an open question. Pb isotopic evidence from ocean island basalts is interpreted by some workers (4) as indicating episodic differentiation of the mantle at least over the period ca. 2500 - 1000 Ma ago. The surface expression of mantle convection at any given time must also be strongly dependent on whether the continental crust was widely dispersed or welded together as one or more "supercontinents." Paleomagnetism forms an essential tool for such investigations.

This brief, qualitative discussion has not really answered the question: "Earth's behavior - how smooth, how spasmodic?" It may have removed some of the confusion that surrounds one aspect of this topic. The confusion is due in part to overgeneralized isotopic age groupings that have sometimes been regarded as globally synchronous. Detailed geochronological comparisons between different regions are still too few to permit firm conclusions to be reached regarding the supposed global synchronicity of regionally episodic crust-forming and orogenic events. On balance the available geochronological evidence supports regional episodicity, but not global synchronicity.

One may hope for a new generation of numerate geologists versed in the recently developed branch of mathematics known as catastrophe theory which is particularly applicable to situations where gradually changing forces lead to abrupt changes in behavior. Every discrete geological event, of whatever magnitude, ultimately represents a "catastrophe" in the mathematical sense.

REFERENCES

(1) Beckinsale, R.D.; Drury, S.A.; and Holt, R.W. 1980. 3360-myr old gneisses from the South Indian craton. Nature 283: 469-470.

(2) Bott, M.H.P. 1964. Convection in the earth's mantle and the mechanism of continental drift. Nature 202: 583-584.

(3) Brookfield, M. 1971. Periodicity of orogeny. Earth Planet. Sci. Lett. 12: 419-424.

(4) Chase, C.G. 1981. Oceanic island Pb: two-stage histories and mantle evolution. Earth Planet. Sci. Lett. 52: 277-284.

(5) Condie, K.C. 1976. Plate Tectonics and Crustal Evolution. New York: Pergamon Press.

(6) Cook, F.A., and Turcotte, D.L. 1981. Parameterised convection and the thermal evolution of the earth. Tectonophys. 75: 1-17.

(7) Dearnley, R. 1965. Orogenic fold-belts and continental drift. Nature 206: 1083-1087.

(8) De Paolo, D.J. 1980. Crustal growth and mantle evolution: inferences from models of elements transport and Nd and Sr isotopes. Geochim. Cosmochim. Acta 44: 1185-1196.

(9) Dewey, J.F., and Windley, B.F. 1981. Growth and differentiation of the continental crust. Phil. Trans. Roy. Soc. Lond. A301: 189-206.

(10) Duyverman, H.J.; Harris, N.B.W.; and Hawkesworth, C.J. 1982. Crustal accretion in the Pan-African - Nd and Sr isotope evidence from the Arabian Shield. Earth Planet. Sci. Lett. 59: 315-326.

(11) Faure, G. 1977. Principles of Isotope Geology. New York: Wiley.

(12) Field, D., and Raheim, A. 1981. Age relationships in the Proterozoic high-grade gneiss regions of southern Norway. Precambrian Res. 14: 261-275.

(13) Gastil, G. 1960. The distribution of mineral dates in time and space. Am. J. Sci. 258: 1-35.

(14) Hamilton, P.J.; Evensen, N.M.; O'Nions, R.K.; and Tarney, J. 1979. Sm-Nd systematics of Lewisian gneisses: implications for the origin of granulites. Nature 277: 25-28.

(15) Hofmann, A.W., and White, W.M. 1982. Mantle plumes from ancient oceanic crust. Earth Planet. Sci. Lett. 57: 421-436.

(16) Jacobsen, S.B., and Wasserburg, G.J. 1979. The mean age of mantle and crustal reservoirs. J. Geophys. Res. 84: 7411-7427.

(17) Lambert, R.St.J. 1980. The thermal history of the earth in the Archaean. Precambrian Res. 11: 199-213.

(18) Langan, R.T., and Sleep, N.H. 1982. A kinematic thermal history of the Earth's mantle. J. Geophy. Res. 87: 9225-9235.

(19) McCulloch, M.T., and Wasserburg, G.J. 1978. Sm-Nd and Rb-Sr chronology of continental crust formation. Science 200: 1003-1011.

(20) Moorbath, S. 1967. Recent advances in the application and interpretation of radiometric age data. Earth-Sci. Rev. 3: 111-133.

(21) Moorbath, S. 1976. Age and isotope constraints for the evolution of Archaean crust. In The Early History of the Earth, ed. B.F. Windley, pp. 351-360. London: Wiley.

(22) Moorbath, S. 1977. Ages, isotopes and evolution of Precambrian continental crust. Chem. Geol. 20: 151-187.

(23) Moorbath, S., and Pankhurst, R.J. 1976. Further Rb-Sr age and isotope evidence for the nature of the late Archaean plutonic event in West Greenland. Nature 262: 124-126.

(24) Moorbath, S., and Taylor, P.N. 1981. Isotopic evidence for continental growth in the Precambrian. In Precambrian Plate Tectonics, ed. A. Kröner, pp. 491-525. Amsterdam: Elsevier.

(25) O'Nions, R.K.; Carter, S.R.; Evensen, N.M.; and Hamilton, P.J. 1979. Geochemical and cosmochemical applications of Nd isotope analysis. Ann. Rev. Earth Planet. Sci. 7: 11-38.

(26) O'Nions, R.K.; Evensen, N.M.; and Hamilton, P.J. 1979. Geochemical modeling of mantle differentiation and crustal growth. J. Geophys. Res. 84: 6091-6101.

(27) O'Nions, R.K., and Hamilton, P.J. 1981. Isotope and trace element models of crustal evolution. Phil. Trans. Roy. Soc. Lond. A301: 473-487.

(28) Ringwood, A.E. 1982. Phase transformations and differentiation in subducted lithosphere: implications for mantle dynamics, basalt petrogenesis and crustal evolution. J. Geol. 90: 611-643.

(29) Runcorn, S.K. 1962. Towards a theory of continental drift. Nature 193: 311-314.

(30) Sutton, J. 1963. Long-term cycles in the evolution of continents. Nature 198: 731-735.

(31) Taylor, P.N.; Moorbath, S.; Goodwin, R.; and Petrykowski, A.C. 1980. Crustal contamination as an indicator of early Archaean continental crust: Pb isotopic evidence from the late Archaean gneisses of West Greenland. Geochim. Cosmochim. Acta 44: 1437-1453.

(32) Van Breemen, O.; Aftalion, M.; and Allaart, J.H. 1974. Isotopic and geochemical studies on granites from the Ketilidian mobile belt of South Greenland. Bull. Geol. Soc. Am. 85: 403-412.

(33) Wells, P.R.A. 1981. Accretion of continental crust: thermal and geochemical consequences. Phil. Trans. Roy. Soc. Lond. A301: 347-357.

Patterns of Change in Earth Evolution, eds. H.D. Holland and A.F. Trendall, pp. 221-242.
Dahlem Konferenzen 1984. Berlin, Heidelberg, New York, Tokyo: Springer-Verlag.

The Archean/Proterozoic Transition: A Sedimentary and Paleobiological Perspective

A.H. Knoll
Biological Laboratories, Harvard University
Cambridge, MA 02138, USA

Abstract. Sedimentary rocks provide sensitive reflections of source terrains and tectonic settings; they thus constitute important sources of information about crustal processes on the early Earth. The preserved stratigraphic record, if read literally, suggests that Lower Proterozoic sediments as a group differ significantly from those characteristic of the Archean. Lower Proterozoic successions include widespread passive margin continental shelf and epicratonic sequences dominated by mature detrital rocks and carbonates. The composition, sedimentary structures, and geometry of these sedimentary packages indicate that large stable continents existed at the time of their deposition. In low-grade Archean terrains, the predominant supracrustal successions are those of greenstone belts. Sedimentary rocks intercalated among thick volcanic sequences in the lower parts of greenstone successions, particularly older ones, provide little evidence for sialic source terrains, but overlying detrital sequences reflect the unroofing of tonalitic/trondhjemitic intrusions and resemble sedimentary packages accumulating along modern, tectonically active continental margins. Extensive epicratonic or passive margin shelf sequences are notably limited in low-grade terrains, although their presence in several localities provides evidence that continental crustal stabilization occurred on at least a regional scale 3000 Ma ago or earlier. Some Archean high-grade terrains contain what appear to be severely metamorphosed greenstone belt packages, but others include probable remnants of ancient shelf deposits. The association of shelf sequences with high-grade terrains reflects deep burial of continental margins, perhaps beneath tectonically emplaced "low-grade" slabs. This highlights a significant problem of Archean geological interpretation, selective preservation. All in all, evident differences between Lower

Proterozoic and Archean stratigraphy appear relatable to the presence of large stable continents during the younger era. This suggests that the late Archean was a period of rapid crustal growth and stabilization, an interpretation corroborated by geochemical evidence. Large-scale cratonization may have had a significant impact on the contemporary biota, especially as regards productivity, and this, in turn, may have influenced atmospheric evolution. This hypothesis highlights many unresolved issues in Precambrian geology.

INTRODUCTION

The division of Precambrian rocks into Archean and Proterozoic systems differentiated on the basis of their characteristic supracrustal successions, metamorphic patterns, and field relationships has long been acknowledged in the stratigraphic literature. Rigorous definition of the boundary between the two systems has proven elusive, in part because the geological events generally associated with the Archean/Proterozoic transition (in a field sense) were not everywhere simultaneous. The IUGS Subcommittee on Precambrian Stratigraphy recently proposed that the boundary be set at 2500 Ma and defined radiometrically (43). Others (e.g., (35)) have suggested that the definition be tied to geological events recognizable in the rocks such as the cooling of the Hartley Complex of the Great Dyke, Zimbabwe (19), or the boundary between the Fortescue and Hamersley groups in Western Australia. In this paper, usage of the terms Archean and Proterozoic follows the recommendation of the IUGS Subcommittee.

Many, if not most, stratigraphic boundaries are in some sense arbitrarily drawn. For example, the Silurian/Devonian boundary is defined by the first appearance of the conodont species Monograptus uniformis in a stratotype section near Klonk, Czechoslovakia (31). In terms of Earth history, this must surely be regarded as an arbitrary definition. Yet, the concepts of Silurian and Devonian are useful, and agreement on a boundary ensures uniform usage of the terms. A second type of boundary is illustrated by that between the Precambrian and the Cambrian. The evolution of mineralized skeletons had significant biological and geological consequences, but different phyla achieved this capability at different times; the choice of a boundary-defining event is therefore somewhat arbitrary. The boundary represents a paleontological transition that took place over a time span of tens of millions of years. But again, designation of a boundary standardizes usage, facilitating communication and providing a fixed point from which changes in the Earth's sedimentary record can be measured.

In this light, it can be seen that the choice of an Archean/Proterozoic boundary is not nearly so important as an understanding of what it represents in terms of Earth history. Does it mark a fundamental geological discontinuity, an arbitrarily designated point along a continuum of crustal development, or a point representative of a period of geological transition? The question can be phrased more explicitly in terms of supracrustal packages. It is clear that the stratigraphic succession of the 3400-3500 Ma Swaziland Supergroup differs significantly from those of the Lower Proterozoic Kaniapiskau or Transvaal Supergroups. However, it is also evident that "Proterozoic-style" sequences were deposited at least as early as 3100-2900 Ma - the Pongola Supergroup of southern Africa is the outstanding example of such a sequence (49, 52) - and that greenstone belt-like successions accumulated in northern South America, in parts of equatorial west Africa, and elsewhere during the Proterozoic Eon (45). This effectively rules out a worldwide synchronous discontinuity in geological pattern or process between 2000 and 4000 Ma ago, and suggests that a choice between the other two possibilities must be made on the basis of data collected from all continents and evalutated in a global context.

The Archean/Proterozoic transition has been associated with the large-scale growth and stabilization of continental crust, accompanied by concurrent changes in tectonic style (e.g., (55)). Sedimentary sequences provide sensitive reflections of source terrains and tectonic settings; therefore, I will concentrate on stratigraphic and sedimentological data in my brief review of Archean and Early Proterozoic geology. This line of evidence strongly favors the "significant transition" view of Earth history, in accord with geochemical and geochronological data. This does not discount the possibility that Archean mantle processes closely resembled those of the Proterozoic, or that if changes occurred, they took place more or less continuously through geological time. Nor does this conclusion require that processes of crustal generation changed markedly at the close of the Archean. Some form of plate tectonic processes may be responsible for the crustal and supracrustal features of the Archean Eon, just as they are for those of younger epochs. It is patterns of Earth surface geology - the geometry, distribution, composition, and deformation features of supracrustal rock packages - that differentiate the preserved records of the Archean and Proterozoic eons.

ARCHEAN STRATIGRAPHY

Low-grade terrains

Sedimentological studies of Archean rocks have been concentrated in weakly metamorphosed, or low-grade, areas where supracrustal packages can be impressively well preserved. The predominant supracrustal successions in these terrains are those of greenstone belts. Forty years ago, Pettijohn (37) compared greenstone belt sequences to Phanerozoic eugeosynclinal packages. He based this comparison on the abundance of greywackes, shales, and conglomerates and the concomitant dearth of mature sandstones and carbonates in both settings. In the past decade, Pettijohn's pioneering observations have been extended and refined by a series of detailed sedimentological analyses of greenstone belt successions on several continents (26).

Two major intervals of greenstone belt development are apparent in southern Africa: an older period dated at 3500–3400 Ma and represented by the Swaziland Supergroup (South Africa, Swaziland) and the Sebakwian Group (Zimbabwe), and a younger period (2900–2600 Ma) best represented in Zimbabwe and adjacent areas. The older sequences can be subdivided into three stratigraphic units. In the Swaziland Supergroup, the lowermost sequence is the Onverwacht Group, a thick pile of ultramific lavas with subordinate amounts of more felsic volcanic rocks. Onverwacht sedimentary units are a minor component of the total succession; they consist predominantly of pyroclastic and volcaniclastic rocks formed during penecontemporaneous volcanism (27). Orthochemical sediments, including chert, carbonates, and local evaporites, often cap volcanogenic detrital sequences. Sedimentary structures in the chemical rocks indicate shallow water deposition, presumably on the top and flanks of relatively high standing volcanic piles. Although sialic crustal blocks existed 3500–3400 Ma ago, there is no evidence that they served as source areas for any Onverwacht sediments.

Overlying the Onverwacht succession are the Fig Tree and Moodies groups, thick sequences of predominantly detrital sediments. Fig Tree rocks include strata that were deposited subaerially in braided stream alluvial settings, as well as turbidites that accumulated below wave base, but there is very little evidence for the presence of intervening shallow shelf environments (12). Clasts in lower Fig Tree sediments are derived principally from preexisting Onverwacht units, but younger Fig Tree units contain Onverwacht, Fig Tree, <u>and</u> sialic igneous materials. The increase in sialic components toward the top of the section records the progessive unroofing of a tonalitic/trondhjemitic terrain (12). Moodies sedimentary units include alluvial conglomerates and relatively mature sandstones deposited in beach to shallow shelf environments. The transition from Onverwacht to Fig Tree and Moodies depositional conditions may reflect tectonic uplift associated with the emplacement of tonalitic/trondhjemitic intrusions (subsequently unroofed) that differentiated during the deep burial, and possibly the tectonic stacking, of Onverwacht lavas.

Approximately contemporaneous greenstone belts in Zimbabwe (53, 54) and Australia (18, 27) have a similar stratigraphic and sedimentological character. The present sinuous to ovoid geometry of these belts is tectonic in origin and does not conform to the distribution of basin margins at the time of deposition. In Zimbabwe and Australia, distinctive sedimentary sequences have been correlated across several adjacent belts (27, 53).

Younger greenstone belts, well developed in Canada, Zimbabwe, and the Yilgarn Block of western Australia, share many features with their older counterparts, but they show a more obvious sialic influence (26). In some cases, direct field evidence indicates that greenstones developed on a sialic basement (e.g., (2)). Stratigraphically lower volcanic sequences contain a higher proportion of detrital sediments, and in some cases these rocks were derived from sialic source areas (46, 51). In at least one case, the Abram Group of northwestern Ontario, sediments associated with basal basalts show no evidence of sialic provenance, but immediately overlying conglomerates contain tonalitic cobbles, suggesting either that granitoid intrusions were emplaced after the initial extrusion of basalts or that sialic basement older than the greenstone belt existed but was not exposed to erosion near preserved sites of sedimentation. As in older greenstone belts, alluvial deposits alternate with turbidites and there is little record of shallow shelf sedimentation.

The general developmental pattern of greenstone belts, then, includes basal mafic to ultramafic lavas overlain by volcanic piles flanked by volcanogenic and chemical sedimentary accumulations. Subsequent or contemporaneous uplift produced thick sequences of compositionally immature sediments deposited predominantly on tectonically unstable alluvial plains or in relatively deep ocean basins. Late stage detrital sediments often include more mature lithologies and contain sedimentary structures indicative of shallow shelf deposition. The presence or absence of sialic basement certainly influenced sediment composition, but apparently made little difference in the overall tectonic development of Archean greenstone belts.

Greenstone belt successions are not the only sedimentary packages found in Archean low-grade terrains. In southern Africa an extensive sequence of predominantly texturally and compositionally mature sandstones, shales, conglomerates, and carbonates was deposited 3100-2900 Ma ago. This succession, the Mozaan Group of the Pongola Supergroup, contains abundant sedimentary structures indicative of fluvial to shallow shelf deposition, including herringbone crossbeds and stromatolites (49, 52). The Mozaan Group strongly indicates that at least small areas of sialic crust had become stabilized by mid-Archean times. The late Archean Witwatersrand and Ventersdorp Supergroups (perhaps in part Proterozoic; better age control is needed) further attest to early stabilization of continental crust in this region. Interestingly, no extensive carbonates are found in any of these successions. Pongola to Ventersdorp deposition took place contemporaneously with the formation of the younger greenstone belts in Zimbabwe.

Mid-Archean continental stabilization apparently also took place in the Pilbara Block of northwestern Australia. Late Archean greenstone belts are not found in this region, although they are widespread in the Yilgarn Block to the south. The 2800 Ma Fortescue Group, located near the southern margin of the Pilbara Block, has characteristics reminiscent of both greenstone belts and epicratonic sequences. The group consists largely of mafic volcanics and tuffs, but it also contains sial-derived clastic sediments and locally well developed shallow water carbonates (15). Like the slightly younger but geologically similar Ventersdorp succession in South Africa, the Fortescue Group may be the consequence of greenstone belt-generating processes constrained by a stiffened and stabilized continental crust. A similar explanation has been advanced for the intrusives of the approximately 2500 Ma old Great Dyke of Zimbabwe (8, 54).

High-grade Terrains

Weakly metamorphosed sequences are attractive because they tend to be scientifically accessible; however, they constitute only about one half of the preserved Archean record. The other half consists of rocks metamorphosed to granulite or amphibolite facies under conditions of deep burial. Units of undoubted sedimentary origin are included within these high-grade terrains (55). Some high-grade metasediments appear to be severely metamorphosed greenstone belt successions, but others resemble metamorphosed passive margin shelf accumulations (9, 13). The high-grade terrains thus indicate that the picture of Archean stratigraphy developed from studies of low-grade terrains is incomplete and that continental shelf deposits were more widespread on the early Earth than one would be led to believe from the examination of weakly metamorphosed regions.

Few passive margin sedimentary accumulations escaped metamorphism under conditions of deep burial (20 km or more (9, 42)), perhaps due to tectonic emplacement of low-grade marginal basin slabs on top of continental margins. One possible explanation for this would be that most Archean continents were not extensive (and/or buoyant) enough to preserve large areas of shelf and epicratonic sediments in an unmetamorphosed state.

LOWER PROTEROZOIC STRATIGRAPHY

In contrast to the Archean, Proterozoic greenstone-like accumulations are limited in geographic extent, whereas epicratonic and miogeosynclinal successions are widespread. Excellent reviews of Lower Proterozoic stratigraphy exist in the literature (6, 22, 44), and they uniformly indicate the distinctly different and modern nature of these packages, relative to those of the Archean. The geometry, distribution, and composition of Lower Proterozoic sedimentary rocks all show the strong influence of large continents. For example, thick carbonates of the Transvaal Supergroup, South Africa, had an original depositional extent of at least 500,000 km^2.

In the Canadian Shield there appear to have been two major tectonic settings for Early Proterozoic sedimentation. Continental margin sequences such as that found in the Wopmay Orogen (10, 17) are quite similar to Phanerozoic geosynclinal successions. Indeed, the entire tectonic development of the Wopmay Orogen compares closely with the Phanerozoic evolution of the North American Cordillera and demands the presence of a reasonably large stable continent. Other sedimentary basins appear to have been initiated by rifting along lines of crustal weakness within continents. The boundary zones between older Archean cratons and areas stabilized 2600-2500 Ma ago appear to have been preferential sites for basin development (examples include the Huronian, Animikie, and Menominee groups of the Lake Superior area and the Ramah Group of Labrador). For want of a better term, I will refer to these rift-initiated basins as "intracratonic ocean basins," although the term is in some sense oxymoronic. These basins exhibit a sedimentary and structural development similar to that of Phanerozoic ocean margin geosynclines, but they are located between areas of Archean continental crust. Available paleomagnetic evidence suggests that any oceans which may have opened up were not wide. Perhaps these regions represent the opening and closing of relatively narrow oceans between distinct crustal regions of a single continent. The apparent restriction of Early Proterozoic iron formations to "intracratonic ocean basins" and their general development near the axes of these basins should be taken into account in hypotheses for the origin of iron formations.

Some sedimentary/volcanic sequences within the Churchill Province are suggestive of collisional suturing between two unrelated continental blocks, in accord with geomagnetic data indicating that prior to 1850 Ma (or a bit earlier) the Slave and Superior cratons moved independently, and that since then they have moved together (25, 30).

In summary, the Lower Proterozoic stratigraphy of the Canadian Shield indicates the presence of large cratonic blocks and plate tectonic processes comparable to those of the Phanerozoic. Lower Proterozoic sequences in southern Africa and Australia exhibit a similar development. Since extensive linear, coupled miogeosynclinal and eugeosynclinal supracrustal packages earlier than the Proterozoic are not known, and since the areal extent of Archean epicratonic sedimentary sequences is quite limited, one can reasonably conclude that the Archean/Proterozoic transition reflects an episode of continental crustal stabilization far larger in scale than those of earlier periods. The estimated original depositional extent of the Pongola Supergroup is 50,000 km^2. In comparison, as mentioned above, the Lower Proterozoic Transvaal Supergroup covered an area of at least 500,000 km^2. Similarly, the Australian Pilbara Block today crops out over an area of 56,000 km^2 (an equally large area is covered by younger rocks), while the exposed area of the Yilgarn Block (stabilized 2600-2500 Ma) is an order-of-magnitude larger (18). This suggests that the time interval between 3000 Ma and the Archean/Proterozoic boundary was a period of very significant crustal growth.

GEOCHEMICAL AND GEOCHRONOLOGICAL CONSIDERATIONS

Geochemical and geochronological data that bear on the nature of the Archean/Proterozoic transition have been evaluated in a number of publications (e.g., (32, 47, 55)). Because of the model dependency of conclusions drawn from these data, widely divergent interpretations of early crustal history have been published. Nonetheless, a brief discussion of the data is appropriate for comparison with the stratigraphic picture developed in preceding paragraphs.

Isotopic analysis of ancient igneous and metamorphic rocks is potentially the most direct means of ascertaining the growth curve for continental crust. Moorbath (33) suggested on the basis of initial $^{87}Sr/^{86}Sr$ values that most of the crustal material added to continents during the Archean Eon was newly derived from the mantle and not recycled from earlier continents. Rb/Sr, U/Pb, and Sm/Nd data have all been interpreted to indicate that while sialic crust certainly existed early in Earth history, its areal extent was limited prior to 3000 Ma (29, 34, 36). In contrast, crustal rocks that yield dates of 2900-2500 Ma are widespread, suggesting that this was a period of rapid continental crust formation. O'Nions and Pankhurst (36) estimated that "perhaps 50% of the present continental crust was stabilized subsequent to an episode of intense magmatic and metamorphic activity unparalleled at any other time in earth history," a view that enjoys wide support among geologists (however, see (1) for a dissenting opinion).

Most of the continental crust generated during this period consists of tonalites and trondhjemites, along with greenstone belt volcanics and sedimentary rocks caught up in the intrusions. Crustal stabilization, as indicated by the development of persistent epicratonic sedimentary sequences containing mature detrital rocks and thick, laterally extensive carbonates, accompanied the differentiation of K-rich granitic plutons - as early as 3000 Ma ago in southern Africa and northwestern Australia, but 2600-2500 Ma ago over larger shield areas.

The geochemistry of Archean and Lower Proterozoic sedimentary rocks is consistent with the stratigraphic and igneous data. Engel et al. (11) noted a significant increase in the K_2O/Na_2O value of detrital rocks across the Archean/Proterozoic boundary. A similar change was documented by Cameron and Garrels (5) in a compositional analysis of several hundred Archean and Lower Proterozoic shales from the Canadian Shield. Both sets of authors concluded that the shift in this ratio reflects both a change in the average composition of source rocks (becoming much more potassic following the terminal Archean granitic intrusions) and the more mature weathering state achieved by Lower Proterozoic sediments deposited on stable cratons. McLennan and Taylor (32) have also discussed a temporal shift in sedimentary REE patterns across the Archean/Proterozoic boundary, attributing the change to the widespread unroofing of granitic plutons in Early Proterozoic times. Enthusiasm for these results must be tempered by the data of Dymek and his colleagues (3, 9) who found that some metasediments from high-grade Archean terrains have REE patterns different from those of greenstone belt shales. The variability of REE patterns in Archean sedimentary rocks suggests that the patterns reflect the nature of source terrains. This interpretation is consistent with stratigraphic and geochronological evidence which indicates that some sialic crustal blocks were differentiated early in Earth history, well before the advent of large Early Proterozoic cratons.

One additional perspective on the Archean/Proterozoic boundary is provided by analysis of $^{87}Sr/^{86}Sr$ in early carbonates. Veizer (47, 48) has reported a significant change in the ratios of strontium isotopes across the boundary, and he has interpreted this to be a reflection of the erosion of large, highly fractionated continents beginning in the Early Proterozoic Era, as well as decreased "mantle" (exhalative, hydrothermal, and low temperature submarine chemical alterations) influences on seawater chemistry.

In general, geochemical and geochronological evidence corroborate the stratigraphic conclusions that the late Archean was an interval of extensive continental crust generation and that the supracrustal patterns characteristic of the Early Proterozoic Era reflect the amalgamation of sialic blocks into large, stable continents. Figure 1 presents a preferred summary of geological history relevant to the Archean/Proterozoic transition.

It is a schematic diagram in which some attempt is made to depict variations in the importance of various supracrustal rock packages and continental crust-generating processes through time. This acknowledges that although quantification is difficult, it is the relative abundance and distribution of rock types that is of primary importance, rather than their presence or absence in a given time period.

Limited areas of tonalitic/trondhjemitic (with incorporated mafic and ultramafic material) crust existed as early as 3700 Ma and perhaps earlier. Passive margin shelf sequences developed on the flanks of these blocks were subsequently metamorphosed under conditions of deep burial. Rifting of (ensimatic?) crust, perhaps along the margins of continental blocks, produced thick lower greenstone belt volcanic piles. Subsequent intrusion

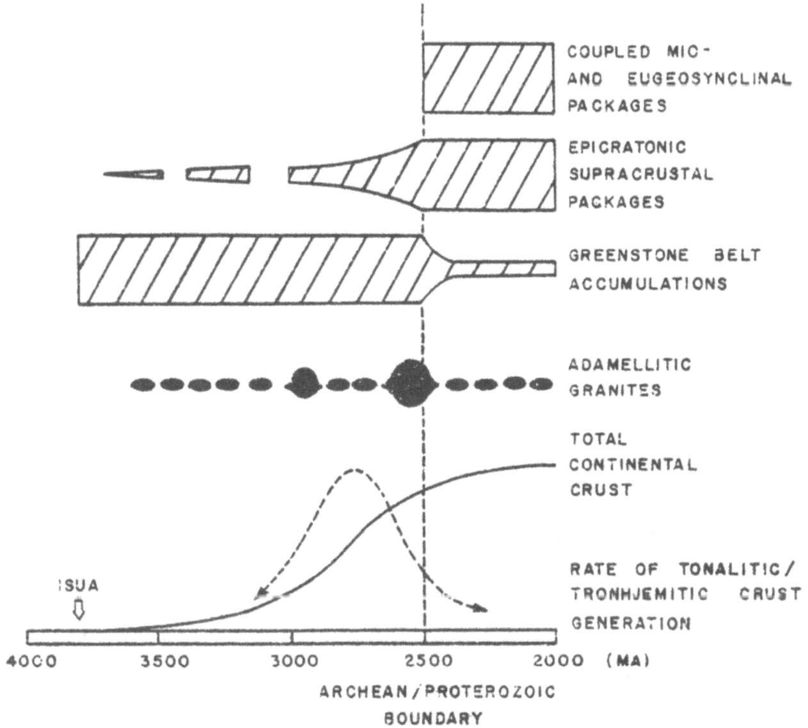

FIG 1 Stylized summary diagram of the geological record documenting the Archean and Early Proterozoic eras. The sigmoidal curve (solid line) above the time scale represents an extremely smoothed picture of continental growth through this interval. The accompanying dotted line indicates rates of tonalitic/trondhjemitic crust generation through time (first derivative of solid curve). Neither curve is intended to be strictly quantitative; each is drawn to convey an indication of the general pattern of crustal evolution. These curves are based on continental growth arguments of Moorbath and others (cited in text). Also shown are generalized patterns of adamellitic granite generation, greenstone belts, epicratonic sedimentary and volcanic sequences, and extensive coupled miogeosynclinal and eugeosynclinal packages through time. Again, general patterns are meant to be conveyed, not quantitative estimates of rock volume. Highly metamorphosed miogeosynclinal and eugeosynclinal rock packages may be present in some high-grade Archean terrains.

of new tonalitic magmas preceded uplift and exposure of land masses whose erosion provided source materials for greenstone belt sediments.

Emplacement of adamellitic granites stabilized regional areas of continental crust as early as 3000 Ma ago. These blocks subsequently accumulated epicratonic sedimentary sequences of mature detrital sediments and carbonates, as well as thick sequences of basalt flows. In other areas, greenstone belt formation continued until the close of the Archean. Indeed, the period 2900-2500 Ma was one of massive additions to the crust. At the close of this period, large areas of continental crust were stabilized, resulting in the initial appearance of large stable continents. The effect of these large continental blocks is evident in Early Proterozoic supracrustal rock packages (Fig. 1).

The nettlesome problem of preservation remains, and it is not likely to disappear in the near future. Is the preserved Archean geological record really representative of the Archean Earth, or have biases introduced by recycling skewed the record? The answers are not known with any certainty; however, if the existing record of the Archean Earth has been shaped largely by preservational biases, then the nature of the tectonic processes that determine crustal and sedimentary preservation must have changed markedly 2600-2500 Ma ago. Perhaps some comfort can be drawn from the fact that several different continents exhibit generally similar Precambrian geological records.

BIOLOGICAL CONSIDERATIONS

On the present-day Earth, physical and biological processes are intimately related via a complex series of mutually dependent biogeochemical cycles that collectively define the biosphere. Presumably, this also held true early in Earth history. It is therefore of interest to ask what might have been the biological consequences of massive continental growth and stabilization at the close of the Archean Eon. Fuller considerations of this question have been presented elsewhere (23, 24), but a brief discussion is relevant to the goals of this workshop.

A small but increasing collection of Archean stromatolites and microfossils demonstrates that prokaryotic life existed as early as 3500-3400 Ma ago; however, these remains do little to constrain hypotheses concerning the metabolic capabilites of early life. $^{13}C/^{12}C$ analyses strongly suggest that some form of photosynthesis (either aerobic or anaerobic) fueled ecosystems at this time, and internal features of the carbon isotope record have prompted the suggestion that aerobic photoautotrophs (cyanobacteria) existed at least 2900-2800 Ma ago (20, 40). The recent discovery of filamentous microfossils in stromatolitic cherts of the 2800 Ma Fortescue Group, Australia, is consistent with this hypothesis. These fossils are morphologically indistinguishable from species of modern oscillatorian blue-greens (41). They are not morphologically similar to extant purple and green photosynthetic bacteria, but given the fact that some living cyanobacteria can use H_2S in photosynthesis - a trait presumably characteristic of cyanobacterial ancestors - this morphological evidence does not demonstrate unequivocally that O_2-producers were present in the Archean Eon. Nevertheless, I believe that reasonable conjecture on the Earth's biota during late Archean times must include a wide variety of anaerobic autotrophic and heterotrophic bacteria, as well as blue-greens capable of "green plant" photosynthesis.

How might the Archean/Proterozoic transition have affected productivity, particularly primary production by cyanobacteria? Besides nutrient supply (considered below), total primary production by photoautotrophic microbenthos is influenced by sedimentary conditions (for example, turbulence) and total area available for colonization. On the modern Earth, the productivity of microbenthos per unit area can differ by as much as two to three orders-of-magnitude between shelves inundated by clastic sediment and quiet water microbial mat environments (23). The transition from an Archean world characterized by limited cratonic areas and tectonic instability to an Early Proterozoic Earth featuring extensive cratons with broad areas of shallow marine platforms and shelves could thus have been accompanied by a significant increase in benthic primary production. An increase of several orders-of-magnitude is not inconceivable.

The effect of crustal growth and stabilization on plankton productivity is more problematical, because we know very little about Archean planktonic photoautotrophs and their importance relative to benthic primary producers. However, the distribution of organic carbon in Archean shales, cytological specializations in various living bacteria, and the presence of several species of cyanobacteria and other prokaryotes in the modern oceans suggest that the oceans were populated early in Earth history. Although some marine ecologists argue that the availability of nitrogen limits primary production in the oceans of today, it appears likely that in an oxygen-poor Archean ocean populated by prokaryotes capable of nitrogen fixation, PO_4^{-3} must have been the limiting nutrient. Again, one can consider the effects of cratonization. Weathering and erosion of large, stable, emergent continents would increase the phosphate flux into the oceans, and large continents would alter the bathymetry of ocean basins in such a way as to promote increased upwelling and to speed the return of nutrients to the photic zone (7). The result would be an increase in global primary production by planktonic photoautotrophs.

A quantitative treatment of the productivity of Precambrian oceans is impossible, but there are good reasons to believe that levels of primary production in the Early Proterozoic Era were significantly higher than in the Archean. Is there any geochemical evidence that might be marshalled in support of this hypothesis? Perhaps. The temporal distribution of detrital uraninite has often been cited in support of an earliest Proterozoic increase in atmospheric oxygen levels. Detrital UO_2 is unknown from rocks younger than about 2300 Ma (although grains have been found in modern sediments of the Indus River), while paleosurface-controlled, vein-like uranium deposits consisting mostly of pitchblende and alteration products in a hematite gangue first appear in weathering horizons developed after this date (38). Redbeds also became prominent for the first time 2300 Ma ago.

Weathering profiles of Archean granitoids indicate that at least small concentrations of O_2 existed in the atmosphere well before the Proterozoic Eon; however, Grandstaff (16) has suggested that uraninite is likely to survive surface exposure, weathering, and erosional transport only if atmospheric oxygen levels are less than approximately 1% PAL. If so, then the nearly simultaneous disappearance of detrital uraninite from the sedimentary columns of at least five continents approximately 2300 Ma ago is most parsimoniously explained by the growth of atmospheric O_2 from low concentrations to levels above 1% PAL. Support for this interpretation is provided by secular trends in the S isotopic record of sedimentary sulfides, which Cameron (4) has interpreted as indicating an increase in seawater sulfate levels about 2350 Ma. (In South Africa, the S isotope transition occurs in beds directly overlying the units that contain the youngest known detrital uraninites.)

The evolution of aerobic photoautotrophs at least 500-400 Ma before the earliest evidence of atmospheric O_2 enrichment and the coincidence of atmospheric transition with the first appearance of widespread "Proterozoic-style" shelf and epicontinental sea deposits permit the hypothesis that global tectonics and atmospheric composition were linked through cyanobacterial productivity. In the Archean, the physical condition of the Earth kept rates of photosynthetic oxygen production below those of O_2 consumption via the oxidation of minerals, organic matter, and reduced volcanic gases. With the terminal Archean stabilization of large cratons, productivity rose, and with it atmospheric oxygen concentrations.* Additionally, Veizer (47) has suggested that tectonic changes accompanying the Archean/Proterozoic transition would have resulted in decreased rates of oxygen consumption by reduced minerals and gases. If true, this would amplify O_2 concentration increases.

A biologically induced increase in atmospheric oxygen levels above 1% PAL would have had further biological consequences. For many prokaryotes, oxygen is toxic; anaerobic bacteria must therefore have become limited to anoxic water bodies and subsurface sedimentary environments below the level of effective O_2 penetration. Other bacteria evolved mechanisms for using oxygen in energetically efficient aerobic respiration. Oligonucleotide cataloging of 16S ribosomal RNA's in a wide variety of prokaryotes suggests that aerobic respiration evolved independently in a number of different bacterial lineages (14).

It should be noted that the Precambrian history of atmospheric oxygen levels remains poorly known. The detrital uraninite arguments, for example, have been attacked on the grounds that the UO_2 is not detrital and that the time distribution of these bodies may reflect a time bounded distribution of suitable source bodies. Holland (21) has recently summarized the available evidence on uraninite emplacement and has shown that, although diagenetic mobilization may well have occurred, at least some of the late Archean and earliest Proterozoic sedimentary uraninites are detrital. The second objection is mitigated by the fact that potential source igneous rocks for detrital uraninites continued to be emplaced after the Early Proterozoic. Uranium-bearing alaskites emplaced in Namibia during the Paleozoic Era closely resemble postulated source igneous bodies for the uraniferous conglomerates of the Huronian Supergroup, Canada (39).

* Walker (50) has suggested that in the absence of an effective ozone shield, lethal ultraviolet radiation would have prevented occupation of the top several meters of the ocean by phytoplankton. If such a situation obtained in the Archean, the Early Proterozoic productivity increase would have been enhanced by the occupation of ocean surface waters as atmospheric oxygen and, hence, ozone levels increased. Whether or not the surface layers of Archean oceans were populated remains a subject for debate.

Holland (21) has also proposed that paleosols developed on mafic bed rocks may place limits on early atmospheric O2 concentrations. Available data are few and their interpretation is complicated by the variable influences of topography, drainage, and diagenesis on soil formation; however, preliminary analyses suggest that atmospheric oxygen levels may have remained very low until mid-Proterozoic times. Clearly all currently available arguments on Early Proterozoic oxygen concentrations still hang by a slender thread; more data and fresh approaches will be required to resolve this question.

CONCLUSIONS AND QUESTIONS

Two main arguments have been advocated in this paper. The first is that the Archean/Proterozoic transition reflects the appearance of large stable continents following a period of uniquely high rates of tonalitic/trondhjemitic crust generation. The second is that these fundamental crustal changes significantly affected the Earth's biota, increasing productivity to the point where metabolically significant concentrations of oxygen accumulated in the atmosphere. The shift in the composition of the atmosphere had further biological consequences, including the reorganization of ecosystems and the evolution of aerobic respiration.

This hypothesis is attractive in that it fits well a large and rather disparate body of geological and biological data and does not depend on special considerations or coincidences. It may have value as a heuristic model of biosphere development through time. Explanations for historically unique events are notoriously difficult to test, yet the importance of historical explanation in geology cannot reasonably be denied (28). The arguments advanced here at least serve to highlight several areas that one would like to understand better. A short list of questions and desiderata is presented for consideration.

1. New methods of estimating the concentration of oxygen in the Precambrian atmosphere are clearly needed. The analysis of paleosols may provide one promising new avenue (21). Other geochemical tests must be developed.

2. Better paleontological and geochemical techniques for determining the biological composition of early biotas are needed. We still know relatively little about Archean life.

3. How did crustal areas now preserved as deeply eroded high-grade metamorphic terrains behave in Archean times? Were they cratonic? What types and amount of sediments accumulated on and about these areas? Metasediments preserved in high-grade terrains hold important clues to early crustal development. A better understanding of these rocks and of processes that may have led to the selective preservation of some supracrustal materials in the Archean is desirable.

4. Did continental stabilization occur throughout the late Archean period of rapid continental growth, or did it occur episodically, particularly during a relatively brief period at the end of the eon? That is, to what extent are the addition of sialic crust and the stabilization of cratons distinct processes? If they are distinct, what processes can account for the widespread generation of K-rich granites 2600-2500 Ma ago?

5. Could a temporary decrease in atmospheric CO_2 concentrations occasioned by Early Proterozoic increases in primary production have initiated the Gowganda continental glaciations?

In conclusion, it seems that insofar as they highlight possible relationships among disparate data sets, integrative, multidisciplinary approaches to Archean and Proterozoic history have much to contribute to studies of Earth evolution.

Acknowledgements. I thank R. Dymek, H.D. Holland, R. Siever, and A. Trendall for many helpful criticisms and E. Burkhardt for her skillful preparation of the manuscript. My research on problems of Precambrian biology has been supported in part by NSF Grants DEB 80-04290 and DEB 82-13682.

REFERENCES

(1) Armstrong, R.L. 1981. Radiogenic isotopes: the case for crustal recycling on a near-steady-state no-continental-growth Earth. Phil. Trans. R. Soc. London A301: 443-472.

(2) Bickle, M.J.; Martin, A.; and Nisbet, E.G. 1975. Basaltic and peridotitic komatiites and stromatolites above a basal unconformity in the Belingwe greenstone belt, Rhodesia. Earth Planet. Sci. Lett. 27: 155-162.

(3) Boak, J.L.; Dymek, R.F.; and Gromet, L.P. 1982. Early crustal evolution: constraints from variable REE patterns in metasedimentary rocks from the 3800 Ma Isua supracrustal belt, West Greenland. Lunar Planet Sci. 13: 51-52.

(4) Cameron, E.M. 1982. Sulphate and sulphate reduction in early Precambrian oceans. Nature 296: 145-148.

(5) Cameron, E.M., and Garrels, R.M. 1980. Geochemical compositions of some Precambrian shales from the Canadian Shield. Chem. Geol. 28: 181-197.

(6) Campbell, F.H.A., ed. 1981. Proterozoic Basins of Canada. Geol. Sur. Can. Paper 81-10: 444.

(7) Chamberlain, W.M., and Marland, G. 1977. Precambrian evolution in a stratified global sea. Nature 265: 135-136.

(8) Coward, M.P.; Lintern, B.C.; and Wright, L. 1976. The pre-cleavage deformation of the sediments and gneisses of the northern part of the Limpopo belt. In The Early History of the Earth, ed. B.F. Windley, pp. 323-330. New York: Wiley.

(9) Dymek, R.F.; Weed, R.; and Gromet, L.P. 1983. The Malene metasedimentary rocks of Rypeø and their relationship to Amîtsoq gneisses. Geol. Unders. Grønland Bull., in press.

(10) Easton, R.M. 1981. Stratigraphy of the Akaitcho Group and the development of an Early Proterozoic continental margin, Wopmay Orogen, Northwest Territories. Geol. Sur. Can. Paper 81-10: 79-96.

(11) Engel, A.E.J.; et al. 1974. Crustal evolution and global tectonics: a petrogenic view. Geol. Soc. Am. Bull. 85: 843-858.

(12) Eriksson, K.A. 1980. Transitional sedimentation styles in the Moodies and Fig Tree Groups, Barberton Mountain Land, South Africa: evidence favoring an Archean continental margin. Precambrian Res. 12: 141-160.

(13) Erselv, E.A. 1981. Petrology and structure of the Precambrian metamorphic rocks of the southern Madison Range, Southwestern Montana. Ph.D. Dissertation, Harvard University, Cambridge, MA.

(14) Fox, G.; et al. 1981. The phylogeny of prokaryotes. Science 209: 457-463.

(15) Goode, A.D.T. 1981. Proterozoic geology of Western Australia. In Precambrian of the Southern Hemisphere, ed. D.R. Hunter, pp. 105-204. Amsterdam: Elsevier.

(16) Grandstaff, D.E. 1980. Origin of uraniferous conglomerates at Elliot Lake, Canada and Witwatersrand, South Africa: Implications for oxygen in the Precambrian atmosphere. Precambrian Res. 13: 1-26.

(17) Hoffman, P.F. 1980. A Wilson Cycle of Early Proterozoic age in the northwest of the Canadian Shield. Geol. Assoc. Can. Spec. Paper 20: 523-549.

(18) Hallberg, J.A., and Glikson, A.Y. 1981. Archean granite-greenstone terrains of Western Australia. In Precambrian of the Southern Hemisphere, ed. D.R. Hunter, pp. 33-104. Amsterdam: Elsevier.

(19) Hamilton, P.J. 1977. Great Dyke and Bushveld mafic phase. J. Petrol. 18: 24-52.

(20) Hayes, J.M. 1983. Geochemical evidence bearing on the origin of aerobiosis, a speculative interpretation. In Origin and Evolution of the Earth's Earliest Biosphere: An Interdisciplinary Study, ed. J.W. Schopf, pp. 291-301. Princeton: Princeton University Press.

(21) Holland, H.D. 1984. The Chemical Evolution of the Atmosphere and Oceans. Princeton: Princeton University Press, in press.

(22) Hunter, D.R., ed. 1981. Precambrian of the Southern Hemisphere. Amsterdam: Elsevier.

(23) Knoll, A.H. 1979. Archean photoautotrophy: some alternatives and limits. Origins Life 9: 313-327.

(24) Knoll, A.H. 1982. Tectonics, productivity, and ecosystems on the early earth. In Proceedings of the 3rd North American Paleontological Convention, vol. 2, pp. 307-311, Montreal, Canada.

(25) Lewry, J.F. 1981. Lower Proterozoic arc-microcontinent collisional tectonics in the western Churchill Province. Nature 294: 69-72.

(26) Lowe, D.R. 1980. Archean sedimentation. Ann. Rev. Earth Planet. Sci. 8: 145-167.

(27) Lowe, D.R. 1982. Comparative sedimentology of the principal volcanic sequences of Archean greenstone belts in South Africa, Western Australia and Canada: implications for crustal evolution. Precambrian Res. 17: 1-29.

(28) Mayr, E. 1982. The Growth of Biological Thought. Cambridge: Belknap Press of Harvard University Press.

(29) McCulloch, M.Y., and Wasserburg, G.J. 1978. Sm-Nd and Rb-Sr chronology of continental crust formation. Science 200: 1003-1011.

(30) McGlynn, J.C., and Irving, E. 1981. Horizontal motions and rotations in the Cambrian Shield during the Early Proterozoic. Geol. Surv. Can. Paper 81-10: 183-190.

(31) McLaren, D.J. 1977. The Silurian-Devonian boundary committee; a final report. The Silurian-Devonian Boundary, IUGS Ser. A5: 1-34.

(32) McLennan, S.M., and Taylor, S.R. 1982. Geochemical constraints on the growth of the continental crust. J. Geol. 90: 347-361.

(33) Moorbath, S. 1975. Evolution of Precambrian crust from strontium isotopic evidence. Nature 254: 395-398.

(34) Moorbath, S. 1977. The oldest rocks and the growth of continents. Sci. Am. 240 (March): 92-104.

(35) Nisbet, E.G. 1982. Definition of 'Archean' - comment and a proposal on the recommendations of the International Subcommission on Precambrian Stratigraphy. Precambrian Res. 19: 111-118.

(36) O'Nions, R.K., and Pankhurst, R.J. 1978. Early Archean rocks and geochemical evolution of the earth's crust. Earth Planet. Sci. Lett. 38: 211-236.

(37) Pettijohn, F.J. 1943. Archean sedimentation. Geol. Soc. Am. Bull. 54: 925-972.

(38) Robertson, D.S.; Tilsley, J.E.; and Hogg, G.M. 1978. The time-bound character of uranium deposits. Econ. Geol. 78: 1409-1419.

(39) Robinson, A., and Spooner, E.T.C. 1982. Source of the detrital components of uraniferous conglomerates, Quirke ore zone, Elliot Lake, Ontario, Canada. Nature 299: 622-624.

(40) Schoell, M., and Wellmer, F.W. 1981. Anomalous ^{13}C depletion in early Precambrian graphites from Superior Province, Canada. Nature 290: 696-699.

(41) Schopf, J.W. 1980. Evidences of early Precambrian (Archean) life. Abstract, p. 356. 5th International Palynological Conference, Cambridge.

(42) Shackleton, R.M. 1976. Shallow and deep-level exposures of Archaean crust in India and Africa. In The Early History of the Earth, ed. B.F. Windley, pp. 317-322. New York: Wiley.

(43) Sims, P.K. 1980. Subdivision of the Proterozoic and Archaean ions: recommendations and suggestions by the International Subcommission on Precambrian Stratigraphy. Precambrian Res. 13: 379-380.

(44) Tankard, A.J.; et al. 1982. Crustal Evolution of Southern Africa. Heidelberg: Springer.

(45) Tarney, J., and Windley, B.F. 1981. Marginal basins through geological time. Phil. Trans. R. Soc. London A301: 217-232.

(46) Turner, C.C., and Walker, R.G. 1973. Sedimentology, stratigraphy, and crustal evolution of the Archean greenstone belt near Sioux Lookout, Ontario. Can. J. Earth Sci. 10: 817-845.

(47) Veizer, J. 1983. Geologic evolution of the Archean-Early Proterozoic earth. In Origin and Evolution of the Earth's Earliest Biosphere: An Interdisciplinary Study, ed. J.W. Schopf, pp. 240-259. Princeton: Princeton University press.

(48) Veizer, J., and Compston, W. 1976. $^{87}Sr/^{86}Sr$ in Precambrian carbonates as an index of crustal evolution. Geochim. Cosmochim. Acta 40: 905-914.

(49) Von Brunn, V., and Mason, T.R. 1977. Siliciclastic-carbonate tidal deposits from the 3000 m.y. Pongola Supergroup, South Africa. Sed. Geol. 18: 245-255.

(50) Walker, J.C.G. 1978. The early history of oxygen and ozone in the atmosphere. Pure Appl. Geophys. 116: 222-231.

(51) Walker, R.G:, and Pettijohn, F.J. 1971. Archaean sedimentation: analysis of the Minnitaki Basin, Northwestern Ontario, Canada. Geol. Soc. Am. Bull. 82: 2099-2130.

(52) Watchorn, M.B. 1980. Fluvial and tidal sedimentation in the 3000 Ma Mozaan Basin, South Africa. Precambrian Res. 13: 27-42.

(53) Wilson, J.F. 1979. A preliminary reappraisal of the Rhodesian basement complex. Spec. Publ. Geol. Soc. S. Afr. 5: 1-23.

(54) Wilson, J.F.; et al. 1978. Granite-greenstone terrains of the Rhodesian Archean craton. Nature 271: 23-27.

(55) Windley, B.F. 1977. The Evolving Continents. New York: Wiley.

Patterns of Change in Earth Evolution, eds. H.D. Holland and A.F. Trendall, pp. 243-259.
Dahlem Konferenzen 1984. Berlin, Heidelberg, New York, Tokyo: Springer-Verlag.

The Archean/Proterozoic Transition As a Geological Event – A View from Australian Evidence

A.F. Trendall
Geological Survey of Western Australia
Perth, W.A. 6000, Australia

Abstract. The concept of an Archean-Proterozoic transition has its roots in a nineteenth century preconception that the development of the Earth during the Precambrian was naturally divisible into two stages; the legacy of this assumption persists not only in the continued (but usually undefined) use of the two names, but also in a widespread reluctance to shake off the burden of established belief which their application implies. An objective assessment of the Archean-Proterozoic transition involves an answer to the general question: "Is it possible to identify a brief period of early Earth history such that the rocks that formed for relatively long periods before and after it have consistently differing and individually distinctive characteristics, other than their relative ages?" Within Australia the geology of the area generally accepted as best displaying the Archean-Proterozoic boundary is complex. This complexity was built up by the interaction of many related processes operating more or less continually within a period of at least 1200 Ma, which began 3500 Ma ago and which probably concluded about 2300 Ma. Although there were many pauses in the processes operating, resulting in discontinuities in the rock record, there does not seem to be a justification for any arbitrary division of this complex history into two parts. Furthermore, any arbitrary distinction between "Archean style" and "Proterozoic style" rocks would require a time boundary at about 2800 Ma, significantly older than in the "type" Archean-Proterozoic unconformity of Canada, while a boundary at the end of the whole history, at about 2300 Ma, would be anomalously young. Evidence from Australia thus gives no support to the concept of a worldwide natural two-stage subdivision of the Precambrian.

INTRODUCTION

Preiss (11) has described a current proposal of the International Subcommission on Precambrian Stratigraphy to assign an age of 2500 Ma to the Archean-Proterozoic boundary as a piece of legislation, "nothing to do with stratigraphy, nor with science at all." It is not the purpose of this paper to examine the Archean-Proterozoic boundary as thus arbitrarily designated, but to discuss the status of the Archean-Proterozoic transition as a significant event in Earth history. This means, in effect, to attempt an answer to the general question: "Is it possible to identify a brief period of early Earth history such that the rocks that formed for relatively long periods before and after it have consistently differing and individually distinctive characteristics, other than their simple relative ages?"

The objectives here are to see to what extent an answer valid for Australia can be derived from Australian evidence, and to discuss briefly what significance such an answer has for the rest of the Earth.

ARCHEAN AND PROTEROZOIC - DEFINITIONS AND HISTORICAL BACKGROUND

Although major international symposia on Archean rocks took place in 1970 and 1980, and several important volumes with "Archean" in their titles have been published during this period, there are no agreed definitions of the terms Archean and Proterozoic, and current usages of the words can only be understood with some appreciation of their historical development.

Rankama (12) reviewed the origin and early usages of both words. Although the term Archean was initially applied to the whole of the period now called Precambrian, a subsequent twofold division of the Precambrian time into Archean and Proterozoic was already established in the nineteenth century, before the advent of isotopic ages. The main basis of this subdivision was the local recognition of flat-lying sedimentary rocks unconformably resting on older gneisses, a situation most clearly displayed in the area north of Lake Huron, in Canada. As structurally and stratigraphically identical situations became known elsewhere, the names Archean and Proterozoic were applied to the rocks respectively below and above the unconformities. Partly because of this nomenclatural consistency, the belief gradually arose that such unconformities were time-equivalent, and by an argument of complete circularity ((8), p. 170), it came to be proposed that because Archean rocks were everywhere strongly deformed and altered, in contrast to the Proterozoic rocks overlying them, the Archean-Proterozoic boundary marked a global event of unique significance.

Despite the fact that the gradual accumulation of isotopic age determinations of increasing reliability has shown unequivocally that major unconformities within the Precambrian vary widely in age, there is still a widespread acceptance that the transition from the Archean to the Proterozoic corresponds to an abrupt and worldwide event in Earth history, and that this took place, based on isotope geochronology from the original Canadian type area, about 2500 Ma ago. However, such an acceptance is by no means universal, and the spectrum of opinion currently held by experienced and competent Precambrian geologists ranges from subscription to this view to a position of outright disbelief in the existence of any abrupt discontinuity in the course of Precambrian crustal development. Before considering Australian evidence, it would be appropriate to review the basis for some of the intermediate beliefs held, and to ask what evidence is required for a resolution of the present differences of opinion.

REQUIREMENTS FOR A DEFINITION OF THE ARCHEAN-PROTEROZOIC TRANSITION

It is not now generally believed that the occurrence of any single rock type, or combination of rock types, is restricted to either the Archean or the Proterozoic, although the relative abundance or some specific attribute of some rocks (e.g., komatiite, BIF) are sometimes argued to be diagnostic. An example of a supposed time-diagnostic attribute is the REE pattern of Archean and Proterozoic sediments proposed by Taylor and McLennan (15). Most cases for the distinctive identity of Archean and Proterozoic rocks are based on complex combinations of compositional, structural, and other geological characters than on simple rock type identity. An example is the frequent description of "granite-greenstone terrain" as characteristically Archean; although this concept is not unreservedly accepted (14), it is a useful one to accept for the purposes of present discussion. Such judgements have in common the concept that during crustal development there have been abrupt changes in the rates at which various geological processes have operated.

The nineteenth century concept that the Archean-Proterozoic boundary was a global event can be represented diagrammatically as in Fig. 1A, where the line represents some index of the rate at which deformation processes took place. Alternatively, the concept in many minds may have been closer to Fig. 1B, or, since the classical basal Huronian unconformity of Canada is sometimes described as marking a widespread depositional hiatus, this might be shown by Fig. 1C, where the line represents a rate of deposition. In all these examples the Archean-Proterozoic transition is represented by the interval T_1-T_2. In reality the range of concepts for Archean-Proterozoic differences is far more complex, as in the hypothetical example of Fig. 1D. In this diagram different but coupled changes in the rates of three processes combine to define the transition. A large number of such diagrams could be constructed to illustrate as many hypotheses. It is their two common elements which are emphasized for the purposes of this paper:

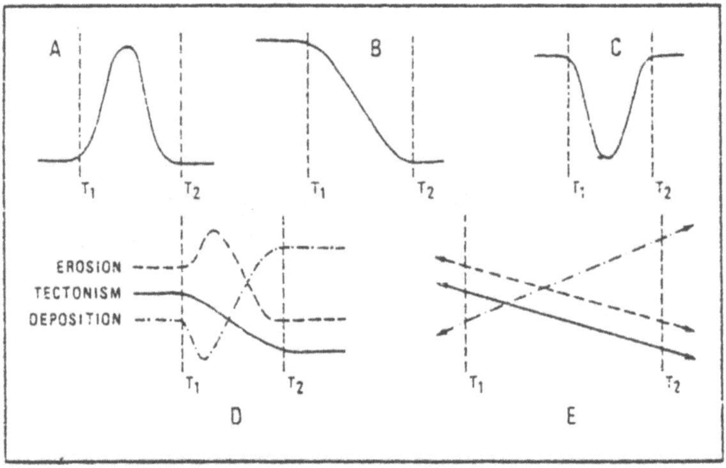

FIG. 1 - Diagrams illustrating concepts of an "Archean-Proterozoic transition"; in all diagrams time flows from left to right and the lines (other than vertical) indicate variations in the intensity (rate) of some geological process. A. A peak of tectonism. B. A decrease in level of tectonism. C. A hiatus in deposition. D. A period of varying interrelated processes between two periods in which their interrelationship is stable, but different. E. Steady change in the interrelationship of processes, not justifying the designation of time-boundaries; the arrows indicate that the lines continue indefinitely without change of slope.

1. a quantitatively different combination of processes operating stably before and after the transition, to produce the resultant Archean or Proterozoic style of crustal structure believed to be typical of each.

2. two abrupt changes in the rates of one or more of these processes at T_1 and T_2 to define the transition.

Note that if the change in rate of these processes is not kinked, then, in spite of the fact that the beginning and ending stages of the change are different, there is no natural break in the total process which it would be appropriate to name. This is illustrated by Figure 1E.

A VIEW FROM AUSTRALIA

Archean and Proterozoic: The "Accepted" View

The largest and most significant outcrop areas in the Australian continent of rocks traditionally assigned to the Archean lie in Western Australia, where they form two ancient stable "blocks" (Fig. 2). The southern of these, the Yilgarn Block, has an area of about 650 000 km^2. It has been subdivided into a number of tectonic units (5), but of these, only the "western gneiss terrain" has been indicated in Fig. 2. The major eastern part of the Yilgarn Block consists of classical granite-greenstone terrain; the granitoid plutons yield ages of about 2700 Ma (2), while the greenstone belts are about 100 Ma older (4, 10). The western gneiss terrain is significantly older (1). The Yilgarn Block is of subordinate interest to the present paper and is referred to again only in the final discussion.

The smaller Pilbara Block to the north, which also consists of granite-greenstone terrain, is of more interest here; its geochronology is presented below. The southern edge of the Pilbara Block is marked by the northern margin of the Hamersley Basin. This unconformity, between the base of the Mount Bruce Supergroup and the underlying Pilbara Block, has been proposed as the Archean-Proterozoic boundary for Australia by Dunn et al. (3). They selected it as one of the "oldest confirmed major unconformities" in Australia, and they were clearly influenced in their choice by the precedent of the Huronian unconformity of Canada. If a major unconformity were to be selected to mark the Archean-Proterozoic boundary in Australia, this one would be the natural choice, and it is towards the relationship between the Pilbara Block and the Hamersley Basin that the following sections of this paper are directed.

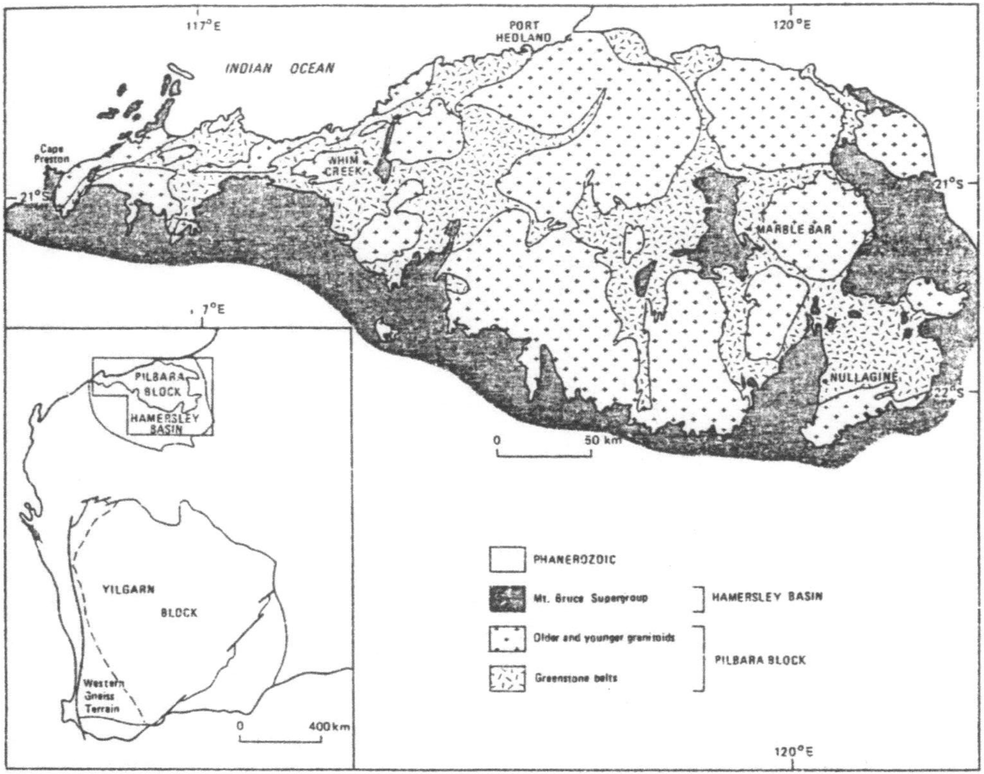

FIG. 2 - Maps showing distribution of "Archean" and "Proterozoic" rocks in the western part of the Australian continent, and the geology of the Pilbara Block and the adjacent part of the Hamersley Basin.

The ("Archean") Pilbara Block

It has already been noted that this block consists of granite-greenstone terrain; the structural pattern defined by the greenstone belts and the generally rounded granitoid plutons to which they are marginal are shown in Fig. 2. Hickman (6, 7) has summarized the structure, stratigraphy, and evolution of the block.

The greenstone belts (or "layered succession" of Fig. 3) are crudely synclinorial, and their stratigraphic succession, with a total thickness of over 30 km, is broadly correlatable over the whole extent of the block.

The basal Warrawoona Group has a thickness of nearly 16 km and consists mainly of mafic to felsic volcanics, together with some ultramafic and sedimentary units. The succeeding Gorge Creek Group, with a total thickness of 12.5 km, also contains volcanic rocks but is dominated by thick granite-derived sediments. The uppermost Whim Creek Group and higher units have a thickness of about 2 km and consist largely of volcanics ranging in composition from felsic to ultramafic. The complete succession is never present in a single transect of any of the belts, and Hickman (6) recognizes three regional and five local unconformities within the sequence. Hickman (7) estimates that the succession may have an actual thickness of up to 15 km in the North Pole area.

The granitoid plutons are complex and internally variable. For present purposes it is enough to note that they can be broadly divided into two components: the "older granitoids" which form the bulk of the present outcrop area and consist mainly of migmatite, gneissic granodiorite, and adamellite, and the "younger granitoids" which are poorly foliated adamellites and granites. Both have either tectonic or intrusive relationships with the greenstone belts.

The geochronological evidence available from the Pilbara Block is displayed in Fig. 3. It includes results from both the older and the younger granitoids, as well as from the layered succession of the greenstone belts. Most of the age determinations labelled "layered succession" in Fig. 3 come from the eastern parts of the Pilbara Block, from the greenstone belts west and southwest of Marble Bar (Fig. 2). Those labelled "Whim Creek" come from the stratigraphically highest units of the succession at Whim Creek, also shown in Fig. 2.

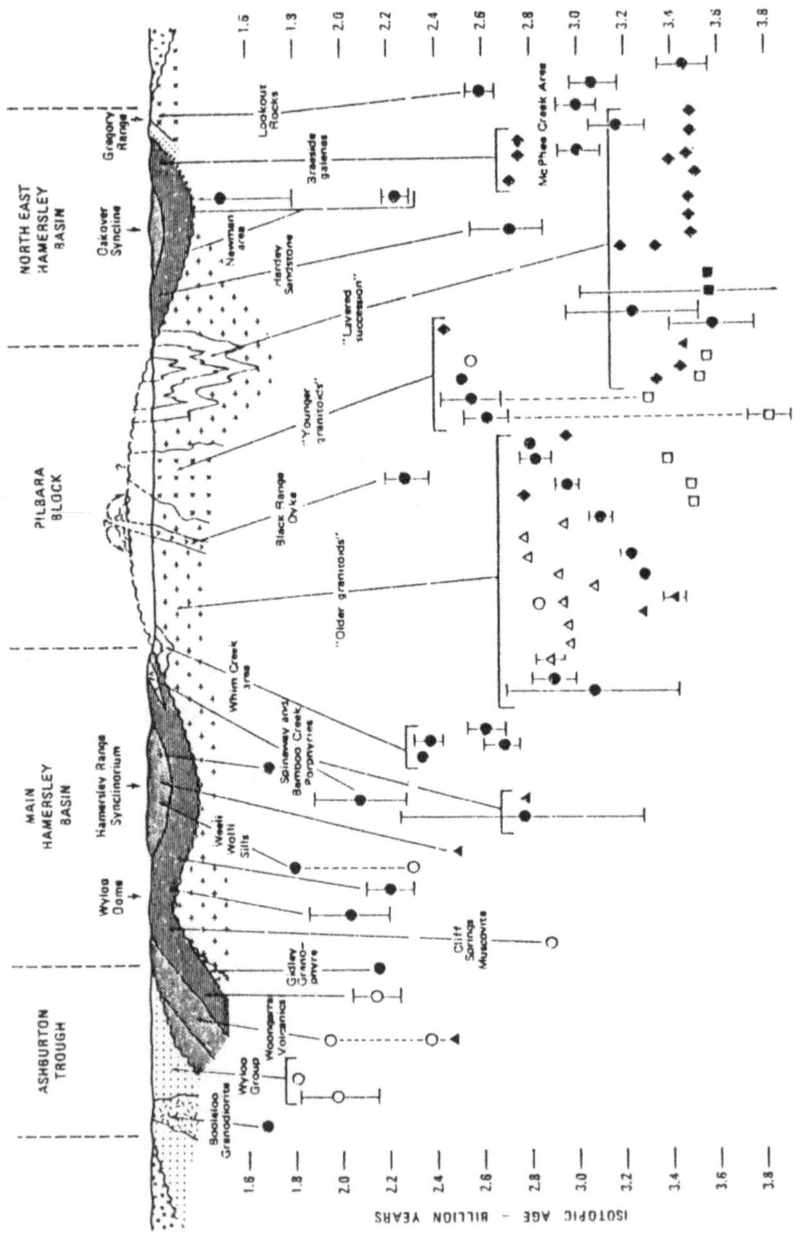
NORTH EAST HAMERSLEY BASIN
PILBARA BLOCK
MAIN HAMERSLEY BASIN
ASHBURTON TROUGH
Gregory Range
Oakover Syncline
Hamersley Range Synclinorium
Wyloo Dome
Lookout Rocks
Braeside galenas
McPhee Creek Area
Newman area
Hardey Sandstone
"Layered succession"
"Younger granitoids"
Black Range Dyke
"Older granitoids"
Whim Creek area
Spinaway and Bamboo Creek Porphyries
Weeli Wolli Sills
Cliff Springs Muscovite
Gidley Granophyre
Woongarra Volcanics
Wyloo Group
Boolaloo Granodiorite
1.6
1.8
2.0
2.2
2.4
2.6
2.8
3.0
3.2
3.4
3.6
3.8
ISOTOPIC AGE - BILLION YEARS

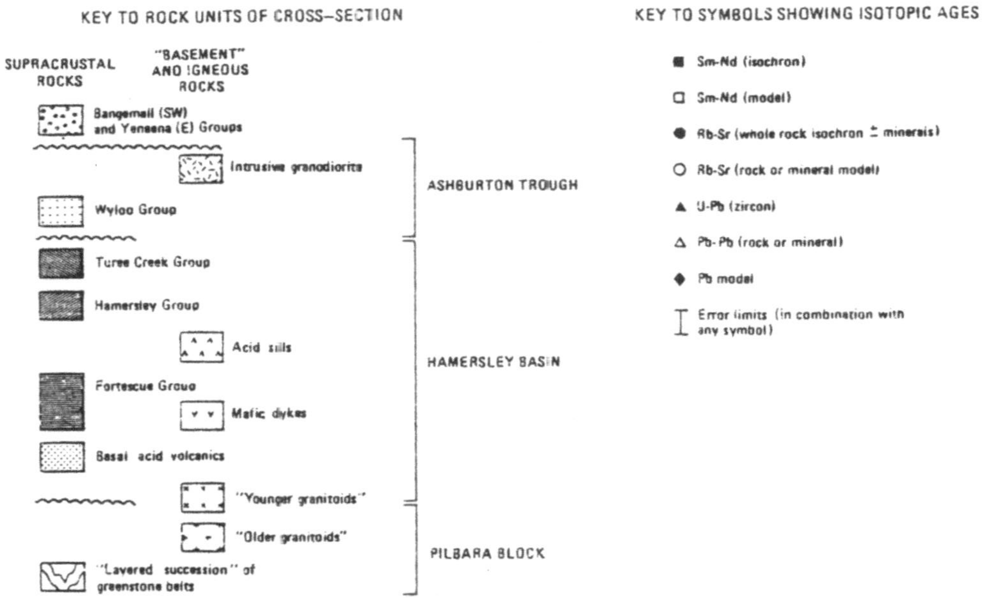

FIG. 3 - Summary of geochronological results from the Pilbara Block and Hamersley Basin. All points are identified and referenced in Table 3-V of (16).

The ("Proterozoic") Hamersley Basin

The Hamersley Basin first became clearly established as a tectono-stratigraphic entity during systematic regional mapping at a scale of 1:250 000 by geologists of the Geological Survey of Western Australia. A recent review is available (16), from which this brief summary is condensed. The present outcrop area of the sedimentary and volcanic rocks of the Hamersley Basin - the Mount Bruce Supergroup - appears in Fig. 2. The original extent and configuration of the depositional basin are uncertain, but it must have covered a substantial part of the Pilbara Block to the north; possibly it did not extend much beyond the limits of the present outcrop to the west, south, or east. The stratigraphy of the Mount Bruce Supergroup is well established and shows little lateral variation over its outcrop area. In much of the northern part of its outcrop it is remarkably undeformed, with a very gentle southerly dip off the unconformably underlying Pilbara Block to the north. Close to the western southern and eastern limits of its outcrop area the Mount Bruce Supergroup has suffered significant deformation.

The Mount Bruce Supergroup consists of three component groups. The basal Fortescue Group, with a maximum thickness in excess of 4 km, consists mainly of an alternation of mafic lavas and mafic-derived pyroclastics, with subordinate terrigenous sediments. An important basal or near-basal unit is the Hardey Sandstone, consisting mainly of granitoid-derived debris. The conformably overlying Hamersley Group, with a thickness of about 2.5 km, is characterized by thick BIF units. These alternate with shale and a little carbonate, and an important felsic sill - the Woongarra Volcanics - occurs near the top. Above the Hamersley Group follows the conformable Turee Creek Group, consisting of poorly sorted terrigenous clastic sediments, with some carbonates and abundant dolerite sills. Its thickest recorded section is in excess of 4 km; the Turee Creek Group is unlikely to have had a thickness of this magnitude over the entire basin.

Age determinations from rocks of the Hamersley Basin are also displayed in Fig. 3. These include results directly from the Mount Bruce Supergroup and from prophyry sills that intrude the Fortescue Group at or near its base.

Relationship Between the Pilbara Block and Hamersley Basin

Early interpretations of the relationship between the Pilbara Block and the Hamersley Basin (17) accepted the conventional wisdom that a spectacular regional unconformity, like that along the northern edge of the main Mount Bruce Supergroup outcrop, was likely to represent a substantial interval of time. The first isotopic results (9) appeared to support this, but the wealth of data obtained subsequently (Fig. 3) has shown this appearance to be misleading. It is now evident that the lowest lavas of the Fortescue Group are at least 2700 Ma old, while "emplacement" of the younger granitoids of the Pilbara Block was substantially more recent than this.

While refinement of some points in Fig. 3, and particularly the reinterpretation of Rb-Sr isochrons with high initial $^{87}Sr/^{86}Sr$ ratios as updates, will no doubt take place, the data as a whole point clearly towards an anomalous absence of any time gap across the "Archean-Proterozoic" unconformity and the existence instead of a conceptually impossible time overlap. This has led to a reinterpretation of regional geological relationships. Much of the main regional unconformity is marked by a stratigraphically low, but not the basal, unit of the Fortescue Group resting unconformably upon an older granitoid. When those parts of the regional unconformity which overlie greenstone belts are critically examined, and especially the unconformity in outlying Fortescue Group areas such as that immediately west of Marble Bar, the stratigraphic and structural relationship of the components of the Pilbara Block and Hamersley Basin appears as shown diagrammatically in Fig. 4.

This relationship, coupled with the geochronological evidence of Fig. 3, leads to the following synoptic view of crustal evolution in the Pilbara-Hamersley region. About 3500 Ma ago extrusion of the Warrawoona Group commenced as a continuous sheet over a primitive crust of unknown composition. As deposition continued, the accumulating volcanics depressed the crust in a random pattern of sinuous "welts" ancestral to the greenstone belts. Sialic material developed, or was emplaced, at the basal part of the succession as the descent of the intervening areas, ancestral to the present plutons, slowed. This pattern of structural differentiation became intensified with time, so that accumulation of the "layered succession" of the greenstone belts became increasingly restricted to their axial areas. By about 3200 Ma the crests of the rising granitoid plutons became breached, so that they supplied material for the Gorge Creek Group as fresh sialic material was added below. This complementary differential ascent and descent of plutons and greenstone belts continued until about 2800 Ma, when the whole crust became rigid, and deposition gradually again became regional, rather than restricted,

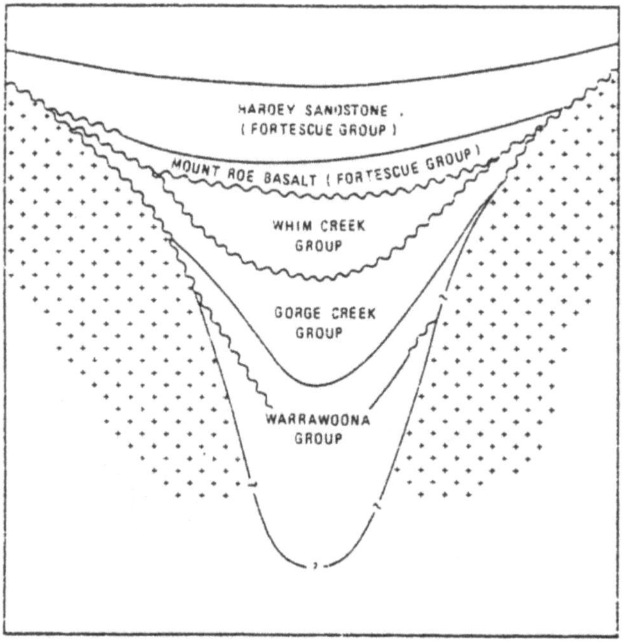

FIG. 4 - Diagrammatic representation of the structural and stratigraphic relationship of the Pilbara Block and Hamersley Basin in greenstone belt areas.

and formed the Mount Bruce Supergroup. This continued until the completion of deposition of the Turee Creek Group, possibly at about 2300 Ma.

During this entire period many interrelated processes - irruption of mafic lavas, growth of granitoid plutons, periodic outflow of felsic volcanics, and clastic and chemical sedimentation - took place intermittently and more or less contemporaneously and continually; the chronology of the development of the area cannot be accommodated within a model involving long and temporally separate periods of granite generation, erosion, and sedimentation.

DISCUSSION AND CONCLUSIONS

From the preceding account, summarized in Fig. 5, there is no obvious justification for dividing the crustal development of the Pilbara-Hamersley

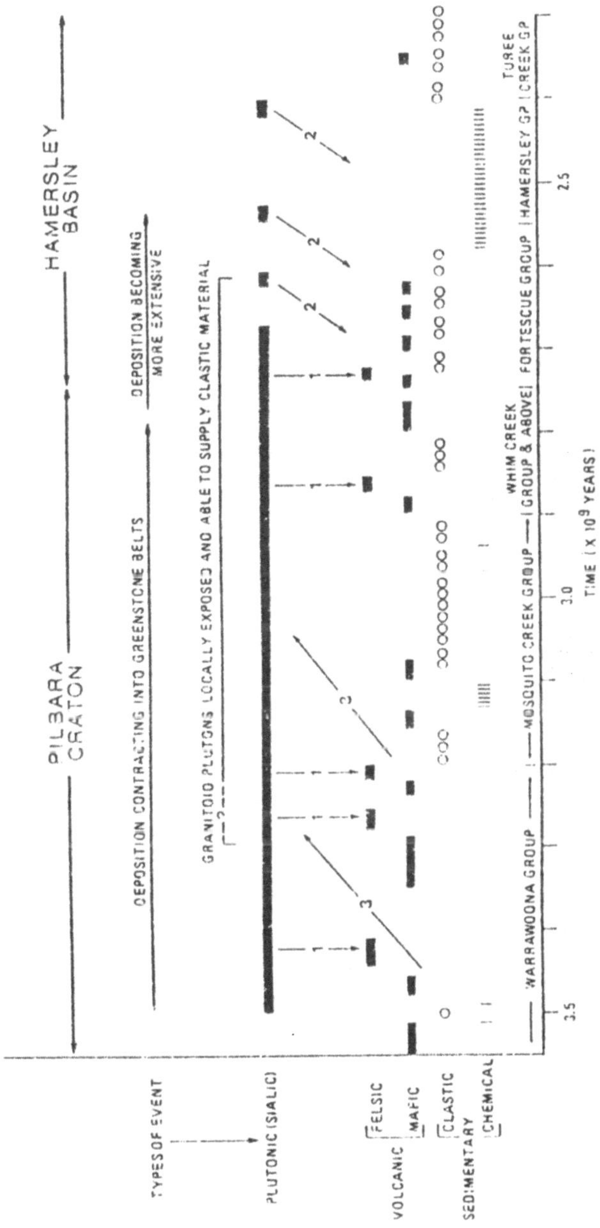

FIG. 5 – Summary of the interrelationship of processes during development of the Pilbara Block and Hamersley Basin.

region before 2300 Ma ago into two distinct and different periods. The processes that were in operation 3500 Ma ago continued until 2300 Ma b.p. Although the relative intensities of these processes varied with time, these variations seem to have been gradational, and none of the processes appears to have stopped or started abruptly. The first natural break in the crustal evolution of the area occurred when the formation of new sialic material ceased about 2300 Ma ago; since that time the only geological processes operating have been regional uplift and associated erosion.

If, following conventional practice, a single time boundary separating rocks of "Archean aspect" from those of "Proterozoic aspect" had to be identified, 2800 Ma would probably be most appropriate. There is no clear period of Archean-Proterozoic transition in the sense discussed in this paper, although the situation at 2900 Ma was very different from that at 2500 Ma, and it is conceivable that the interval 2900 Ma to 2700 Ma could be a complex transition of the conceptual type represented in Fig. 1D. The first unroofing of the plutons, probably at about 3200 Ma (but poorly fixed), represents a sharp break in surface conditions, but not in tectonic evolution. Not one of these possible boundaries has the unique regional significance needed to justify its designation as "Archean-Proterozoic"; each is arbitrary and artificial.

It is, nevertheless, true that there does exist in the Pilbara-Hamersley region a distinctive area of "granite-greenstone terrain" conveniently designated as the Pilbara Block, and an equally distinctive area of flat-lying sediments and volcanics conveniently referred to as the Hamersley Basin. These could be referred to as "Archean" and "Proterozoic" if those terms are regarded as local chronotectonic ones with a sequential rather than an absolute chronological significance, and loosely designating the early and late stages in the continuous complex evolution of that part of the Earth's crust.

It is when the terms are extended outside an area where their meaning is clearly specifiable that problems arise. Even within Western Australia not all parts of the crust share a common history of development. The granite-greenstone terrain of the Yilgarn Block, in almost every aspect not easily distinguishable from that of the Pilbara Block, is broadly contemporaneous with the early development of the Hamersley Basin in the period 2800-2600 Ma. The western gneiss terrain, on the other hand, is broadly contemporaneous (3500-3000 Ma) with much of the Pilbara Block, but it is not granite-greenstone terrain. Thus there is no evidence that the Pilbara-Hamersley region and the Yilgarn Block share a common sequence of synchronous "events" in their development. The Archean-Proterozoic transition, viewed from Australian evidence, does not exist as an identifiable period of time.

In a paper of this length it is not possible to review the equivalent situation in other continents. However, faced with such widely disparate chronotectonic development in two nearby areas of a single continent, it is clearly inappropriate to allot any special status to the classical Huronian unconformity of Canada as a standard by which to assess the worldwide synchroneity or otherwise of Precambrian crustal evolution. Although there has been less geochronological work in the Huronian area than in the Pilbara-Hamersley region, it does appear that a substantial time elapsed between the latest thermal event in the granitoids of the Superior Province, at about 2600 Ma, and the initiation of deposition of the Huronian Supergroup. This is not well established, but the available data suggest a time of about 2300 Ma (13). During this time interval no processes are known to have operated other than, presumably, uplift and erosion. Any "Archean-Proterozoic" transition represented by this local quiescent interval has no demonstrable equivalence in the evolution of the Pilbara-Hamersley region.

Acknowledgements. Among the very many geologists to whom I am indebted for discussion which helped to develop the views argued in this paper, R.D. Gee and A.H. Hickman deserve special mention. M.J. Bickle and H.D. Holland also made comments resulting in the improvement of the final paper. Elsevier Scientific Publishing Company are thanked for permitting the use of Figs. 3 and 5, which originally appeared in essentially the same form in (16).

REFERENCES

(1) de Laeter, J.R.; Fletcher, I.R.; Rosman, K.J.R.; Williams, I.R., Gee, R.D.; and Libby, W.G. 1981. Early Archaean gneisses from the Yilgarn Block, Western Australia. Nature 292: 322-324.

(2) de Laeter, J.R.; Libby, W.G.; and Trendall, A.F. 1981. The older Precambrian geochronology of Western Australia. Geol. Soc. Australia, Spec. Paper 7: 145-157.

(3) Dunn, P.R.; Plumb, K.A.; and Roberts, H.G. 1966. A proposal for time-stratigraphic subdivision of the Australian Precambrian. Geol. Soc. Australia J. 13: 593-608.

(4) Fletcher, I.R.; Rosman, K.J.R.; Trendall, A.F.; and de Laeter, J.R. 1982. Variability of ε^1_{Nd} in greenstone belts in the Archean of Western Australia. Abstracts, pp. 100-101. Fifth International Conference on Geochronological and Cosmochronological Isotope Geology, Nikko, Japan.

(5) Gee, R.D. 1979. Structure and tectonic style of the Western Australian Shield. Tectonophysics 58: 329-369.

(6) Hickman, A.H. 1981. Crustal evolution of the Pilbara Block, Western Australia. Geol. Soc. Australia, Spec. Paper 7: 57-69.

(7) Hickman, A.H. 1983. Archean diapirism in the Pilbara Block, Western Australia. In Precambrian Tectonics Illustrated, eds. A. Kroner and R. Greeling. Stuttgart: Schweizerbart'se Verlagsbuchhandlung, in press.

(8) James, H.L. 1960. Problems of stratigraphy and correlation of Precambrian rocks with particular reference to Lake Superior region. Am. J. Sci. (Bradley Volume) 258A: 104-114.

(9) Leggo, P.J.; Compston, W.; and Trendall, A.F. 1965. Radiometric ages of some Precambrian rocks from the Northwest Division of Western Australia. Geol. Soc. Australia J. 12: 53-65.

(10) McCulloch, M.T., and Compston, W. 1981. Sm-Nd age of Kambalda and Kanowna greenstones and heterogeneity in the Archaean mantle. Nature 294: 322-327.

(11) Preiss, W.V. 1982. Untitled letter to Editor. The Australian Geologist. Geol. Soc. Australia Newsletter 42: 8-9.

(12) Rankama, K. 1970. Proterozoic, Archean, and other weeds in the Precambrian rock garden. Bull. Geol. Soc. Finland 42: 211-222.

(13) Roscoe, S.M. 1973. The Huronian Supergroup, a Paleoaphebian succession showing evidence of atmospheric evolution. Geol. Asso. Can. Spec. Paper 12: 31-47.

(14) Tarney, J.; Dalziel, I.W.D.; and de Wit, M.J. 1976. Marginal basin 'Rocas Verdes' complex from South Chile: A model from Archaean greenstone belt formation. In The Early History of the Earth, ed. B.F. Windley. New York: John Wiley & Sons.

(15) Taylor, S.R., and McLennan, S.M. 1981. The composition and evolution of the continental crust: rare earth element evidence from sedimentary rocks. Phil. Trans. Roy. Soc. Lond. A301: 381-399.

(16) Trendall, A.F. 1983. The Hamersley Basin. In Iron-formation: Facts and Problems, eds. A.F. Trendall and R.C. Morris, pp. 69-129. Amsterdam: Elsevier.

(17) Trendall, A.F., and Blockley, J.G. 1970. The iron formations of the Precambrian Hamersley Group, Western Australia, with special reference to the associated crocidolite. West. Australia Geol. Surv. Bull. 119: 366.

Standing, left to right:
Hans-Ulrich Schmincke, Andy Knoll, Mike Steckler, Gerhard Wörner, Gerhard Brey.

Seated, left to right:
Jim Wilson, Kenneth Deffeyes, Alec Trendall, Stephen Moorbath.

Patterns of Change in Earth Evolution, eds. H.D. Holland and A.F. Trendall, pp. 261-270.
Dahlem Konferenzen 1984. Berlin, Heidelberg, New York, Tokyo: Springer-Verlag.

Events on a Time Scale of 10^7 to 10^9 Years Controlled by Tectonism or Volcanism
Group Report

K.S. Deffeyes, Rapporteur
G.P. Brey
A.H. Knoll
S. Moorbath
H.-U. Schmincke
M. Steckler
A.F. Trendall
J.F. Wilson
G. Wörner

INTRODUCTION

On the 100 my time scale we are able to identify changes in sea level and dramatic changes in the percentages of newly formed rock types. Equally important are phenomena which demonstrably do not exhibit changes on this time scale: examples are the major element compositions of continental tholeiites and of calc-alkaline plutonic rocks. Some problems arising from these observations seem solvable with additional work, but we wish to express our frustration with the realization that other problems cannot be solved with existing techniques.

ARE THERE RELATIONS AMONG SEA LEVEL CHANGES, OCEANIC THERMAL OUTPUT, AND VOLCANICITY?

At first it might seem that interconnections among the stand of sea level, heat flow, and volcanic output would be speculative. It turns out that the most prominent hypotheses for changing sea level have in common a requirement that the average thermal output from the ocean changes. For the last 150 my, for which there is a readable seafloor record, tests among the various hypotheses are possible.

Sea Level

The presently available techniques for observing sea level with respect to the continents consist of mapping the areal and vertical distribution of marine/nonmarine rocks (which is beset by problems of preservation) and by studying continental margins, which have theoretically predictable subsidence rates, using either seismic or borehole data. Although observations of the second type are, as yet, few in number, existing estimates agree reasonably well that there were two major high stands of sea level during the Phanerozoic: one in the Ordovician and the other in the Cretaceous. Particularly low stands of sea level occurred during the early Cambrian, during the Triassic, and in Recent time. The amplitude of the excursions seems to have been on the order of 200 meters.

Many causes of sea level change have been cataloged; most of these can be shown to result in only modest long-term changes of sea level. The major determinant of sea level at any given time is very probably the average thermal age of the oceanic lithosphere. The thermal "age" is either the time since a given segment of seafloor was generated or a thermal time that has been reset to a younger age by hot spot thinning of the lithosphere.

A wide range of plausible effects could change the average thermal age of the ocean floors: a) faster seafloor spreading rates; b) subduction of younger oceanic crust by installing more mid-ocean trenches, such as the Tonga-Kermadec trench; c) moving segments of the ridge from mid-oceanic to subcontinental rifting locations; or d) moving hot spots from oceanic to continental areas. Because the heat flow also depends on the average age of the ocean floor, all these effects also change the oceanic heat output.

Parsons (5) proposed that a 25 percent increase or decrease in the oceanic heat output is necessary to cause a 150 meter change in sea level. On the very longest time scales, the heat production has to equal the surface heat flow. However, on the time scale of the Cretaceous sea level excursion (about 50 my), a 25 percent change in heat flow would result in a mantle temperature change of roughly two degrees, which is modest enough to be plausible. The relative importance of the various causes (spreading rate, age at subduction, or hot spot activity) could be examined for the last 150 to 200 my, but for older times we do not have methods for discriminating among the various causes.

Extending the sea level observations back into the Precambrian requires techniques both to tell time and to distinguish marine from nonmarine rocks. Age-measuring techniques with 50 my resolution have recently become available for the late Precambrian through the study of planktonic microfossils (3). Supplementary help is available from magnetostratigraphy and from radiometric ages (particularly on zircons). The Cretaceous and Ordovician high sea level stands were initially inferred on the basis of widespread carbonates deposited during these periods. Although comparable work on Proterozoic sequences is still in its infancy, abundant late Precambrian carbonates with ages of about 700 to 800 my suggest that high sea levels existed just prior to the Early Cambrian/latest Precambrian sea level low that marks the beginning of the Phanerozoic.

Volcanism

A sensitive test of the uniformity of both the chemical composition of the upper mantle and of the partial melting process comes from the composition of basalt. We are aware of the obvious point that the chemical composition of a "basalt" cannot change too drastically; otherwise the resulting rock would no longer be classified as a basalt. A considerable range of basalts are known today: continental tholeiites, alkali basalts, and the various mid-ocean ridge basalts. Theoleiites persist throughout all of the geologic record without a discernible change in chemical composition.

Although the chemical and trace element composition of individual basalt types has remained constant, the proportions of the several basalt species has changed. About 10 percent of the Archean lava flows are komatiites, rocks so rich in MgO that they are not formally classified as basalts. The other apparent major change with time is the increase in the proportion of alkali basalts and their derivatives relative to tholeiites in Proterozoic and younger rocks.

It seems probable that the major effect controlling basalt chemistry is the degree of partial melting and not changes of mantle composition. Either lower heat production or, more likely, increased thickness of crust and/or lithosphere selects for lower degrees of partial melting with time. The extreme product at very low degrees of partial melting is probably kimberlite. One would not expect to find conventional kimberlite pipes in terrains that do not have stable, thick lithosphere of the Proterozoic and later type.

A second suite of igneous rocks which shows exceedingly small variations with time is the calc-alkaline suite. The incompatible element content of Precambrian calc-alkaline rocks is indistinguishable from that of modern rocks associated with subduction zones, except for a depletion in heavy rare earth elements and a slight enrichment in Ni and Cr compared to their young equivalents. It is still not clear whether the early Precambrian calc-alkaline rocks should be attributed to the same subduction process that produces their modern equivalents. Although recent petrogenetic experiments have done much to clarify the origins of calc-alkaline rocks, not everyone is satisfied that all of the trace element data are fully explained by the current petrogenetic models.

WHAT ROCK TYPES, TECTONIC STYLES, AND SEDIMENTS CHARACTERIZE EARLY EARTH HISTORY?

We wish to begin with an analogy. Towards the end of the Archean there was a period of maturation. Cratons became stabilized and mature sediments developed. A specific time (the cooling of the mafic phase of the Hartley Complex of the Great Dyke) has been suggested as a date to mark this passage. The analogy is to human development in which puberty occurs at various ages in different individuals; it occurs over a span of time in any one individual, and its effects are readily recognized in the field. In contrast, the age of legal majority is a precisely defined time, useful for certain purposes, but bearing little relationship to the actual process of maturation. We choose to bypass the precise legal definition of the Archean/Proterozoic boundary and to proceed to discuss the Earth's puberty.

The best known areas (Australia, southern Africa, and Canada) all have provinces which made an early transition from greenstone to shelf sedimentation; they also harbor areas which made the same transition considerably later, after the "official" end of the Archean. However, it is not clear what the successors are to the greenstone belts. Their later equivalents may be intercratonic rifts, marginal basins, entire ocean basins, or they may have no later analogs at all.

Some terrains progressed through the greenstone stage and started receiving shelf sediments as early as 3200 my; however, none of these terrains has proved to be as areally extensive as the terrains that passed through this transition about 2600 my ago. Most of the panel members felt that the transition is best chronicled by the generation of tonalites. Because of their intermediate position in the transition from basaltic to granitic rocks, and because their petrogenesis is suspected to be simpler, it may be rewarding to focus geochronologic efforts on the tonalites rather than on the later granites.

In addition to the well-known shelf sedimentation that characterizes the Proterozoic, there are a number of anorogenic igneous complexes which are characteristic of the Proterozoic. These rock types include anorthosite massifs, syenite-alkali granite-rapikivi granite complexes, and mangerites. Anorthosite emplacement is no longer regarded to have taken place during a unique brief episode in Earth history, but the bulk of these large, central, anorogenic complexes have ages between 2000 and 1000 my.

Models for the origin of greenstone belts cover a spectrum that extends from the deformation of preexisting flag-lying strata by rising granites, to rifting that produces basins in which the volcanics and sediments were deposited. In Australia and Zimbabwe we note that in the younger greenstone belts, major lithostratigraphic units have been traced from belt to belt across the intervening granites (albeit with differing degrees of reliability). What these correlations mean in terms of the nature and evolution of the greenstone belts has yet to be established. More sedimentological studies, particularly those including sediment transport directions and variations in thickness, would assist in clarifying the genesis of greenstone belts.

Accumulation rate measurements of the volcanics and sediments also place constraints on models of greenstone belt development. In principle, these rates can be determined by radioactive dating. In Australia, Rb/Sr ages from volcanics at the base and top of greenstone belt sequences suggest slow accumulation rates. In contrast, U/Pb ages on zircons from similar sequences in Canada suggest much faster rates. Accumulation rates are important constraints on models of the Archean Earth, and strong efforts should continue to determine accumulation rates in a variety of greenstone belts.

Although the early results are apparently in conflict, both as to mechanism and as to accumulation rate, we should not expect that all greenstone belts have the same characteristics. Part of the reward of further studies might be the recognition of two or more genetically different units which we currently lump together as "greenstone belts."

DEVELOPMENT OF CRATONIC AREAS THROUGH GEOLOGIC TIME

Despite the fact that "cratonization" may be a loaded word (an area is a "craton" only until some orogeny catches up with it), the word may usefully symbolize a very real stabilization process involving continental crust. In most cratonic areas, the last tectonic event is the emplacement of granites that cut across all other structure and which are themselves undeformed.

This has led many field geologists to state informally that the late-stage granites "stiffen" the crust, but we discuss below the possibility that the post-tectonic granites are an important mechanism for removing heat-producing elements from the lower crust. The upward redistribution of radioactive elements reduces the average geothermal gradient and thereby establishes lower temperatures which then result in a stabilized crust. The primary readable signature of a stabilized continental block is a sequence of thin-bedded, laterally correlatable sedimentary rocks deposited in shallow water. In Paleozoic rocks, known rates of sedimentation correspond rather closely to the subsidence rates (with isostasy) expected from cooling of a thermal boundary layer which makes up the mantle lithosphere beneath the continent. For about ten years, Jordan (2) has been presenting teleseismic evidence for a cooled subcontinental lithosphere; this has been confirmed recently by evidence based on lithospheric flexure. If accumulation rates of Proterozoic shelf sediments were determined, comparison with theoretical cooling models could test the hypothesis that these sediments recorded the growth of subcontinental lithosphere.

Where major rearrangements take place during the collision of two previously cooled continental blocks, self-heating due to the included radioactivity ought to give rise to post-tectonic granites. On the other hand, crust generated at subducting continental margins is born hot; therefore, the process of segregating the heat-producing elements into the upper part of the crust can begin at once. Regardless of whether a tectonic event begins hot or cold, there ought to be an upper bound on the time by which post-tectonic granites can postdate crustal rearrangement. Specifically, the thermal conduction time for heat to escape to the surface from the base of the continental crust (about 40 my) ought to be the maximum time. Yet many older cratons apparently contain young, undeformed, cross-cutting granites that are about 200 my younger than the previous metamorphic, magmatic, and tectonic events. This is a ripe opportunity for revising the model and/or redetermining the age dates.

Integrating the studies of new crustal formation of various ages ought to improve our perspective on the processes that are involved. For instance, we speak of establishing 2700-my-old crust by accumulating greenstone belts, of building the 900 to 500-my-old Nubian shield by accreting island arcs, and of generating the 40-my-old Oregon Coast Range by accreting a hot-spot-generated oceanic plateau. It would be surprising if these processes did not have some features in common; an exchange of scholars and data might bring out some commonality in processes.

A few years ago it seemed that the peaks of age dates on histograms reflected a fundamental fact about historical geology. From our present perspective we have serious reason to doubt, not the validity of the dates, but the interpretation that the peaks have a fundamental significance. Through plate tectonics, we realize that a single continent-continent collision can reset geochronometers in a very large area. In addition, we now know that something more complex than just "age" is being measured: there are whole-rock isochrons, cooling ages for individual minerals, and partially overprinted ages. A reexamination of the entire question may not be feasible. However, there are a number of reasons to make a beginning. A first step would be an international data bank for radiogenic isotopic observations. Both for mapping and for global studies a data bank of published and privately supplied data would be valuable.

DO MORE RECENT GEOLOGIC PROCESSES INVOLVE SMALLER VOLUMES AND/OR SHORTER TIMES?

Most geologists realize that we might have lumped together ancient groups of unrelated events because we did not have the resolving power to separate them in time. A slightly more subtle problem is self-similarity: distributions which look the same at any magnification. Statisticians have identified a few geological spatial distributions which are well-described by self-similarity, but as yet we have no proof for self-similar patterns in time. Because we are aware that a temporal illusion may exist, we turned to those phenomena which could be examined with high resolving power even in very old rocks.

Foremost among these phenomena are the reversals of the Earth's magnetic field. If an arbitrarily old sedimentary sequence is sufficiently unmetamorphosed to retain magnetic remanence, magnetic reversals a few thousand years apart are potentially observable. What has been observed so far is a Phanerozoic pattern of roughly 30 my blocks of time, each characterized by rapid reversals, or by less frequent reversals, or by magnetic quiet without reversals. As far as the magnetic polarity record has currently been read back into the late Precambrian, the same behavior persists. In this one instance, admittedly extending only partially into the Precambrian, we see that short-term changes, from periods of no reversals to periods of fast reversals, completely dominate any long-term trends that may exist in the behavior of the magnetic field.

We would like to see more long-term records full of detail. The stable-isotope records in marine sediments are potentially another data set that could be searched for secular trends. Our individual wish lists are quite varied: a way of dating abundant dike swarms, observations of the Na content of greenstones, or a catalog of isotopic ages of newly formed continental crust.

For many of the phenomena which we have discussed, such as the frequency of magnetic reversals or the composition of basalts, a strictly uniformitarian view is consistent with our present knowledge. There is one circumstance that would render a uniformitarian view terribly incorrect: if there is a short-term excursion happening today either in the existence of a process or in the rate of a process, then in one way the present-day Earth is not typical of most of geologic time. We nominate the following example: most ancient shield areas contain some lightly metamorphosed rocks which have never been deeply buried. These rocks set an upper bound on the total depth of erosion that could have occurred in the shield. Monitoring the chemical and detrital load in rivers gives a present-day erosion rate for these shields: typically 5cm/1000 years. If the upper limit to the allowable depth of erosion is 5 km, then 100 my is the total allowable erosion time, 1/30th of the age of the shield. Even if the shield surface were to oscillate upwards and downwards in time, the present erosion rate is typical only of the brief periods of maximum uplift. The constraints are often even tighter; one line of reasoning suggests that western Australia was at its present depth of erosion 1800 my ago!

The concern with erosional depths is not just an academic exercise. The economically valuable levels of diamond pipes are quite shallow; deep erosion following their emplacement leaves only a small, barren scrap. Similarly, uranium prospecting is hopeless in a granulite terrain that has been systematically stripped of its uranium. The evidence seems to be telling us that shield areas spend most of their history under flat-lying sedimentary cover. When we want a historical picture of a typical cratonic setting, we should think of the Russian platform or the American Midcontinent rather than the exposed Canadian shield.

CONCLUSION

The multitude of processes by which igneous, sedimentary, and metamorphic rocks form are essentially the same throughout geologic history. The rate of growth of crust and of stable continents, however, has not been constant through time. We therefore see certain types of rocks and rock associations dominating the early part of Earth history. Instead of regarding the Archean granite-greenstone terrains as the result of a unique mode of crustal accretion, we suspect that they will eventually be seen as an ancient process that gradually evolved into the accretion processes that we observe today.

Our greatest frustration concerns our continuing inability to choose between some mutually incompatible hypotheses. Specifically, we cannot now choose between: a) steady-state models in which continental crust segregates early and recycles thereafter with decreasing efficiency (1), and b) growth models in which irreversible chemical differentiation of the continents from the mantle takes place throughout geologic time (4). Not only are we unable to choose between these models, we are not yet able to propose definitive tests which would cause one model or the other to be rejected. Many of the available observations are only intuitively linked to one hypothesis or the other. The constancy of basalt composition with time, which seems to support the steady-state model, is avoided in the growth model by proposing that each mass of mantle contributes one time only to continental tholeiites. The very clear isotopic evidence that most of the continental crust carries no isotopic signature of the preexisting crust is explained away in the steady-state model by postulating efficient isotopic homogenization with an extensive mantle reservoir.

Depending on your intuition, one or the other of these hypotheses seems terribly wrong. Neither hypothesis is objectively disprovable. A compromise between the two seems unlikely because one postulates perfect mantle mixing and the other excludes any mixing at all. Detailed comparisons between the predictions of various continental evolution models and observations of radiogenic daughter isotopes may eliminate some models, but we suspect that new techniques may be needed before one of the two major hypotheses is proved to be incorrect.

Our final impression and our concluding recommendation is that the quality of many of the isotopic and geochronological observations is limited by the need for very extensive and detailed field-mapping programs. Unless the samples are from well characterized, thoroughly deciphered settings, the precision of modern isotopic techniques is largely wasted. In each box of samples shipped to the mass spectrometer lab there should be included a worn-out pair of field boots.

REFERENCES

(1) Armstrong, R.L. 1981. Radiogenic isotopes, the case for crustal recycling on a near-steady-state no-continental-growth earth. Phil. Trans. Roy. Soc. London A301: 443-472.

(2) Jordan, T.H. 1978. Composition and development of the continental tectosphere. Nature 274: 544-548.

(3) Knoll, A.H. 1981. Paleoecology of late Precambrian microbial assemblages. In Paleobotany, Paleoecology, and Evolution, ed. K.J. Niklas, vol. 1, pp. 17-54. New York: Praeger.

(4) O'Nions, R.K., and Hamilton, P.J. 1981. Isotope and trace element models of crustal evolution. Phil. Trans. Roy. Soc. London A301: 473-487.

(5) Parsons, B. 1982. Causes and consequences of the relation between the area and age of the ocean floor. J. Geophys. Res. B87: 289-302.

Patterns of Change in Earth Evolution, eds. H.D. Holland and A.F. Trendall, pp. 271-289.
Dahlem Konferenzen 1984. Berlin, Heidelberg, New York, Tokyo: Springer-Verlag.

Time and Space Scales of Mantle Convection

F.M. Richter
Dept. of Geophysical Sciences, University of Chicago
Chicago, Il 60637, USA

Abstract. Aside from the obvious mantle flow associated with the creation and subduction of plates, geophysical observations suggest a spectrum of other scales of motion ranging in size from about 100 km to many thousand kilometers. A classification of those "other" motions in terms of their origin and measurable consequences is proposed. Three time scales are then discussed: the overturn time ($\sim 10^9$ years for the plate scale and $\sim 10^8$ years for "other" scales), the thermal adjustment time ($\sim 10^9$ years for models with a uniform viscosity of 10^{22} poise and $\sim 10^8$ years for models with temperature-dependent viscosity), and the time scale for the dispersal of chemical heterogeneities in the mantle.

INTRODUCTION

To the extent that mantle convection is the principal driving force of tectonics, it is appropriate to review what we know or expect of its spatial and temporal scales.

THE PLATE TECTONIC SCALE OF MOTION

Plate tectonics implies a mantle flow on the same scale. Figure 1 illustrates this in terms of a simple two-dimensional model. The flow under the moving, subducting plate consists of two parts: a shear driven flow of mantle material being dragged towards the trench by the moving boundary and a pressure-driven flow that returns the mass flux of plate plus shear-driven flow back to the ridge. The downgoing slab also induces a local circulation which is best seen under the stationary non-subducting plate. The relative importance of the shear-driven and pressure-driven

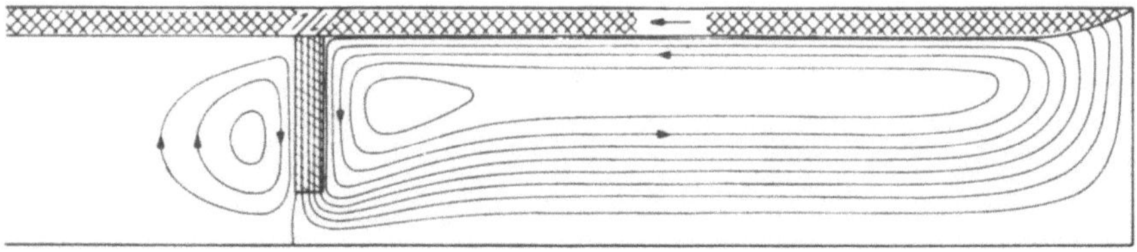

FIG. 1 - Two-dimensional idealization of the scale of motion associated with plate tectonics.

components of flow depends on the detailed viscosity structure of the mantle, and in the extreme case of a pronounced low viscosity zone at the base of the plates, the shear-driven flow will be almost entirely absent (16).

The distinction between the shear-driven flow and the pressure-driven return flow becomes very important once one considers the more realistic case of a spherical shell representing the mantle (or upper mantle). While the shear-driven flow is always in the direction of plate motions, the pressure-driven flow need not be antiparallel to the plate motions (4, 9). For example, it is quite likely that the North Mid-Atlantic ridge receives via the pole a significant part of its mass flux from subduction around the Pacific.

The uncertainties surrounding the mantle flow associated with the plates are compounded once one begins to consider the extent of its depth. If we accept the relatively uniform viscosity of the mantle inferred from the study of glacial rebound (2, 24), the alternatives seem to be a single circulation from the surface down to the core or a layered mantle with convection throughout but confined to discrete and separate levels. The traditional seismic arguments for a separate convecting layer in the upper mantle, based on the state of stress of downgoing slabs and the absence of any seismicity below 700 km, are inconclusive (26). Jordan (14, 15) argues for mantle-wide convection on the basis of near source velocity anomalies associated with deep seismicity; these are considered to be evidence for the penetration of slabs below 700 km. Many geochemical models (7, 11, 21), on the other hand, suggest a depleted and relatively isolated upper mantle which favors the layered model. Arguments based on the thermal structure of the mantle, which in the case of layers would result in a hotter lower mantle, have been used to favor layered convection (13) and to argue against it (34). Clearly the depth extent remains an outstanding issue.

To summarize, there are three principal reasons why we cannot at present go beyond a conceptual framework to specify in detail the mantle flow associated directly with plate tectonics: a) we do not know the appropriate frame of reference in which to measure the plate velocities; b) the viscosity structure of the mantle is not well enough known to be able to specify the relative importance of the shear-driven and pressure-driven components of motion; and c) we do not know the depth extent. Seismology will contribute to reducing the uncertainties as the velocity and anisotropy structure of the mantle are mapped in greater detail. Geochemistry should refine (or cause us to abandon) the arguments suggesting a mantle composed of various relatively isolated geochemical reservoirs. Geochemical measurements on intraplate volcanics have the added potential of mapping the net sublithospheric flow if subducted material that has contaminated the mantle can be used as a tracer (see (31) for an illustration of this approach).

OTHER SCALES OF MOTION

There are both theoretical and observational grounds for expecting motions in the mantle other than the simple kinematic flow discussed above. One approach for establishing the existence of such "other" scales of motions is to model the geophysical observables of the kinematic plate scale flow and ask if it can account for the actual observations.

The most relevant observations for present purposes involve oceanic bathymetry and the sea surface geoid, these being sensitive in different ways to the thermal structure of the plates and upper mantle. The thermal structure associated with the kinematic flow is well represented by an error function temperature solution (see (23) for a detailed discussion).

$$T(z,t) = T_m \operatorname{erf}\left(\frac{z}{2\sqrt{\kappa t}}\right) , \qquad (1)$$

where T is temperature, T_m the mantle temperature, z the depth, t the age, and κ the thermal diffusivity. Such a continuously cooling thermal structure results in the prediction of ocean depth increasing as $t^{\frac{1}{2}}$ and a geoid anomaly that decreases linearly with age. Both ocean depth (23) and the geoid variations (3, 8, 32) follow the continuously cooling relationships at young ages, but at older ages depart significantly in the sense of a shallower ocean depth and a more slowly varying geoid anomaly with age (Fig. 2). The departure of ocean depth from the square root of age relationship can be explained in a variety of ways: a) a convective heat flux by some small-scale flow to the base of the cooling plate (23), b) a tilting of the plate by the pressure gradient (28), c) heating

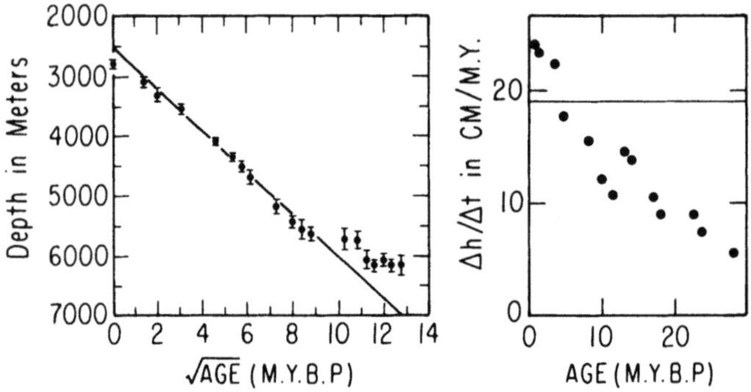

FIG. 2 - Depth as a function of $\sqrt{age}$ for the North Pacific (23) and change in geoid height divided by age difference across fracture zones as a function of age using data by Cazenave et al. (3). The solid straight lines are what one expects for simple cooling given by Eq. 1.

of the plate due to viscous dissipation (33), or d) the inappropriateness of Eq. 1 when whole mantle convection heated mainly from below is considered (12). The reason why so many different mechanisms are in principle possible is that bathymetry data, while described as a function of age, come from measurements made in different places. It is the spatial distribution of the observations that allows the pressure gradient effect, for example, to be a possible mechanism. The geoid data are therefore especially significant when taken across fracture zones, since the age differences are then not spatially dispersed. For this reason the geoid data cannot be explained by the last three mechanisms, and a very local convective instability (and associated transfer of heat) seems the only likely explanation.

A second important feature of the thermal structure of the kinematic flow is that it predicts geophysical observables to be a function of age alone. This, of course, is not the case, as both bathymetry and the geoid vary significantly around the "average" value for a given plate age. Figure 3 shows this for the sea surface geoid over a portion of the Pacific. The magnitude of these variations are well above the "noise level" and are another clear indication that something besides simple cooling is taking place. There are also geoid (and gravity variations) on even larger scales, and these are not always obviously associated with plate boundaries. Their explanation must involve large-scale dynamical processes that are undoubtedly convective in origin, since the Earth is not sufficiently "strong" to support such variation in any static way.

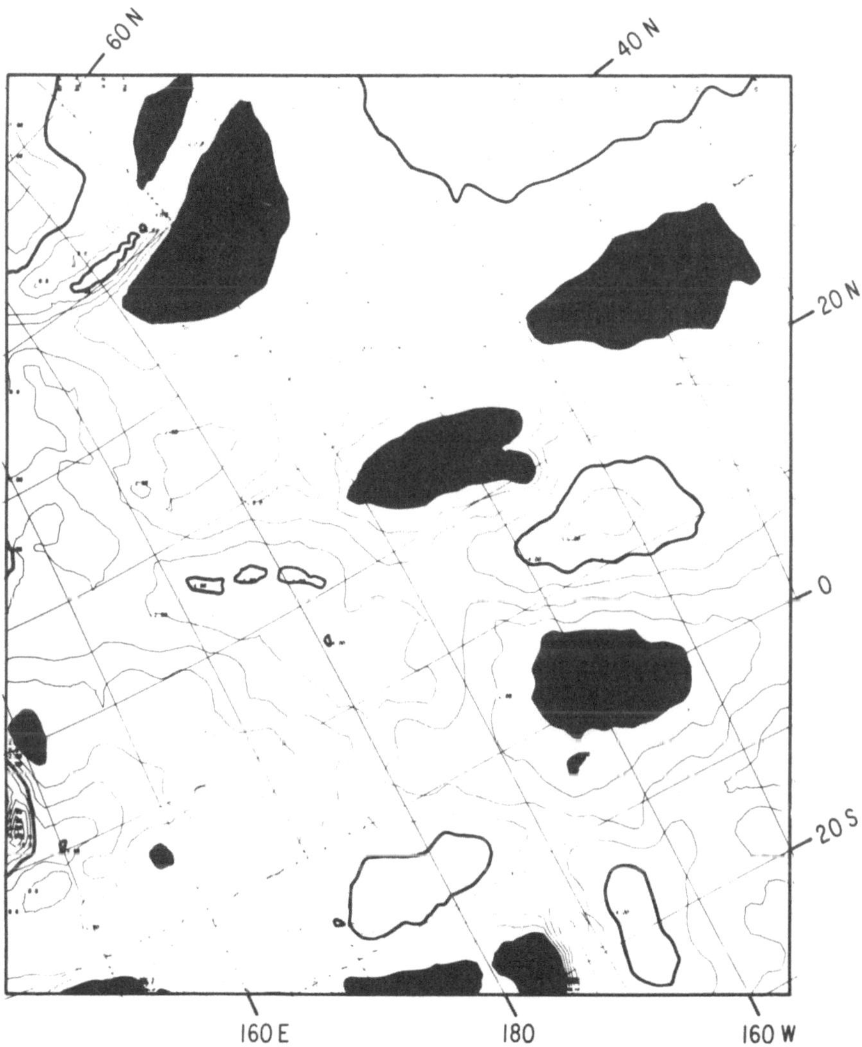

FIG. 3 - The sea-surface geoid in the Pacific, with a reference geoid (GEM7, spherical harmonic degree and order 10) subtracted and filtered to remove features less than about 100 km (19). The black areas exceed +4 m and the heavy contour encloses areas of less than -4 m.

Bathymetry and geoid data give the clearest indication that motions other than the kinematic flow associated with the lithospheric plates do in fact exist (19). In the past these "other motions" have been referred to as the small-scale flow (20, 25, 30), but it is becoming increasingly clear that the situation is richer in scales than originally conceived. One should really consider a spectrum of scales; a classification based on specific mechanisms and specific observations is given below to serve as a framework for discussion.

1. **Surface boundary layer instabilities:** The near surface thermal structure, which is probably not very different from Eq. 1 for young oceanic plates, may (depending on the details of the rheology of mantle materials) become unstable and give rise to a convective flow on a very small scale (22). Because of the way oceanic plates are generated, this flow would represent the initial stages of instability and therefore might very well be characterized by a scale as small as 100 km, reflecting the length scale of the portion of the boundary layer that is becoming unstable. A convective flow with such a small characteristic scale seems to be required by the variations of the geoid across fracture zones, which are very localized. In continental areas the surface thermal structure is old, and we are no longer concerned with the onset of boundary layer instabilities; therefore the upper boundary will not lead to a distinctively small scale as it might in oceanic regions.

2. **Lower boundary layer instabilities:** The lower boundary layer lies within the mantle in the case of layered mantle convection or at the core mantle boundary; it is a permanent feature, and thus flows resulting from buoyancy fluxes across it will have a characteristic scale determined by the layer depth and boundary conditions. The existence of this type of flow seems to be required by the geoid anomalies shown in Fig. 3. It would seem at first sight that the scale of the geoid anomalies should resolve the question of the depth extent of convection, but the issue is quite complicated. The horizontal scale of convection depends not only on the depth but is also affected (in the sense of making the horizontal scale larger than the depth) by the details of the thermal boundary conditions (10). The relation between "observed" horizontal scale and depth extent remains ambiguous.
3. **Large-scale thermal flows:** Plate tectonics is itself a case of large, variable scale thermal flow, since much of the driving forces are

directly associated with temperature differences. This need not be the only large-scale flow, and, indeed, the observations of large-scale gravity and geoid anomalies not correlated with plate boundaries are evidence of this.

TIME SCALES

The discussion of convective time scales is inevitably more theoretical since most geophysical measurements reflect the present Earth structure, while in the case of geochemistry, where the time element is more directly involved, the data are still somewhat sparse. Three distinct time scales are discussed in detail: the convective overturn time, the thermal adjustment time, and the chemical dispersal time.

The Overturn Time

The overturn time can be characterized by the ratio of the typical dimensions to some representative convective velocity. For the "kinematic" large-scale flow associated with the plates themselves, we again come up against the question of a low viscosity zone at the base of the lithosphere. In the absence of such a zone of decoupling, the overturn time will be relatively short, only several times larger than the typical age of plates at subduction. A reasonable guess would be something on the order of 5×10^8 years. If, on the other hand, there is a low viscosity zone, the return flow will be slower and the time longer. In the limit of complete decoupling (no shear-driven component of flow) the turnover time will be approximately the age of subducted material times the ratio of layer depth to plate thickness: about 10^9 years for flow restricted to the upper mantle and approaching the age of the Earth for the mantle-wide case.

The other scales of mantle convection are expected to be more closely related to flows that have been studied numerically, and we can make use of this experience in estimating reasonable turnover times. The characteristic length scale is the depth, d, while the velocities scale as (18)

$$u = C(\kappa/d)\ Ra^{\frac{1}{2}}\ , \qquad (2)$$

where C is a constant, κ is the thermal diffusivity, and Ra is the Rayleigh number defined as

$$Ra = \frac{g\alpha Fd^4}{\rho c_p \kappa^2 \nu} \quad . \tag{3}$$

Here g is the gravitational acceleration, α is the coefficient of thermal expansion, F is the heat flux per unit area at the surface, ρ is density, c_p is specific heat, and ν is the kinematic viscosity. The overturn time, τ_{ot}, is order d/u, which by combining Eqs. 2 and 3 is

$$\tau_{ot} = (d^2/C\kappa)Ra^{-\frac{1}{2}} = C^{-1}\left(\frac{\rho c_p \nu}{g\alpha F}\right)^{\frac{1}{2}} \quad . \tag{4}$$

The constant, C (≈0.12), can be found from numerical experiments covering a range of Rayleigh numbers (5, 18). If the values from Table 1 are used, the present-day overturn time is on the order of 10^8 years.

A time scale of 10^8 years is probably the shortest time scale associated with mantle convection; this must be kept in mind when appealing to mantle processes for an explanation of particular geological events. The overturn time was undoubtedly shorter during the earlier part of Earth history, because the heat flux was larger, and the higher mantle temperatures were associated with a lower mantle viscosity. We cannot say how much shorter the time scale of mantle was, because of uncertainties in the actual value of the viscosity. Just as an illustration one might consider an Archean mantle 200°C hotter (resulting in a viscosity lower than today's by about two orders-of-magnitude) and twice

TABLE 1 - Values assumed when evaluating relations given in text.

gravity:	$g = 9.8\ ms^{-2}$	thermal diffusivity:	$\kappa = 10^{-6} m^2 s^{-1}$
thermal expansion:	$\alpha = 3x10^{-5}\,^\circ C^{-1}$	density:	$\rho = 3.3x10^3 Kg\ m^{-3}$
mean heat flux:	$F_o = 7x10^{-2}\ Wm^{-2}$	specific heat:	$c_p = 1.2x10^3 JKg^{-1}\,^\circ C^{-1}$
depth:	$d = 7x10^5$ or $2x10^6$ m	viscosity:	$\nu_o = 3x10^{17}\ m^2 s^{-1}$
thermal conductivity:	$k = 4\ Wm^{-1}\,^\circ C^{-1}$		

the present heat flux. The hypothetical Archean overturn time would then have been about 10^7 years.

Thermal Adjustment Time

This is the time scale appropriate to thermal evolution, for which we can write the governing equation (see (5)) as

$$\rho c_p d \frac{\partial \langle T \rangle}{\partial t} = H - F , \tag{5}$$

where ρ and c_p are suitably averaged density and specific heat, d is a depth scale such that when multiplied by the surface area it equals the total volume, $\langle T \rangle$ is the average temperature, H is the rate of radiogenic heating per unit surface area, and F is the surface flux of heat per unit area. Even if the rate of heating is assumed (based on some bulk composition model), a second relationship between $\langle T \rangle$ is needed to close the problem. Traditionally this is given in the form of a Nusselt number–Rayleigh number relation:

$$Nu = \gamma Ra^{\beta} . \tag{6}$$

The Nusselt number is simply the heat flux when convecting normalized by some appropriate conductive flux:

$$Nu = \frac{F}{k \langle T \rangle / d} , \tag{7}$$

where k is the thermal conductivity ($= \rho c_p \kappa$). With the Rayleigh number given by Eq. 3, Eq. 6 can be converted into the explicit relationship between $\langle T \rangle$ and F that we need:

$$\langle T \rangle = \frac{F^{3/4}}{\gamma k^{3/4} \left(\frac{ga}{\kappa \upsilon}\right)^{1/4}} . \tag{8}$$

I have chosen $\beta = 1/4$ which is the value found in numerical experiments when the Rayleigh number is defined in terms of the surface flux.

A special feature of mantle convection is that the ease of deformation is very temperature-dependent. Despite laboratory experiments with variable viscosity fluids (1, 29), it is not yet clear how to incorporate this variation into Eq. 8 once we are faced with the extreme behavior of mantle materials. We will proceed by asserting that the relevant viscosity for Eq. 8 is that corresponding to the temperature $\langle T \rangle$. If we assume a viscosity low

$$\frac{\nu}{\nu_o} = \left(\frac{T}{T_o}\right)^{-n} , \tag{9}$$

which Davies (6) shows is a reasonably good approximation (with $n \approx 30$) for the rheology of mantle materials, we obtain the relation

$$\frac{\langle T \rangle}{F_o} = (F/F_o)^{3/n+4} \; (d/\gamma\kappa) \; R_o^{-1/4} , \tag{10}$$

where

$$R_o = \frac{g\alpha F_o d^4}{\rho c_p \kappa^2 \nu_o} ,$$

and F_o is a reference flux that satisfies the Nusselt number-Rayleigh number relation Eq. 6 with $R_a = R_o$ and $\langle T \rangle = T_o$.

Combining Eqs. 10 and 5 we get an evolution equation for the surface flux:

$$(d^2/\gamma\kappa) \quad R_o^{-1/4} \quad \frac{\partial}{\partial t} \quad (F/F_o)^{3/n+4} = \frac{H}{F_o} - \frac{F}{F_o} \quad . \tag{11}$$

Since Eq. 11 is nonlinear even for uniform viscosity ($n = 0$), I will give only approximate solutions, so that they can be put in closed form; these are, in fact, very accurate when evaluated for the present day. The nature of the approximation is to write Eq. 11 as

$$(d^2/\gamma\kappa) \; R_o^{-1/4} \; (3/n+4) \, (F/F_o)^{\frac{3}{n+4}-1} \; \frac{\partial}{\partial t} (F/F_o) = H/F_o - F/F_o , \tag{12}$$

and since $F/F_o = 1$ at present, let $(F/F_o)^{\frac{3}{n+4}-1}$ equal 1 for all time. This results in an equation that we can write as

$$\frac{\partial}{\partial \tau}\left(\frac{F}{H_o}\right) = H/H_o - F/H_o , \tag{13}$$

where H_o is the present rate of heating (by radioactive decay), and where time has been nondimensionalized by

$$\tau = \frac{\gamma\kappa R_o^{1/4}}{d^2\,(3/n+4)}\, t . \tag{14}$$

In order to bring out the innate time scale of the system, we can consider first a constant heating rate, H_o, but with $F/H_o = F_i$ ($\neq 1$) at $\tau = 0$. The solution is

$$F/H_o = F_i e^{-\tau} + 1 - e^{-\tau}. \tag{15}$$

The system forgets its initial conditions, and surface flux comes into balance with heat production on a time scale $\tau \sim 1$. In dimensional terms the adjustment time is

$$\tau \sim \frac{d^2\;3/(n+4)}{\gamma\kappa\; R_o^{1/4}} . \tag{16}$$

The appropriate choice of d is 2000 km if we want to consider that the entire planetary temperature can change. The value of γ is uncertain: 0.67 is found in numerical experiments and 0.2 results from Eq. 8 when typical values for the Earth are used. Estimates of this adjustment time are then

(a) uniform viscosity (n=0): $\begin{cases} 1.1 \times 10^9 \text{ yrs } (\gamma=.67) \\ 3.5 \times 10^9 \text{ yrs } (\gamma=.2) \end{cases}$

(b) variable viscosity (n~30): $\begin{cases} 1.2 \times 10^8 \text{ yrs } (\gamma=.67) \\ 4.1 \times 10^8 \text{ yrs } (\gamma=.2) \end{cases}$

A second important question concerns the adjustment of the system to a decreasing rate of heat input, as is the case with radiogenic heating in the Earth. If we set $H = H_o e^{-\lambda (t-t_p)}$, where t_p is the present time and again consider $F/H_o = F_i$ at $t = 0$, then the solution evaluated at the present time (when it is most accurate) is

$$F/H_o = F_i e^{-\tau_p} + \frac{1}{1-\lambda'}\,[1-e^{(\lambda'-1)\tau_p}] , \tag{17}$$

where

$$\lambda' = \frac{\lambda d^2}{\gamma\kappa Ra_o^{1/4}} \quad 3/(n+4) ,$$

and

$$\tau_p = \frac{t_p \gamma\kappa R_o^{1/4}}{d^2 \quad 3/(n+4)} \quad .$$

The appropriate choice of λ is about 2×10^{-10} yrs^{-1}, corresponding to a present-day half-life of 3.5×10^9 yrs (35).

The variable viscosity cases are in the limit $\lambda' \ll 1$, $\tau_p \gg 1$, therefore

$$F/H_o \approx 1 + \lambda' \approx 1.1, \tag{18}$$

so that heat flow and heat production are very nearly equal. This is typical of many variable viscosity thermal evolution models (see (35)). The results given by Davies (6) are consistent with Eq. 18, but he estimates λ' to be order 1, thus $F/H_o \sim 2$. His λ' is unreasonably large, because he uses an average λ instead of a present-day value, and his adjustment time is too large by a factor of about 3.

The uniform viscosity limit ($\lambda' \sim 1$, $\tau_p \sim 1$) produces values of F/H_o closer to 2, consistent with the results of Daly (5) and McKenzie and Richter (17), but these models would generally be viewed as unrealistic since they ignore a key feature of planetary convection, namely, that the viscosity is a very sensitive function of temperature.

The reason for discussing the thermal adjustment time in such detail is that it controls the ratio of heat flow to contemporaneous heat production, and this ratio is something for which we have separate estimates based on bulk chemical composition models. But the bulk composition models typically estimate $F/H_o \sim 2$ (see summary in (17)), while the variable viscosity mantle-wide convection models result in a value much closer to 1. The simplest way to reconcile the results of convective thermal evolution models with bulk composition estimates is to assume a layered mantle. Layering can result in $F/H_o \sim 2$.

Chemical Dispersal Time

The recognition of long-lived isotopic heterogeneities in the mantle raises the question of the fate of "marked" volumes of material in an actively convecting system such as the mantle. Such a marked parcel might represent the depleted residue of partial melting. To begin with an example, we can consider a numerical calculation of an internally heated convecting system at a Rayleigh number of 1.4×10^6 which is appropriate to flow restricted to the upper mantle with a surface heat flux equal to the present-day value (Fig. 4). The flow is time-dependent in that cold sinking sheets are continuously becoming detached from the surface boundary layer. The characteristic time scale on which new cold sheets are formed is of the same order as the overturn time discussed earlier. The effect of such a flow on a "marked" volume of fluid is shown in Fig. 5. The initial volume is drawn out into ever more complicated streaks, and the region containing "marked" material increases with time. Both the mind and present computers are too limited to follow this behavior for long times, and so an alternative representation is necessary. One such representation is to focus on the rate at which the region containing marked material increases with time and to describe this process of dispersal in terms of an effective eddy diffusivity.

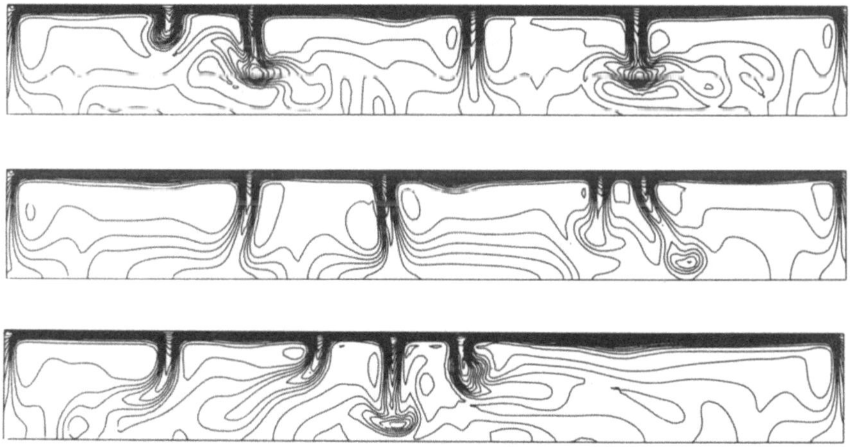

FIG. 4 - Isotherms of internally heated convection at different times (2×10^8 years apart) showing the time dependence of the cold sinking regions. Ra = 1.4×10^6.

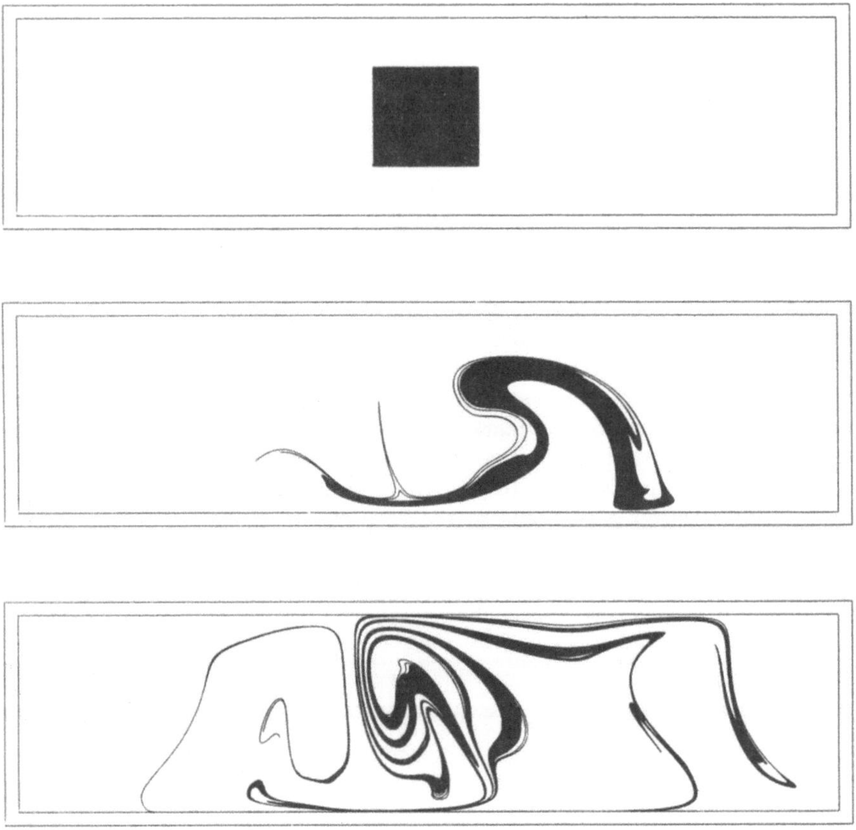

FIG. 5 - The evolution of a marked volume in the flow shown in Fig. 4. The time increment is 2×10^8 years for depth equal to 700 km. This calculation is by N. Hoffman, Cambridge University (personal communication).

To do this we can repeat the experiment using a discrete set of particles and follow their dispersal (Fig. 6). The second moment of the particle distribution increases more or less linearly with time; the process can be described in terms of an effective diffusity κ_e of 3×10^{-5} m^2/sec (27). The dispersal time for spreading over a distance ℓ is then

$$\tau_d = \ell^2/\kappa_e \ . \qquad (19)$$

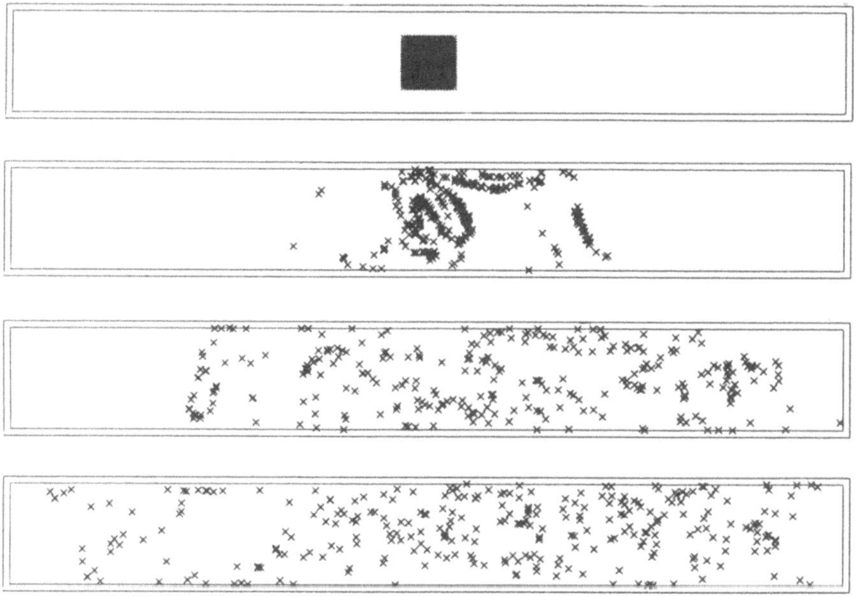

FIG. 6 - The dispersal of a number of marked particles by the flow shown in Fig. 4. The time increment is 4 x 10^8 years when the depth equals 700 km (again by N. Hoffman, personal communication).

The time scale is not unreasonable for convective flows that do not involve the plates themselves. One should keep in mind that the plate scale return flow will also tend to disperse heterogeneities but on a time scale that depends on its own overturn time and only in the direction of the return flow.

The key question for the dispersal of chemical heterogeneities is whether the lavas represent "samples" from a small volume of the mantle and are then measures of individual streaks or whether they are collected from a sufficiently large volume so that their composition reflects the streak density, which is characterized by diffusive behavior. Observationally the difference is that the streaks represent a signal with increasing high spatial frequency, while the diffusive representation damps the high frequencies leaving only the very long spatial scales of variation.

Some Outstanding Questions

Several key issues can be abstracted from the preceding discussion.

1. Is the mantle layered, with separate levels of convection and isolated geochemical reservoirs?
2. Is there a low viscosity zone at the base of the lithosphere affecting the direction and overturn time of the plate scale of motion?
3. Can the evidence for "other" scales of motion be made quantitative?
4. Is the difference in heat production inferred from thermal evolution models vs. bulk composition models significant, and if so can they be reconciled?
5. Can geochemical data distinguish between the streak vs. diffusive representation for the evolution of chemical heterogeneities in the mantle?

Acknowledgements. I thank N. Hoffman and D. McKenzie for discussions and for the figures used in this paper. The research was supported by The Royal Society and The National Science Foundation grant NSF-EAR 82-00003.

REFERENCES

(1) Booker, J.R. 1976. Thermal convection with strongly temperature-dependent viscosity. J. Fluid Mech. 76: 741-754.

(2) Cathles, L.M. 1975. The Viscosity of the Earth's Mantle. Princeton: Princeton University Press.

(3) Cazenave, A.; Dominh, K.; and Lago, B. 1982. Thermal parameters of the oceanic lithosphere estimated from geoid height data. J. Geophys. Res., in press.

(4) Chase, C.G. 1979. Asthenospheric counterflow: a kinematic model. Geophy. J.R.A.S. 56: 1-18.

(5) Daly, S.F. 1980. Convection with decaying heat sources: constant viscosity. Geophy. J.R.A.S. 61: 519.

(6) Davies, G.F. 1980. Thermal histories of convective earth models and constraints on radiogenic heat production in the Earth. J. Geophys. Res. 85: 2517-2530.

(7) DePaolo, D.J., and Wasserburg, G.J. 1976. Inferences about magma sources and mantle structure from variations of $^{143}Nd/^{144}Nd$. Geophys. Res. Lett. 3: 743-746.

(8) Detrick, R.S. 1982. An analysis of geoid anomalies across the Mendocino Fracture Zone: implications for thermal models of the lithosphere. J. Geophys. Res. 86: 11751-11762.

(9) Hager, B.H., and O'Connell, R.J. 1979. Kinematic models of large scale flow in the earth driven by the moving plates. J. Geophys. Res. 84: 1031-1048.

(10) Hewitt, J.M.; McKenzie, D.P.; and Weiss, N.O. 1980. Large aspect ratio cells in two-dimensional thermal convection. Earth Planet. Sci. Lett. 51: 370-380.

(11) Jacobsen, S.B., and Wasserburg, G.J. 1979. The mean age of crustal and mantle reservoirs. J. Geophys. Res. 84: 7411.

(12) Jarvis, G.T., and Peltier, W.R. 1980. Oceanic bathymetry profiles flattened by radiogenic heating in a convective mantle. Nature 285: 649-651.

(13) Jeanloz, R., and Richter, F.M. 1979. Convection, composition, and the thermal state of the lower mantle. J. Geophys. Res. 84: 5497-5504.

(14) Jordan, T.H. 1977. Lithospheric slab penetration into the lower mantle beneath the Sea of Okhotsk. J. Geophys. 43: 473-496.

(15) Jordan, T.H., and Creager, K.C. 1982. Slab penetration into lower mantle. Terra Cognita. 2: 146-147.

(16) McKenzie, D.P., and Richter, F.M. 1976. Convection currents in the earth's mantle. Sci. Am. 235: 72-89.

(17) McKenzie, D.P., and Richter, F.M. 1981. Parameterized thermal convection in a layered region and the thermal history of the Earth. J. Geophys. Res. 86: 11667-11680.

(18) McKenzie, D.P.; Roberts, J.M.; Weiss, N.O. 1974. Convection in the earth's mantle: towards a numerical simulation. J. Fluid Mech. 62: 465-538.

(19) McKenzie, D.P.; Watts, A.; Parsons, B.; and Roufosse, M. 1980. Planform of mantle convection beneath the Pacific Ocean. Nature 288: 442-446.

(20) McKenzie, D.P., and Weiss, N.O. 1975. Speculations on the thermal and tectonic history of the Earth. Geophys. J.R.A.S. 42: 131-174.

(21) O'Nions, R.K.; Hamilton, P.J.; and Evensen, N.M. 1977. Variations in $^{143}Nd/^{144}Nd$ and $^{87}Sr/^{86}Sr$ ratios in oceanic basalts. Earth Planet. Sci. Lett. 34: 13-22.

(23) Parsons, B., and McKenzie, D.P. 1978. Mantle convection and the thermal structure of the plates. J. Geophys. Res. 83: 4485-4496.

(23) Parsons, B., and Sclater, J.G. 1977. An analysis of the variations of ocean floor bathymetry and heat flow with age. J. Geophys. Res. 82: 803-827.

(24) Peltier, W.R., and Andrews, J.T. 1976. Glacial-isostatic adjustment, I. The foreward problem. Geophys. J.R.A.S. 46: 605-646.

(25) Richter, F.M. 1973. Convection and the large scale circulation of the mantle. J. Geophys. Res. 78: 8735-8745.

(26) Richter, F.M. 1979. Focal mechanisms and seismic energy release of deep and intermediate earthquakes in the Tonga-Kermadec region and their bearing on the depth extent of mantle flow. J. Geophys. Res. 84: 6783-6795.

(27) Richter, F.M.; Daly, S.F.; and Nataf, H.-C. 1982. A parameterized model for the evolution of isotopic heterogeneities in a convecting system. Earth Planet. Sci. Lett. 60: 178-194.

(28) Richter, F.M., and McKenzie, D.P. 1978. Simple plate models of mantle convection. J. Geophys. 44: 441-471.

(29) Richter, F.M.; Nataf, H.-C.; and Daly, S.F. 1983. Heat transfer and horizontally averaged temperature of convection with large viscosity variations. J. Fluid. Mech., in press.

(30) Richter, F.M., and Parsons, B. 1975. On the interaction of two scales of convection in the mantle. J. Geophys. Res. 80: 2529-2541.

(31) Richter, F.M., and Ribe, N.M. 1979. On the importance of advection in determining the local isotopic composition of the mantle. Earth Planet. Sci. Lett. 43: 212-222.

(32) Sandwell, D., and Schubert, G. 1980. Geoid height versus age for symmetric spreading ridges. J. Geophys. Res. 85: 7235-7241.

(33) Schubert, G.; Yuen, D.A.; Froidevaux, C.; Fleitout, L.; and Soriau, M. 1978. Mantle circulation with partial shallow return flow: effects of stresses in oceanic plates and topography of the sea floor. J. Geophys. Res. 83: 745-759.

(34) Spohn, T., and Schubert, G. 1982. Modes of mantle convection and the removal of heat from the earth's interior. J. Geophys. Res. 87: 4682-4696.

(35) Turcotte, D.L. 1980. On the thermal evolution of the earth. Earth Planet. Sci. Lett. 48: 53-58.

Patterns of Change in Earth Evolution, eds. H.D. Holland and A.F. Trendall, pp. 291-302.
Dahlem Konferenzen 1984. Berlin, Heidelberg, New York, Tokyo: Springer-Verlag.

Isotopic Evolution of the Crust and Mantle

R.K. O'Nions
Dept. of Earth Sciences
University of Cambridge, Cambridge CB2 3EQ, England

INTRODUCTION

The age and chemical evolution of the continents together with the nature and locale of their depleted mantle complement have been investigated using naturally occurring isotope tracers. The approach adopted has its roots in the pioneering studies of the 1960s, particularly following the successful development of the ^{87}Rb-^{87}Sr system, and has matured to the present situation where a variety of natural radiogenic and non-radiogenic tracers are available in the geochemist's armamentarium (see (2)).

Broadly speaking, the isotopic tracers that have been successfully exploited are divisible into two categories according to their general geochemical behavior. In classical geochemical terminology one of these groups of tracers is termed lithophile and includes those elements that exhibit a preference for silicate phases within the terrestrial planets. Rb, Sr, Sm, Nd, U, Th, and Pb are all lithophile elements. The second group includes those elements that are gases under normal conditions and therefore show an affinity for the atmosphere rather than the Earth's interior: these are the atmophile elements and include the rare gases He, Ne, Ar, Kr, and Xe. The lithophile and atmophile elements listed above are also characterized by a low abundance within the Earth. They are all trace elements and have concentrations below 1000 ppm.

The prime utility of the several natural isotope tracers derives from their basic geochemical preferences. For example, during the differentiation of the continental crust the heat-producing elements have been strongly concentrated into the outer part of the Earth and there has been a marked fractionation of Rb, Sr, Sm, Nd, Lu, and Hf between the mantle and the continental crust. Consequently the isotopic composition of Sr, Nd, Hf, Pb, etc., has evolved in a different manner in these two parts of the Earth during the course of geological history.

The basic problem, then, is to document the isotopic evolution of the daughter products of long- and short-lived parent isotopes in the continents, in the atmosphere, and in the mantle, and to place constraints on the time-dependence of the transport processes responsible for the evolution of these reservoirs.

BRIEF REVIEW OF DECAY SCHEMES

This brief review of relevant decay schemes and isotope ratios is far from complete; it is intended merely to provide some necessary, basic reference information.

Rare Gases

Helium ($^3He/^4He$)

Primordial: $^3He/^4He \simeq 10^{-4}$

Radiogenic: $^6Li(n,\alpha)^3He$

$^{235}U \rightarrow {}^{207}Pb + 7\,{}^4He$ ($T_{\frac{1}{2}}$ = 0.70 Ga)

$^{238}U \rightarrow {}^{206}Pb + 8\,{}^4He$ ($T_{\frac{1}{2}}$ = 4.47 Ga)

$^{232}Th \rightarrow {}^{208}Pb + 6\,{}^4He$ ($T_{\frac{1}{2}}$ = 14.0 Ga)

Radiogenic production ratio: $\frac{^3He}{^4He} \simeq 2 \times 10^{-8}$

Atmosphere: $^3He/^4He = 1.4 \times 10^{-6} \equiv$ (Ra)

Mantle: MORB: R/Ra = 6-10

Hot spot: R/Ra = 5-30

Argon ($^{40}Ar/^{36}Ar$)

Primordial: $^{40}Ar/^{36}Ar \simeq 10^{-4}$

Radiogenic: $^{40}K \rightarrow {}^{40}Ar + {}^{-}e$ $\{T_{\frac{1}{2}} = 1.3$ Ga$\}$

$\rightarrow {}^{40}Ca + \beta$

Atmosphere: $^{40}Ar/^{36}Ar$ (Ra) = 295.6

Mantle:	MORB:	$5 < R/Ra < 100$
	Hot spot:	$R/Ra \simeq 1.0$ (?)

Xenon ($^{129}Xe/^{130}Xe$)

Primordial:	$^{129}Xe/^{130}Xe = 6.3$	
Radiogenic:	$^{129}I \rightarrow {}^{129}Xe + \beta$	$\{T_{\frac{1}{2}} = 17Ma\}$
Atmosphere:	$^{129}Xe/^{130}Xe = 6.48$	
Mantle:	MORB:	$^{129}Xe/^{130}Xe = 6.8 \pm .2$
	Hot spot:	$^{129}Xe/^{130}Xe = 6.5$ (?)

Lithophile Elements

Strontium ($^{87}Sr/^{86}Sr$)

Primordial:	$^{87}Sr/^{86}Sr \simeq 0.699$	
Radiogenic:	$^{87}Rb \rightarrow {}^{87}Sr + \beta$	$\{T_{\frac{1}{2}} = 48.8$ Ga$\}$
Bulk Earth:	$\frac{Rb}{Sr} \simeq 0.03$	
	$\frac{^{87}Sr}{^{86}Sr} \simeq 0.7047$	
Mantle:	MORB:	0.702-0.703
	Hot spot:	0.703 -0.706
Average Continent:	$\simeq 0.710$	

Epsilon Notation:

Deviation from bulk Earth value in parts per 10^4, i.e., $\epsilon^{o}_{Sr} = 0$ equivalent to present day $^{87}Sr/^{86}Sr = 0.7047$

Neodymium ($^{143}Nd/^{144}Nd$)*

Primordial:	$^{143}Nd/^{144}Nd = 0.5066$	$\{T = 4.6$ Ga$\}$
Radiogenic:	$^{147}Sm \rightarrow {}^{143}Nd + \alpha$	$\{T_{\frac{1}{2}} = 106$ Ga$\}$
Bulk Earth (today):	(Normalized $\frac{146}{144}$ Nd $= .7219$)	

* In considering $^{143}Nd/^{144}Nd$ ratios note should be taken of different normalization procedures that are in common use.

Bulk Earth (today) : (Normalized $\frac{146}{144}$ Nd = .7219)

$$\frac{^{147}Sm}{^{144}Nd} = 0.1966$$

$$\frac{^{143}Nd}{^{144}Nd} = 0.512638$$

Epsilon Notation:

Deviations from bulk Earth value in parts per 10^4. $\epsilon^{o}_{Nd} = 0$ equivalent to present day $^{143}Nd/^{144}Nd = 0.512638$

Mantle: MORB: $^{143}Nd/^{144}Nd = .5130 \rightarrow 0.5133$

ϵ^{o}_{Nd} = +7 to +14

Hot spot: ϵ^{o}_{Nd} = -5 to +8

Average continents upper crust: $\epsilon^{o}_{Nd} \simeq -12$

Lead

Primordial: $^{208}Pb/^{204}Pb = 29.48$

$^{207}Pb/^{204}Pb = 10.29$

$^{206}Pb/^{204}Pb = 9.31$

Radiogenic: See Helium above.

Mantle: MORB: $^{206}Pb/^{204}Pb \simeq 18\text{–}20$

Hot spot: $^{206}Pb/^{204}Pb = 18\text{–}21$

Continents: Note strong relative enrichment of U in upper crust; this results in upper crust having much higher $^{206}Pb/^{204}Pb$ ratios than those of the lower crust. The latter are depleted in U and Th.

HOW OLD IS THE ATMOSPHERE?

The view is widely held that the atmosphere is secondary and produced by outgassing of the Earth's interior. The main constraint on the timing of volatile release has been provided by Ar-isotope data.

The observations are as follows:

1. The atmosphere contains 1% Ar with $^{40}Ar/^{36}Ar = 295.6$.
2. Ocean floor basalts are relatively depleted in ^{36}Ar with $1000 < {}^{40}Ar/^{36}Ar < 20{,}000$.
3. ^{36}Ar is primordial and nearly all ^{40}Ar is radiogenic.

A large number of degassing schemes will satisfy these observations but all concur in requiring a higher rate of degassing in the early part of Earth history. Degassing models based upon Ar isotopes could be further constrained if the variation of the $^{40}Ar/^{36}Ar$ ratio in the atmosphere were known as a function of time.

The isotopic composition of Xe imposes a tighter constraint. Recently the Paris group (9) have reported excess ^{129}Xe (see systematics above) in mid-ocean ridge basalts; this has been attributed to the decay of now extinct ^{129}I ($T_{\frac{1}{2}}$ = 17 Ma) in the Earth. Removal of a major portion of the Xe inventory into the atmosphere within the first 10^8a of Earth history is required to accommodate this observation if the isotopic composition of Xe was uniform in the early mantle.

Notions concerning the evolution of the atmosphere gained from Ar and Xe isotopes are therefore not at variance. A higher rate of degassing of both elements in the early part of Earth history seems to be required. If the rare gases were accompanied by other volatile species, then the major constituents of the atmosphere (neglecting oxygen) were also probably present prior to 4.0 Ga.

HOW OLD ARE THE CONTINENTS AND HOW HAVE THEY GROWN?

The oldest known continental crust has been dated using Rb-Sr, U-Pb, Sm-Nd, and Lu-Hf geochronometers at 3.7 to 3.8 Ga. At the present time there are no reported remnants of still older continental crust; events during the first 700 to 800 Ma of Earth history remain obscure. The Sm-Nd isotope data for a number of Archean terranes have been summarized in Fig. 1. It should be noted that the Sm/Nd fractionation between continental crust and mantle is about 40% which results in a divergence in their Nd-isotope evolution which should be resolvable within a few hundred million years of the fractionation event. Initial Nd-isotope compositions can therefore betray the incorporation of older material that has formerly resided in the continental crust. The ϵ_{Nd} values of Isua and of other Archean terranes (Fig. 1) fail to reveal evidence for incorporation of preexisting continental material: if it exists, it must itself be comparatively young or present in small quantities. Metasedimentary rocks preserved at Isua similarly fail to reveal any evidence for cannibalism of older sediments.

At this juncture we return to the following long-standing and still unresolved questions:

1. Did the continents commence their formation 3.8 Ga ago? If so, what was the thermal state of the Earth between 4.5 and 3.8 Ga and why did continents wait so long to appear when the atmosphere seems to have formed much earlier?
2. Has the evidence for the existence of earlier continental crust been entirely eradicated?
3. How has the continental mass changed as a function of time?

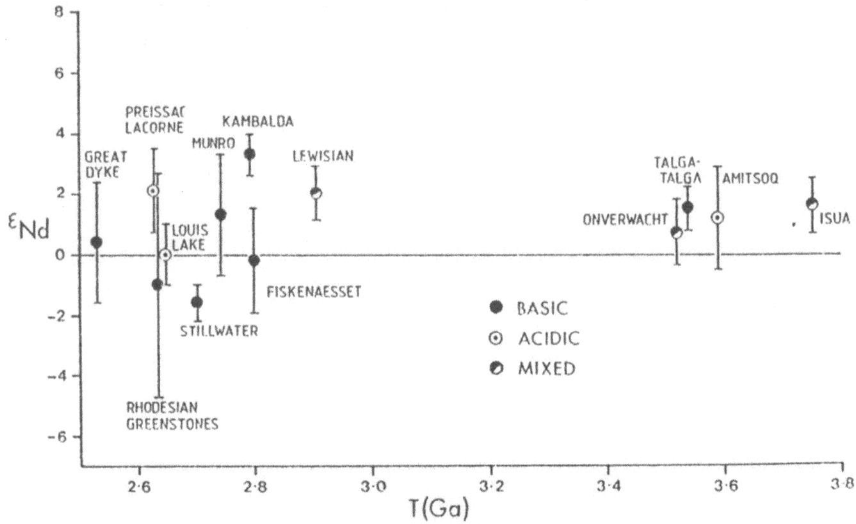

FIG. 1 - Comparison $\epsilon_{Nd}(T)$ values for samples of Archean continental crust. A value of $\epsilon_{Nd}(T) = 0$ corresponds to a system which has a $^{147}Sm/^{144}Nd = 0.1966$ and $^{143}Nd/^{144}Nd = 0.512638$ as measured today ($^{146}Nd/^{144}Nd = 0.7219$). These are the so-called CHUR PARAMETERS. The $^{147}Sm/^{144}Nd = 0.1966$ is close to the average value for chondrite meteorites and is the current best estimate of the bulk Earth value.

Continental crust is formed with a Sm/Nd ratio that is approximately 40% lower than CHUR and evolves with time to less radiogenic Nd (lower ϵ_{Nd} values).

Figures 2 and 3 illustrate the general similarity in form of average isotopic age that can be generated by fundamentally different styles of continent evolution. On the one hand, constant continental mass (no growth or constant mass) models such as championed by Armstrong (1) are compatible with the isotopic patterns in the present-day continents as long as continental material has been recycled into the mantle and balanced by the accretion of new mantle-derived material.

Growth models which do not include the recycling of continental material into the mantle may yield identical patterns of istopic and age distribution within the continents, a point which has been emphasized by Armstrong (1).

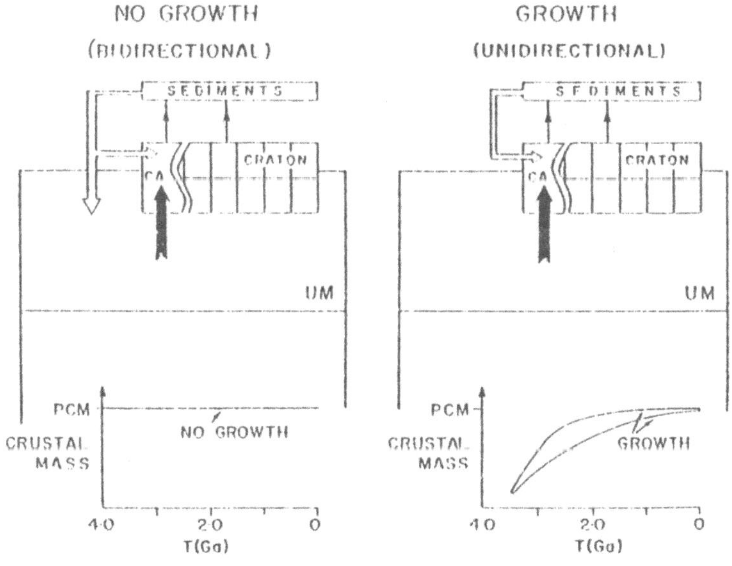

FIG. 2 - Schematic diagrams illustrating the distinction between rival hypotheses for the generation of continental crust. On the one hand the continental mass has remained constant from 4.0 Ga to present. The paucity of radiometric ages from 3.0 to 4.0 Ga ago is explained by recycling of continental material to the mantle, balanced by a return flux from the mantle to the continents. At the other extreme the continents are considered to have grown continuously without a return flux of continental material to the mantle. The paucity of radiometric ages in the interval 4.0 to 3.0 Ga ago is accommodated by an initially slow growth rate of continental crust.

One of the most common objections to the no-growth hypothesis is that the mantle material undergoing accretion must be balanced by a return flux of low-density buoyant continental crust. The amount of material required at the present day, however, is equal to the amount of pelagic clay in the oceans.

What happens to all the pelagic clay that is deposited in the world oceans? It is certainly scarce within the sedimentary record.

It would seem that there are two lines of attack available for the future:
1. The Freeboard argument. An apparently constant freeboard to the continents since the Archean is cited as support for a no-growth hypothesis. This is in need of reevaluation.
2. Calculation of fluxes. Recent calculations by Javoy et al. (5) suggest that the C-flux from the mantle may be sufficient to supply the estimated surface carbon inventory in $\sim 10^8$a. This would require that C is efficiently recycled. If such claims can be substantiated, they would provide major support for the no-growth hypothesis.

CHEMICAL DIFFERENTIATION AND MANTLE CONVECTION

Mid-ocean ridge basalts sample a region of the mantle that has been

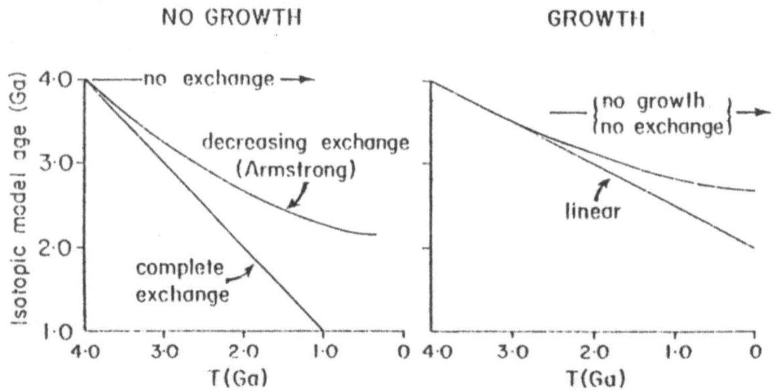

FIG. 3 - Schematic diagram illustrating how the average isotopic age of continental crust varies according to a no-growth model of continental development that includes exchange with the mantle and growth models. Note that similar overall patterns can emerge suggesting that the age distribution within the continents alone will not readily distinguish growth and no-growth models.

depleted in the heat-producing elements and in other trace elements of large ionic radius; these include the natural radiogenic isotope tracers discussed above. The depleted character was indicated by early Rb-Sr isotope measurements which demonstrated that the radiogenic ^{87}Sr in MORB was unsupported by the ^{87}Rb present. This observation suggested that Rb has been preferentially partitioned into the continents.

If the Sm/Nd ratio of the Earth is equal to that of average chondrite, then the Nd and Sr isotope measurements of MORB (Fig. 4) suggest that

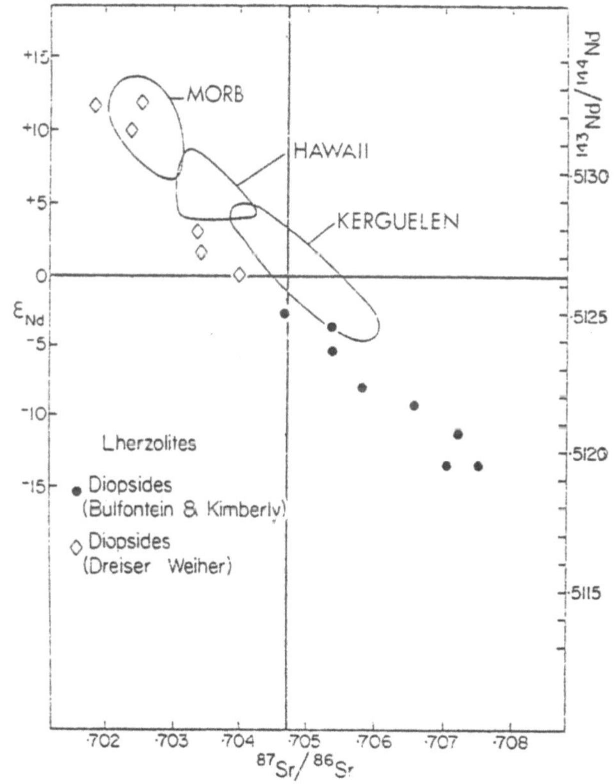

FIG. 4 - The Nd-Sr-isotope mantle array. Note that all mid-ocean ridge basalts and the majority of ocean island basalts have isotopic compositions that are identified with a depleted source region.

all mid-ocean ridge and a majority of ocean island basalts are derived from mantle that has been depleted in Rb and Nd relative to Sr and Sm, respectively. The accepted dogma is that the continental crust is the complement to the depleted portion of the mantle which is sampled by ocean ridge volcanism.

The definition of the amount of mantle that has been depleted to the extent of the material sampled by MORB is a matter of great interest to geochemists and geophysicists alike.

The hope of the geochemists is that geophysicists will contrain the properties and the flow pattern within the mantle to the point where the portion available for depletion can be specified. Needless to say, geophysicists would like a clear message from geochemists on this issue to resolve some of their debates concerning whole mantle versus two-layer mantle convection.

Various attempts were made at the end of the seventies to estimate the proportion of depleted mantle from mass balance considerations. Estimates varied from about 25-50% and concurred in ruling out depletion of the whole mantle (4, 7, 8). Either implicitly or explicitly this result was interpreted to mean that convection in the upper mantle was decoupled from the lower mantle and, furthermore, that upper mantle convection was driven by heating from below. In recent reassessments of the mass balance it has been suggested that earlier claims were conservative and that the available data are compatible with whole mantle convection (e.g., (3)).

Isotopic data on the rare gases are also highly relevant to this problem. For example, ^{3}He in excess of the atmospheric abundance has been identified in MORB and ocean island volcanics (Fig. 5). The average excess found for MORB is about eight times the atmospheric ratio, but in Iceland and Hawaii the $^{3}He/^{4}He$ ratios are up to twenty-five to thirty times the atmospheric value. Mantle ^{3}He is almost entirely primordial and has resided in some portion of the Earth's interior from the time of its formation. This primordial ^{3}He is preferentially sampled by the plumbing system operative beneath Iceland and Hawaii hot spots. It is difficult to see how the ^{3}He could have been stored at shallower levels in the Earth than the lower mantle.

The strongest arguments for two-layer convection within the mantle probably come from considerations of the rare gases rather than the lithophile elements.

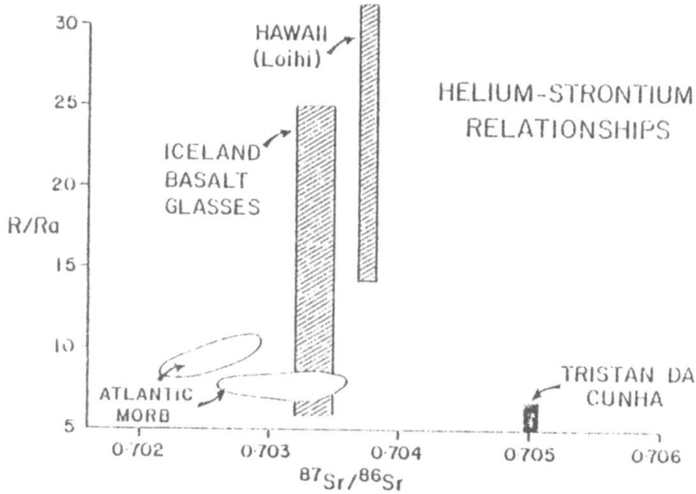

FIG. 5 - Comparison of $^3He/^4He$ and $^{87}Sr/^{86}Sr$ ratios in oceanic basalts. Iceland data from Cambridge, remainder from Kurz (6).

Note: the highest $^3He/^4He$ (R/Ra) ratios are associated with Iceland and Hawaii. All ocean ridge basalts analyzed to date exhibit both lower upper limits for $^3He/^4He$ than oceanic islands and a smaller total range.

Note: the low $^3He/^4He$ ratio of Tristan da Cunha has been cited as evidence for the recycling of He-depleted crust.

OUTSTANDING MAJOR PROBLEMS

1. Has the mass of continental crust remained constant for the last 4.0 Ga, or has it grown continuously in size?
2. Was there any low-density continental crust prior to ca. 3.8 Ga? If so, how was it destroyed?
3. Are pelagic sediments recycled into the mantle at the present day, or do they fail to penetrate the island-arc "filter"?
4. Is the C-cycle at steady state?
5. Does mantle chemistry tell us anything about the scale of mantle convection?

SELECTED REFERENCES

(1) Armstrong, R.L. 1981. Radiogenic isotopes: the case for crustal recycling on a near-steady-state no-continental growth Earth. In Origin and Evolution of the Earth's Continental Crust. Phil. Trans. R. Soc. Lond. A 301: 443-472.

(2) Faure, G. 1977. Principles of Isotope Geology. New York: Wiley.

(3) Davies, G.F. 1981. Earth's neodynium budget and structure and evolution of the mantle. Nature 240: 208.

(4) Jacobsen, S.B., and Wasserburg, G.J. 1979. The mean age of mantle and crustal reservoirs. J. Geophys. Res. 84: 7411.

(5) Javoy, M.; Pineau, M.; and Allègre, C.J. 1982. Carbon geodynamic cycle. Nature 300: 171.

(6) Kurz, M.D.; Jenkins, W.J.; Schilling, J.G.; and Hart, S.R. 1982. Helium isotopic variations in the mantle beneath the central N. Atlantic. Earth Plant. Sci. Lett. 58: 1.

(7) O'Nions, R.K.; Evensen, N.M.; and Hamilton, P.J. 1979. Geochemical modeling of mantle differentiation and crustal growth. J. Geophys. Res. 84: 6091.

(8) O'Nions, R.K., and Hamilton, P.J. 1981. Isotope and trace element models of crustal evolution. Phil. Trans. R. Soc. Lond. A 301: 473-487.

(9) Staudacher, T., and Allègre, C.J. 1982. Terrestrial xenology. Earth Planet. Sci. Lett. 60: 389.

Patterns of Change in Earth Evolution, eds. H.D. Holland and A.F. Trendall, pp. 303-311. Dahlem Konferenzen 1984. Berlin, Heidelberg, New York, Tokyo: Springer-Verlag.

Degassing of the Earth

H.D. Holland
Dept. of Geological Sciences, Harvard University
Cambridge, MA 02138, USA

Abstract. The discovery of ^3He in basaltic glasses and in seawater that has cycled through mid-ocean ridges has confirmed the reality of the release of juvenile gases from the mantle today. The rate of ^3He loss from the mantle is now reasonably well-known. Since the ratio of the concentration of this nuclide to that of ^{20}Ne, ^{36}Ar, ^{84}Kr, and ^{132}Xe has been measured in a number of mantle xenoliths and volcanic rocks, the rate at which the rare gases are released from the mantle can be estimated. The most likely value of the degassing rate of ^{20}Ne and ^{36}Ar is approximately 13% of the mean degassing rate, defined as the total quantity of these nuclides in the atmosphere divided by the age of the Earth. The low value of the present degassing rate is consistent with an early period of intense degassing from the mantle. Most of the CO_2 that is added to the atmosphere today is recycled rather than juvenile. This seems to have been true during much of geologic time. The acid-base balance of crustal rocks has been remarkably constant during the past 3000 m.y.; this indicates that the rate of supply of weathering acids has been matched rather closely by the rate of conversion of igneous and high-grade metamorphic rocks into sedimentary rocks.

PRIMARY DEGASSING

Numerous attempts have been made to reconstruct the degassing history of the Earth since Rubey's (26) classic paper on the subject (see, for instance, (6-10, 27, 29)).

These reconstructions have tended to rely heavily on the present-day distribution of the rare gases, particularly that of ^{40}Ar, between the solid Earth and the atmosphere. It is clear that ^{40}Ar has been generated in the Earth largely by the decay of ^{40}K since the formation of the Earth; hence degassing could not have taken place entirely at or very close to 4.5 b.y. ago. It has, however, proved very difficult to extract a detailed degassing schedule from the ^{40}Ar data, because the construction of such schedules requires more precise data than are currently available for the total potassium content of the Earth, for the distribution of potassium between the Earth's crust and mantle, and for the retentivity of Ar in crustal rocks.

Really convincing evidence for the present-day addition of any primordial gas to the atmosphere was lacking until the discovery of excess ^{3}He in deep waters of the Pacific Ocean by Clarke, Beg, and Craig ((1); see also (5)). The ^{3}He excess which they observed could only be explained by the injection of primordial helium at seafloor spreading centers on mid-ocean ridges. Subsequent measurements have confirmed the presence of excess ^{3}He in deep ocean waters (2, 3, 14, 15) and have demonstrated the presence of excess ^{3}He in rapidly quenched, glassy margins of oceanic basalts (4, 19, 22) as well as in hydrothermal fluids associated with mid-ocean ridges (16, 23, 24). The discovery of large excesses of ^{3}He in basaltic glasses has also shown that other rare gases in these basalts are largely primordial rather than due to atmospheric contamination, and this discovery has opened what promises to be a fruitful approach to the degassing history of the Earth.

The flux of ^{3}He from the mantle can be estimated by multiplying the ^{3}He concentration in hydrothermal solutions in the Galapagos and in other parts of the East Pacific Rise (16, 23, 30, 31) by the total annual flow of such solutions required to account for the heat flow deficit in the vicinity of mid-ocean ridges (32). If one adds to the marine flux the somewhat uncertain ^{3}He flux through the continents, one arrives at a total primordial ^{3}He flux of $(3.0 \pm 1.0) \times 10^{19}$ atoms/sec (i.e., $10^{3.2}$ mol/yr), a rate which is compatible with the estimated rate of escape of ^{3}He from the atmosphere (Jenkins, unpublished manuscript).

If we could obtain values for the average ratio of the primordial nonradiogenic isotopes of the other rare gases to ^{3}He in mantle-derived volatiles, we could combine these with the estimated rate of ^{3}He escape to derive a figure for the present-day release rate of these gases to the atmosphere. Some of the necessary data are available, but their sum total is still small. Kaneoka, Takaoka, and Aoki (18) have determined the concentration of the rare gases in three samples of volcanic rocks from the island of Hawaii; Kaneoka and Takaoka (17) have reported rare gas analyses of samples from Maui and Oahu; and Kyser (20) and Kyser and Rison (21) have analyzed a collection of samples from Oahu, the island of Hawaii, South Africa, the Grand Canyon, and the Massif Central in France.

These data have been used to infer the present degassing rate of the nonradiogenic isotopes ^{20}Ne, ^{36}Ar, ^{84}Kr, and ^{132}Xe ((13), Chapter 3). The results are shown in Table 1 together with the mean degassing rate of these nuclides. The mean degassing rate was defined as the total number of moles of each nuclide in the atmosphere divided by the age of the Earth, 4.5 x 10^9 yr. For both ^{20}Ne and ^{36}Ar the probable present-day degassing rate is nearly an order-of-magnitude less than the mean degassing rate; however, the maximum permitted values of the current degassing rate of these nuclides are indistinguishable from their mean degassing rate. The most probable present-day input rates of ^{84}Kr and ^{132}Xe are, respectively, somewhat less and comparable to the mean input rate of these nuclides as defined above. If, as seems likely, significant quantities of Kr and Xe have been lost to sediments and sedimentary rocks from the atmosphere the calculated mean degassing rates for these gases are considerably less than the true degassing rates. It is likely that the ratio of the present to the mean degassing rate of Kr and Xe corrected for loss to the crust is similar to the ratio of the present to the mean degassing rate of Ne and Ar.

TABLE 1 - Comparison of the estimated current degassing rate of ^{20}Ne, ^{36}Ar, ^{84}Kr, and ^{132}Xe with the mean degassing rate as defined in the text (from Holland (13), Chapter 3).

Nuclide	Present-day Degassing Rate (mol/yr)			Mean Degassing Rate (mol/yr)
	Minimum	Probable	Maximum	
^{20}Ne	$10^{4.6}$	$10^{4.9}$	$10^{5.4}$	$10^{5.81}$
^{36}Ar	$10^{4.6}$	$10^{5.2}$	$10^{6.0}$	$10^{6.10}$
^{84}Kr	$10^{2.9}$	$10^{3.8}$	$10^{4.9}$	$10^{4.41}$
^{132}Xe	$10^{1.9}$	$10^{2.9}$	$10^{4.2}$	$10^{2.96}$

If the data in Table 1 for the ^{20}Ne and the ^{36}Ar flux from the Earth today turn out to be correct, the present flux of these gases is ca. 13% of their mean flux. This implies that if the degassing rate has remained constant during most of Earth history, ca. 87% of ^{20}Ne and ^{36}Ar were released early in Earth history. Such an evolutionary model is rather unlikely. Heat generation in the Earth has decreased with time, and it is likely that the rate of degassing of the Earth has decreased concomitantly, both in response to the decreasing metabolic activity of the Earth and to the decreasing quantity of the nonradiogenic rare gases in the mantle. The functional relationships between the degassing rate, the rate of heat generation, and the rare gas content of the mantle are not known. However, plausible models can be constructed to see whether and under what conditions a period of rapid degassing early in Earth history is required by the rare gas release data. Several models have been constructed and evaluated ((13), Chapter 3).

The results of this analysis show that a substantial early degassing event seems to be required except for degrees of mantle degassing and rehomogenization which are in conflict with present evidence for the evolution of the mantle from the systematics of the Nd-Sm and Rb-Sr decay systems. Staudacher and Allègre (28) have reached the same conclusion on the basis of differences between the isotopic composition of Xe in the atmosphere and in MORB gases. It must, however, be remembered that the data base for the proposed present-day release rate of the nonradiogenic rare gases is still very slim. If present-day release rates turn out to be more than a factor of four greater than the probable rates listed in Table 1, the case for a period of rapid early degassing becomes very weak.

RECYCLING OF VOLATILES

Ne and Ar degassed from the Earth's interior during geologic time have largely accumulated in the atmosphere. The present quantity of these gases in the atmosphere is therefore nearly equal to their integrated degassing rate during the last 4.5 b.y. This is not true for most of the other volatiles. A sizable fraction of the degassed nitrogen and water, most of the sulfur, and nearly all of the carbon have been transferred from the atmosphere to the Earth's crust. Nitrogen is present in the crust largely as a constituent of organic compounds, water as a constituent of hydrous silicates, sulfur as a constituent of pyrite and anhydrite, and carbon as a constituent of organic compounds, graphite, and carbonates of calcium and magnesium. The residence times of carbon and sulfur for burial, or reburial, with sedimentary rocks are geologically short. The residence time of carbon in the oceans is approximately 8 x 10^4 yrs, the residence time of sulfur approximately 8 x 10^6 yrs (12). The chemistry of the ocean-atmosphere system therefore tends to adjust itself on a geologically short time scale, so that the rate of output of dissolved constituents is equal to the rate of input.

The present-day cycling of many volatiles is rapid compared to their primary rate of degassing from the mantle. This is best illustrated by the operation of the geochemical cycle of carbon, for which a great deal of information is now available (see, for instance, (12), pp. 262-283). The present total rate of erosion and of sedimentation is close to 2 x 10^{16} gm/yr (11, 25). The carbon buried with new sediments amounts to some (3.4 $\pm$ 0.6) x 10^{14} gm/yr ((12), Table 6-6). Most of this carbon is derived from the erosion of older sedimentary rocks. Elemental and organic carbon in old sedimentary rocks is oxidized to CO_2 during weathering; carbonate carbon is released by reaction with atmospheric and soil CO_2 and is transported to the oceans, largely as HCO_3^- in river water.

Approximately 75% of recent sediments are derived from the weathering of old sedimentary rocks, the remainder from the weathering of igneous and high-grade metamorphic rocks. Thus, approximately 75% of the carbon in new sediments has been derived from the weathering of old sedimentary rocks. Since igneous and high-grade metamorphic rocks contain very little carbon compared to modern sediments, approximately 25% of the carbon in modern sediments remains to be accounted for. The only reasonable sources for this carbon are the carbon released from sedimentary rocks during metamorphism and juvenile carbon. The missing carbon amounts to some (8 $\pm$ 3) x 10^{13} gm/yr.

This quantity is large compared to the quantity of juvenile carbon that is probably added annually to the atmosphere. The total inventory of carbon in the crust today is approximately 9 x 10^{22} gm. If this quantity of carbon has been released from the mantle at a constant rate, the rate of carbon addition from this source has been 2 x 10^{13} gm/yr, a figure that is almost certainly a strong upper limit for the actual present-day flux of juvenile carbon. The analysis of the present rate of degassing of ^{20}Ne and ^{36}Ar suggested that these gases are being released presently at about 13% of the mean rate calculated over geological time. There is no reason to believe that this rate is equal to the mean rate at which juvenile CO_2 is being released from the mantle; but if CO_2 and the rare gases are behaving roughly similarly, then juvenile carbon is being added at a rate of approximately 0.3 x 10^{13} gm/yr. This accounts for only about 4% of the carbon which must be degassed annually to account for the carbon content of modern sediments. Even if the proposed figure for the degassing rate of juvenile carbon is too low by a factor of four, it still follows that carbon released during the metamorphism of sedimentary rocks accounts for most of the non-anthropogenic CO_2 that is currently being added to the atmosphere. This is probably true also for sulfur and nitrogen. Degassing during the operation of the geological cycle is therefore currently much more important for the geochemistry of these elements than the contribution of juvenile material from the mantle.

The rate of degassing and recycling of volatiles in the past can be estimated, at least roughly, from the acid-base balance of sedimentary rocks formed during the past 3800 m.y. During weathering, rocks are titrated with a variety of acids; among these H_2CO_3, H_2SO_4, and HCl have been and are particularly important. Most of the cations that are released during titration accompanying weathering appear in sedimentary rocks as salts of these acids. The acid-base balance of sedimentary rocks has been defined as the fraction of titratable cations in parent igneous rocks that have been titrated during conversion into sedimentary rocks (for a fuller discussion, see (13), Chapter 5). The acid-base balance in sedimentary rocks of a given time period is determined by the injection of weathering acids, the nature and quantity of igneous and high-grade metamorphic rocks undergoing weathering, and the quantity and acid-base balance of the sedimentary rocks that are weathered and redeposited during the given time period.

The record of the acid-base balance of sedimentary rocks formed during the early parts of Earth history is still fragmentary, but we know that carbonation reactions were taking place during weathering at least as early as 3800 m.y.b.p., and that the intensity of these reactions 3400 m.y. ago was comparable to their intensity today. Perhaps the most remarkable conclusion that can be drawn from the rather large body of analytical data for Proterozoic and Phanerozoic sedimentary rocks is that the acid-base balance of sediments has varied very little during the past 2.5 b.y. ((13), Chapter 5). If the acid-base balance of sedimentary rocks has varied at all during the past 2500 m.y., the effects of these variations cannot be identified with certainty in the available data.

REFERENCES

(1) Clarke, W.B.; Beg, M.A.; and Craig, H. 1969. Excess ^{3}He in the sea: Evidence for terrestrial primordial helium. Earth Planet. Sci. Lett. 6: 213-220.

(2) Clarke, W.B.; Beg, M.A.; and Craig, H. 1970. Excess helium 3 at the North Pacific Geosecs Station. J. Geophys. Res. 75: 7676-7678.

(3) Craig, H.; Clarke, W.B.; and Beg, M.A. 1975. Excess ^{3}He in deep water on the East Pacific Rise. Earth Planet. Sci. Lett. 26: 125-132.

(4) Craig, H., and Lupton, J.E. 1976. Primordial neon, helium, and hydrogen in oceanic basalts. Earth Planet. Sci. Lett. 31: 369-385.

(5) Craig, H., and Lupton, J.E. 1981. Helium-3 and mantle volatiles in the oceans and the oceanic crust. In The Oceanic Lithosphere, vol. 7, of The Sea, Chapter 11. New York: Wiley.

(6) Damon, P.E., and Kulp, J.L. 1958. Inert gases and the evolution of the atmosphere. Geochim. Cosmochim. Acta 13: 280-292.

(7) Fanale, F.P. 1971. A case for catastrophic early degassing of the Earth. Chem. Geol. 8: 75-105.

(8) Fisher, D.E. 1978. Terrestrial potassium and argon abundances as limits to models of atmospheric evolution. In Terrestrial Rare Gases, eds. E.C. Alexander, Jr., and M. Ozima, pp. 173-183. Tokyo: Japan Scientific Societies Press.

(9) Hamano, Y., and Ozima, M. 1978. Earth-atmosphere evolution model based on Ar isotopic data. In Terrestrial Rare Gases, eds. E.C. Alexander, Jr., and M. Ozima, pp. 155-171. Tokyo: Japan Scientific Societies Press.

(10) Hart, R., and Hogan, L. 1978. Earth degassing models and the heterogeneous vs. homogeneous mantle. In Terrestrial Rare Gases, eds. E.C. Alexander, Jr., and M. Ozima, pp. 193-206. Tokyo: Japan Scientific Societies Press.

(11) Holeman, J.N. 1968. The sediment yield of major rivers of the world. Water Resour. Res. 4: 737-747.

(12) Holland, H.D. 1978. The Chemistry of the Atmosphere and Oceans. New York: Wiley.

(13) Holland, H.D. 1984. The Chemical Evolution of the Atmosphere and Oceans. Princeton, NJ: Princeton University Press.

(14) Jenkins, W.J.; Beg, M.A.; Clarke, W.B.; Wangersky, P.J.; and Craig, H. 1972. Excess ^{3}He in the Atlantic Ocean. Earth Planet. Sci. Lett. 16: 122-126.

(15) Jenkins, W.J., and Clarke, W.B. 1976. The distribution of ^{3}He in the western Atlantic Ocean. Deep Sea Res. 23: 481-494.

(16) Jenkins, W.J.; Edmond, J.M.; and Corliss, J.B. 1978. Excess ^{3}He and ^{4}He in Galapagos submarine hydrothermal waters. Nature 272: 156-158.

(17) Kaneoka, I., and Takaoka, N. 1980. Rare gas isotopes in Hawaiian ultramafic nodules and volcanic rocks: Constraint on genetic relationships. Science 208: 1366-1368.

(18) Kaneoka, I.; Takaoka, N.; and Aoki, K-I. 1978. Rare gases in mantle-derived rocks and minerals. In Terrestrial Rare Gases, eds. E.C. Alexander, Jr., and M. Ozima, pp. 71-83. Tokyo: Japan Scientific Societies Press.

(19) Kurz, M.D.; Jenkins, W.J.; Schilling, J.G.; and Hart, S. 1982. Helium isotopic variations in the mantle beneath the central North Atlantic Ocean. Earth Planet. Sci. Lett. 58: 1-14.

(20) Kyser, T.K. 1980. Stable and rare gas isotopes and the genesis of basic lavas and mantle xenoliths. Ph.D. Dissertation, Department of Geology, University of California, Berkeley.

(21) Kyser, T.K., and Rison, W. 1982. Systematics of rare gas isotopes in basic lavas and ultramafic xenoliths. J. Geophys. Res. 87: 5611-5630.

(22) Lupton, J.E., and Craig, H. 1975. Excess ^{3}He in oceanic basalts: Evidence for terrestrial primordial helium. Earth Planet. Sci. Lett. 26: 133-139.

(23) Lupton, J.E.; Weiss, R.F.; and Craig, H. 1977a. Mantle helium in the Red Sea brines. Nature 266: 244-246.

(24) Lupton, J.E.; Weiss, R.F.; and Craig, H. 1977b. Mantle helium in hydrothermal plumes in the Galapagos Rift. Nature 267: 603-604.

(25) Milliman, J.D., and Meade, R.H. 1983. World-wide delivery of river sediment to the oceans. J. Geol. 91: 1-21.

(26) Rubey, W.W. 1951. Geologic history of seawater: An attempt to state the problem. Bull. Geol. Soc. Am. 62: 1111-1147.

(27) Schwartzman, D.W. 1978. On the ambient mantle $^4He/^{40}Ar$ ratio and the coherent model of degassing of the Earth. In Terrestrial Rare Gases, eds. E.C. Alexander, Jr., and M. Ozima, pp. 185-191. Tokyo: Japan Scientific Societies Press.

(28) Staudacher, T., and Allègre, C.J. 1982. Terrestrial xenology. Earth Planet. Sci. Lett. 60: 389-406.

(29) Turekian, K.K. 1959. The terrestrial economy of helium and argon. Geochim. Cosmochim. Acta 17: 37-43.

(30) Turekian, K.K. 1964. Degassing of argon and helium from the Earth. In The Origin and Evolution of Atmospheres and Oceans, eds. P.J. Brancazio and A.G.W. Cameron, pp. 74-82. New York: Wiley.

(31) Welhan, J.A. 1981. Carbon and hydrogen gases in hydrothermal systems: The search for a mantle source. Ph.D. Dissertation, University of California at San Diego.

(32) Wolery, T.J., and Sleep, N.H. 1976. Hydrothermal circulation and geochemical flux at mid-ocean ridges. J. Geol. 84: 249-275.

Patterns of Change in Earth Evolution, eds. H.D. Holland and A.F. Trendall, pp. 313-343.
Dahlem Konferenzen 1984. Berlin, Heidelberg, New York, Tokyo: Springer-Verlag.

Supracrustal Rocks, Polymetamorphism, and Evolution of the SW Greenland Archean Gneiss Complex

R.F. Dymek
Dept. of Geological Sciences, Harvard University
Cambridge, MA 02138, USA

Abstract. The gneiss complex of southern West Greenland contains the world's oldest well characterized rocks, the ~3800 Ma Isua Supracrustal Sequence of metavolcanic amphibolite, and chemical and clastic metasediment, for which there is no known basement. Stratigraphic relationships suggest a broad similarity to upper portions of some Archean greenstone belts, whereas geochemical characteristics of clastic sediments may record progressive unroofing of a volcanic source terrain. The Isua rocks underwent amphibolite-grade metamorphism (T ~ 550°C, P ~ 5 Kb) before ~3600 Ma, following emplacement of ~3700 Ma Amîtsoq orthogneiss.

The ~3000 Ma Malene Supracrustal suite consists of metavolcanic amphibolite, minor carbonate and calc-silicate, cordierite-orthoamphibole gneiss, and clastic metasediment possibly deposited in a continental margin environment on a basement that may have included mafic oceanic crust and older Amîtsoq gneiss. Amphibolite- to granulite-grade metamorphism of the Malenes preserves a kyanite→sillimanite transition, with T ~550-850°C and P ~5-9 Kb, and occurred during major orogenic activity at ~2800 Ma, accompanied by formation of Nûk orthogneiss. Retrogression to greenschist/amphibolite grade at ~2650 Ma appears related to shear zone formation; kyanite formed during this event, suggesting no substantial decrease in P.

The following general conclusions can be applied: mineral assemblages and inferred P-T conditions indicate a "normal" Barrovian-type environment for Isua and Malene main-stage metamorphism; crustal thicknesses >15 km before 3600 Ma and >25 km before 2800 Ma are indicated; P-T relationships are compatible with a metamorphic thermal gradient of ~30-40°C/km, but the relationship between these values and an equilibrium continental geotherm are unclear; widespread retrograde metamorphism may follow a transition from regional flow folding to deformation in discrete zones that provided pathways for fluid migration during final stages of continent stabilization (Nûk gneiss).

Models for Precambrian geological history that invoke a thin, hot sialic crust are not supported by petrological studies on the SW Greenland gneiss complex.

INTRODUCTION

The thermal evolution of the Earth is a topic of widespread interest to geoscientists, as it brings into focus many of the uncertainties regarding the formation and subsequent development of the continents. Hypotheses on the secular nature of tectonic regimes, as well as the time scales and processes of crustal thickening, are linked in some way to the magnitude of terrestrial heat production and mechanisms for its dissipation.

The commonly held opinion that Archean crust was "hot and thin" represents a logical inference from substantially higher heat production early in Earth history (8, 26, 37, 44, 50, 58). The thesis developed here is that this conclusion seems at variance with the record of Archean metamorphism, extending back to the oldest rocks presently available for study.

The purpose of this essay, then, is to summarize information on the lithological characteristics and metamorphic history of supracrustal (= sedimentary and volcanic) rocks from the Archean gneiss complex of southern West Greenland. This region assumes a rather unique position geologically, as it represents the "type example" of Archean high-grade terrains, contains the oldest known rocks on Earth, as alluded to above, and preserves a record of crust-forming and crust-modifying events that spans a period of more than 1300 million years.

The data and the ideas discussed in this paper are based on several seasons of field work and on petrological study of a substantial number of samples. The presentation is largely descriptive and qualitative in nature, but these results are a necessary prelude to the development of meaningful quantitative geochemical and thermal models of the ancient Earth. Future work should focus on understanding better the relationship between observations on regional polymetamorphism and on tectonic events, to causative heat sources and to the underlying mechanical state of the Archean lithosphere.

The text begins with a brief summary of the Archean geology of West Greenland, in order to familiarize the reader with local terminology. This is followed by descriptions of early and later Archean supracrustal sequences and their metamorphism. The paper concludes with a discussion of the implications of these results for Archean crustal development.

GEOLOGICAL OVERVIEW

Extensive field mapping during the past approximately fifteen years by geologists of the Greenland Geological Survey (GGU) has shown that ~75% of the West Greenland Archean craton is composed of polyphase, complexly deformed granitoid gneisses which appear to be derived primarily from calc-alkaline igneous precursors (18, 19, 25, 27, 60, 61). Within the Godthåb District (see Fig. 1), McGregor (59) suggested that the presence or absence of a broken-up mafic dike swarm, the Ameralik Dikes, represented the most reliable field criterion for distinguishing between older and younger components of this gneiss complex. Subsequent Rb-Sr and U-Pb isotopic studies confirmed this field distinction (5, 7, 46, 66, 67), with stratigraphically older Amîtsoq gneiss having ages of formation in the 3600-3700 Ma range, and stratigraphically younger Nûk gneiss having ages of formation in the 2800-3000 Ma range. Known occurrences of Amîtsoq gneiss are confined to the general region of Fig. 1, whereas Nûk-type gneisses represent the most abundant rock group in the entire West Greenland Archean craton (19).

Scattered throughout these gneisses are extensive areas of metasedimentary and metavolcanic rocks, of which two age groups are also recognized. The older Isua Supracrustals crop out in an arcuate belt near the edge of the inland ice, ~120 km northeast of Godthåb. These rocks are intruded by both Ameralik dikes and locally by various phases of Amîtsoq gneiss, and have yielded isotopic ages in the 3700-3800 Ma range (6, 49, 64, 65), with the most precise determination being a U-Pb age of 3769 $\pm^{11}_{8}$ Ma on single zircons (63).

Rocks that are lithologically similar to the Isua Supracrustals occur as enclaves (up to kilometer size) in Amîtsoq gneiss throughout the region, where they are known informally as the Akilia Association (62). Although no isotopic ages have been reported for Akilia units, they are cut by Ameralik dikes and presumably are penecontemporaneous with the Isua sequence.

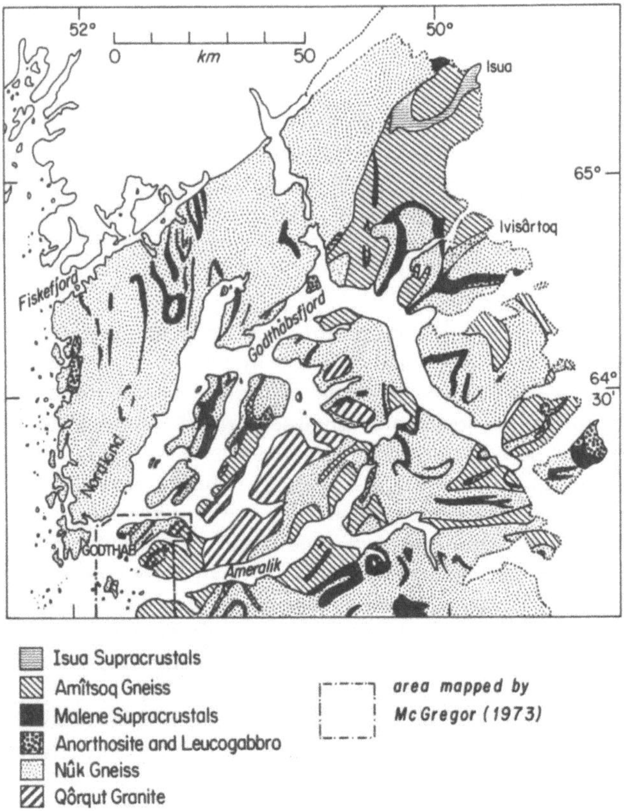

FIG. 1 - Generalized geological map of the Archean gneiss complex in the vicinity of Godthåb (adapted from McGregor (60)).

The younger Malene Supracrustal suite forms continous map units up to tens of kilometers long and several hundred meters thick. These units are intruded by several phases of Nûk gneiss but in most cases are in structural contact with Amîtsoq gneiss, although rare examples of possible depositional contacts have been reported (24, 35, 73). Attempts to determine the age of the Malene Supracrustals by isotopic methods have produced ambiguous results (e.g., (6)), although a recent $^{143}Nd/^{144}Nd$ model age of $\lesssim$3000 Ma has been determined for the source area of one Malene paragneiss unit (49). At present the only apparent constraints on the age of the Malene Supracrustals are that they are younger than

~3600 Ma (the minimum age of Amîtsoq gneiss) and older than ~2800 Ma (the minimum age of Nûk gneiss). Ameralik dikes are not observed in the Malene Supracrustals.

Bodies of layered calcic anorthosite/leucogabbro constitute a minor but widespread lithology throughout West Greenland. These are intrusive into certain Malene Supracrustal units and are themselves intruded by Nûk gneisses. Selected samples of anorthositic rocks have been dated at ~2800 Ma by the Pb-Pb method (45). The youngest significant map unit in the Godthåb District is the ~2550 Ma Qôrqut Granite Complex, which appears to have formed by extensive partial melting of older gneisses, based on its field relationships and isotopic characteristics (7, 22, 68).

The field and isotopic studies reviewed above can be used to construct an internally consistent model for the overall geological history of the Godthåb region. This model provides a framework within which a variety of detailed petrological and geochemical studies can be carried out. Undoubtedly, future work will refine details of the lithostratigraphic chronology.

EARLY ARCHEAN METAMORPHISM: ISUA SUPRACRUSTALS

Geological Setting

The geology of the Isua region has been reviewed by several previous investigators (1, 20, 70) who presented general descriptions of the stratigraphy and major rock types and established the presence of metavolcanic, and clastic and chemical metasedimentary units. More recent detailed mapping (33, 72) has largely confirmed the earlier results and established a coherent stratigraphy for the entire belt (74).

The principal rock types in the Isua Belt include: a) banded green to black metabasaltic amphibolite; b) massive to layered gray-green, Mg- and Al-rich leucoamphibolite, commonly with "garbenschiefer" texture; c) quartz-rich chemical metasediments ranging from banded magnetite ironstone to types with abundant Fe-silicates or carbonate (green quartzitic to calc-silicate gneisses containing Ba-Cr muscovite (13, 34) comprise a minor but widespread variant of this rock type); d) ferruginous "shale" comprising pelitic to semi-pelitic to mafic metasediment (garnet-biotite schist); e) volcaniclastic metagraywacke (muscovite-biotite gneiss); and f) "metaperidotite" (derived from serpentinite through prograde metamorphism), since retrograded to talc/chlorite schist or serpentinite.

Tectonic, Stratigraphic, and Metamorphic Development

Deformation of these units is complex. It involved early episodes of isoclinal folding (and thrusting), which produced an intense LS tectonite fabric, followed by later broad, open folding that was responsible for the arcuate form of the Isua Belt (1, 56, 74). Most supracrustal lithologies show evidence of polymetamorphism, which may correspond to tectonic and thermal events at ca. 3600-3700 Ma, 2800-3000 Ma, 2500-2600 Ma, and 1600-1800 Ma, based on regional geological relationships (cf. (19)). The main stage of metamorphism occurred at amphibolite grade conditions and is associated with early folding events (14, 17, 72). A subsequent greenschist/amphibolite event or events retrograded the supracrustal rocks (manifested principally by growth of chlorite and sericite) and recrystallized Ameralik dikes. Interestingly enough, the Isua Belt appears to have escaped nearly completely the intense later Archean tectonothermal events; a "tectonic front" occurs just a few kilometers south of the Isua Belt, beyond which the older gneisses become thoroughly reworked (72).

A consideration of the stratigraphic relationships at Isua, which involve the association of felsic volcaniclastic graywacke and ferruginous shale with abundant volcanogenic (?) chemical sediment (including ironstone) and mafic volcanic rocks, suggests a possible similarity to the upper portions of some Archean greenstone belts (15, 28). Geochemical variations within the stratigraphy are significant: each clastic metasedimentary unit has a distinct major, minor, and trace element composition; this distinction includes the rare earth elements as well (12, 15-17). The muscovite-biotite gneisses appear to be derived from different monolithologic felsic volcanic sources, whereas garnet-biotite schists were derived from a mixed source of felsic, mafic, and possibly even ultramafic rocks. Collectively, these observations suggest that the clastic sediments at Isua may record the progressive evolution and/or unroofing of a volcanic construct. The nature of the basement to the Isua sediments is unknown, and there is no compelling geological or geochemical evidence for the existence of older sialic crust at Isua.

Petrology

A detailed account of Isua metamorphism, including a discussion of the relationship between mineral assemblages and geological events has been published previously (14, 17), and only a brief summary is presented here. The assemblages staurolite-garnet-biotite and kyanite-biotite characterize Isua main-stage metamorphism (Fig. 2). The lack of more Mg-rich phase assemblages probably reflects a local absence of sediments

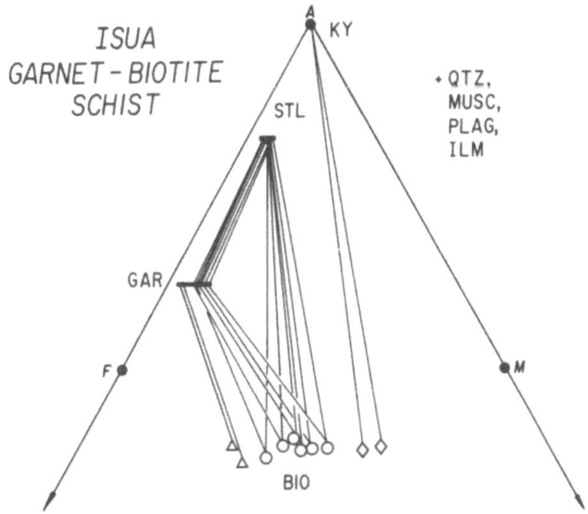

FIG. 2 - Diagnostic mineral assemblages in Isua clastic metasediments (updated from results presented in (14, 17)).

with appropriate bulk compositions. Variations of mineral compositions in the limiting Stl-Garn-Bio assemblages may be due to effects of the non-AFM components and/or partial retrograde equilibration; alternatively, this feature may reveal a small variation in metamorphic conditions along the belt.

Estimated P-T conditions for Isua main-stage metamorphism are illustrated in Fig. 3, in which temperatures (550 $\pm$ 50°C) were calculated from Fe-Mg partitioning between coexisting garnet-biotite pairs (39), and the pressure is constrained to lie within the field of kyanite. Comparison of these values to calculated petrogenetic grids for pelitic compositions (e.g., (81)) suggests that the average temperature may be slightly low (by about 50°C). This is not unreasonable, as most samples show some degree of secondary chloritization and growth of partial, slightly more Fe-rich euhedral rims on garnet; the events that produced these features probably disturbed the Mg-Fe partitioning systematics somewhat.

Detailed petrological studies on other Isua rock types are not yet complete, but preliminary results are generally consistent with those obtained on the garnet-biotite schists. For example, silicate-ironstones (grunerite + hornblende or actinolite ± garnet ± Fe-Ca-Mg carbonates ±magnetite + quartz), "garbenschiefer" amphibolites (anthopyllite ± hornblende ± cummingtonite + plagioclase + quartz), and meta-serpentinites (principally olivine + tremolite ± cummingtonite ± chlorite ± Mg-Ca carbonate) contain coexisting amphiboles with compositions compatible with staurolite-kyanite zone metamorphism (cf. (77)). However, a few Isua meta-serpentinites also contain rare examples of metamorphic orthopyroxene and green spinel (23). These samples yield olivine-spinel equilibration temperatures ranging up to 700°C; this result is problematic and points out the need for additional study.

EARLY ARCHEAN METAMORPHISM: AKILIA ASSOCIATION

The Akilia Association is best known in the vicinity of the small islands south of Godthåb, where it consists predominantly of basaltic (to komatiitic) amphibolite, with minor quartz-banded ironstone and garnet-biotite (-sillimanite) schist (62). These rocks have been extensively deformed and recrystallized during the late Archean (ca. 2800 Ma), and evidence for earlier periods of metamorphism has been nearly completely obliterated. However, in some Akilia amphibolites there are relict occurrences of coexisting garnet + clinopyroxene + orthopyroxene, an assemblage not observed in younger gneisses. Griffin et al. (48) have presented various lines of evidence establishing that this assemblage formed during an early Archean (ca. 3600 Ma) granulite-grade regional metamorphism, which also affected Amîtsoq gneisses and is correlated here with the amphibolite-grade event at Isua. Compositions of appropriate minerals yield T ~650°C and P ~7-8 Kb (48); the inferred range of P-T conditions for Akilia metamorphism is illustrated in Fig. 3. Because petrographic evidence for retrogression is so widespread, these P-T values are probably minimal.

LATE ARCHEAN METAMORPHISM: MALENE SUPRACRUSTALS

Geological Setting

The name Malene Supracrustals was introduced by McGregor (60) for a diverse suite of amphibolites and paragneisses in the vicinity of Godthåb (Fig. 1) that occupy an intermediate stratigraphic position with respect to Amîtsoq and Nûk gneiss. Subsequent mapping has shown that many of the "type units" can be followed well beyond the original map area, and the name Malene Supracrustals has since been applied to all rocks with lithological characteristics and structural setting similar to those described by McGregor (cf. (19)). It is not clear whether all units now called Malene Supracrustals are of the same age or were once part of

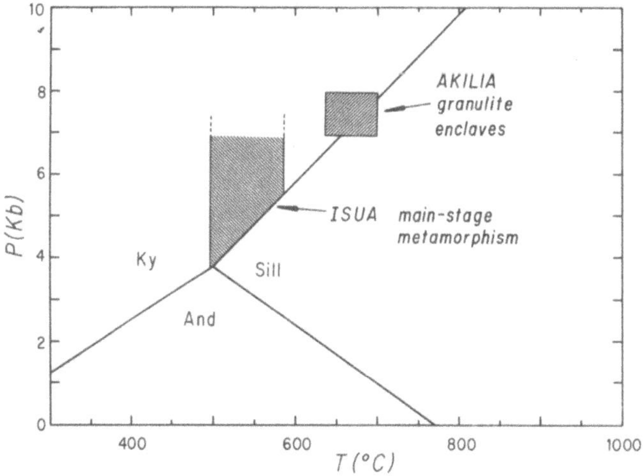

FIG. 3 - Estimated P-T conditions for early Archean metamorphism, West Greenland; Isua results from Boak and Dymek (13, 17); Akilia result modified from Griffin et al. (48), using revised garn-opx geobarometer of Newton and Perkins (71). Aluminosilicate phase boundaries are those of Holdaway (54).

an originally contiguous sequence. Five broad lithological associations can be distinguished on the basis of mineralogy, rock composition, and regional field relatationships:

a) The amphibolite association includes the most common Malene lithology, a banded gray to black amphibolite composed of hornblende + plagioclase + quartz + ilmenite; less abundant varieties contain biotite, garnet, cummingtonite, or anthophyllite in addition. Their higher-grade equivalents, termed pyribolite, contain hornblende + plagioclase + orthopyroxene ± garnet ± clinopyroxene ± quartz ± magnetite + ilmenite. Overall, this group of rocks appears to represent metamorphosed intermediate to mafic (to ultramafic?) flows and sills (cf. (41)), whereas finely laminated types were probably derived from a tuffaceous protolith.

b) The skarn amphibolite association includes rocks with abundant Ca-rich phases. Diopside amphibolites (diopside + hornblende + plagioclase + quartz ± calcite ± epidote) are striking green-black rocks layered on a centimeter to meter scale, which locally contain calc-silicate pods (grossular-diopside-epidote-scapolite-sphene-calcite-quartz); similar pods may also be found in a variety of other Malene units. Thinly layered ($\lesssim$10 cm) diopside-calcite gneiss is particularly abundant on the Lille Narssaq peninsula (see Fig. 4), and on the north coast of Qilangarssuit, a few outcrops of impure marble occur. This marble contains microcline + mica + graphite + zircon in addition to quartz + calcite (plus diopside-clinozoisite-tremolite), suggesting the presence of a minor detrital component. Diopside quartzites also form part of this association, some of which contain Ba-microcline, Ba-Cr muscovite, and Cr-zoisite (13).* Plausible protoliths for this entire group of rocks include "carbonated" mafic tuff, mixed chert-carbonate or pelite-carbonate.

c) The clastic association includes a variety of semipelitic to pelitic gneisses and quartzofeldspathic metasediments containing quartz + plagioclase ± biotite ± sillimanite ± garnet ± cordierite ± muscovite ± K-spar. These range from finely laminated to massive, and are typically brown- to red-weathering sulfide- and graphite-bearing paragneisses, derived from a protolith ranging from shale to arkosic sandstone to quartzite. Locally, rocks of this association contain vivid green muscovite ("fuchsite") that is Cr-bearing but lacks Ba, in contrast to the green micas mentioned above.

On a regional scale, there is a close spatial relationship between green mica-bearing quartzitic and quartzofeldspathic metasediments and Amîtsoq gneiss, which may represent an extensive basement-cover contact (24, 35, 73). However, Hamilton et al. (49) have recently reported a Sm-Nd model age of ~3000 Ma for Malene clastic metasediments from one locality. This result seems to rule out derivation of these sediments from "underlying" 3600-3700 Ma Amîtsoq gneiss and is consistent with a source similar in age to Nûk gneiss. If the Malene metasediments were derived from Nûk gneiss, then the accepted regional geological history may have to be revised. Alternatively, some areas presently mapped as Amîtsoq gneiss may in fact be Nûk gneiss. In either case, additional isotopic studies of the Malenes are sorely needed.

d) Cordierite-orthoamphibole gneiss is a collective term applied to a series of aluminous, ferromagnesian metasediments containing various

* This unusual occurrence of Ba and Cr enrichment is noted for both the Isua and Malene Supracrustal rocks. It is premetamorphic in origin and may be related to some type of hydrothermal alteration process.

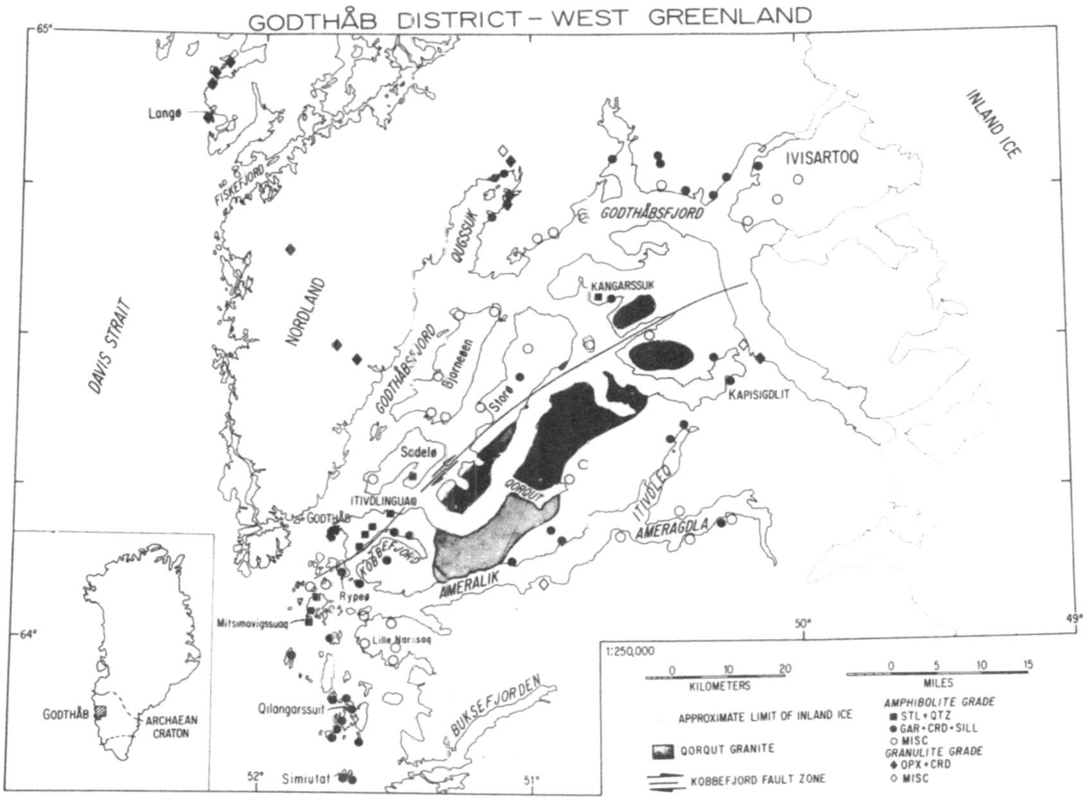

FIG. 4 - Outline map of the Godthåb region showing the distribution of Malene Supracrustal sample localities: symbols correspond to main-stage metamorphic mineral assemblages, as discussed in the text (see Fig. 5); "miscellaneous" refers to localities that lack a diagnostic AFM mineral facies type.

combinations of the minerals: quartz-plagioclase-biotite-talc-anthophyllite-gedrite-orthopyroxene-commungtonite-staurolite-cordierite-garnet-kyanite-sillimanite-corundum. These rocks range from very quartz-rich types (e.g., cordierite quartzite, sillimanite quartzite) to nearly quartz-free, garnet or gedrite gneisses.

The nature of the protolith for this association is unclear, but reworked altered mafic volcanic rocks are plausible candidates, as there is a clear, intimate association of cordierite-orthoamphibole gneisses with Malene amphibolites. A comprehensive regional chemical study of these rocks is not yet available, but preliminary chemical data as well as modal mineralogy indicate a pronounced relative depletion in CaO, Na_2O, and K_2O, and enrichment in MgO and Al_2O_3, compared to all common sediment types. These rocks are also enriched in several incompatible trace elements, as shown by their high content of zircon and niobian rutile (31).

Based on a study of a suite of samples from the islands northwest of the mouth of Buksefjorden (see Fig. 4), Beech and Chadwick (10) suggested that some of these gneisses may have evolved from a protolith that included Mg-rich pelagic clay (palygorskite) and a quartz-rich continental clastic component. Perhaps the palygorskite formed from the hydrothermal alteration or deep-sea weathering of basaltic rocks. Whatever the case, these Malene gneisses are widespread, and the process(es) that led to the development of such unusual sediment compositions operated on a regional scale.

e) Ultramafic rocks occur as widespread isolated "pods" (up to tens of meters long) in Malene amphibolite or adjacent granitoid gneiss. These typically form concentrically zoned bodies with mineralogically variable cores (olivine-orthopyroxene-tremolite/hornblende-diopside-anthophyllite-talc-spinel-magnetite) and rims (chlorite-biotite-actinolite, etc.). Hornblendite inclusions that abound in Amîtsoq and Nûk gneiss may be relics of ultramafic rock. Very little is known about these rocks, but their spatial association with amphibolite is noteworthy.

Metamorphic and Tectonic Development

The Malene Supracrustals preserve textural and mineral-chemical evidence for variable degrees of recrystallization and mineral growth during at least three episodes of metamorphism (29, 30). The earliest main-stage episode of regional metamorphism (M_1) ranged from middle amphibolite to hornblende granulite grade and occurred almost entirely within the sillimanite stability field. This prograde event can be assigned with confidence to the ~2800 Ma orogenic/magnetic activity that affected the entire West Greenland craton. M_1 is well documented as accompanying structural interleaving of Malene and Amîtsoq units concomitant with the syntectonic intrusion of vast quantities of calc-alkaline magma, which were the Nûk gneiss precursors (19, 24, 60, 61, 86). These intrusions were emplaced predominantly as diapiric sheets in a deformation event that culminated in the formation of large, nappe-like recumbent folds (24, 61). Chadwick and Nutman (24) have argued that this regional deformation involved a transition to transcurrent shearing in steep linear zones which deformed the recumbent folds into flattened dome and basin interference structures. These authors argue that certain mineral growth lineations are coaxial with these subconical structures, indicating that M_1 continued through the final stages of deformation. Observations made in the course of the present study indicate that M_1 tended to outlast the major phase of deformation. Petrological aspects of M_1 are considered in more detail below.

The second metamorphism (M_2) is a regional retrograde event that has been interpreted as resulting from renewed crustal heating and hydration rather than cooling from the M_1 thermal peak. M_2 is characterized by the development of distinct new mineral assemblages and occurred within the stability field of kyanite. Simplified examples of M_2 retrograde reactions are listed in Table 1. M_2 displays a wide range of development at any given outcrop down to the scale of a thin section; when it was first proposed as a regionally significant event (30), no source for the water of hydration or thermal input was known, nor was it possible to relate M_2 to a dated or datable event in the history of the region.

TABLE 1 - Malene Supracrustals - M_2 reactions (schematic).

"Pelite"	Garn (or Cord) + Ksp + H_2O — Kyanite + Bio
	Sill + Ksp + H_2O — Musc + Qtz
"Cord-Oamph gneiss"	Cord + Opx + H_2O — Anth (Al-rich) ± Kyanite
	Cord (or Garn) + Bio (Ti-rich) + H_2O — Stl + Chl + Bio (Ti-poor) ± Kyanite
Amphibolite and Pyribolite	Pyx + Plag + H_2O — Hbl (Na-, Al-, Ti-poor) + Qtz
	Anth + H_2O — Chl + Qtz
	Hbl + Plag + H_2O — Ep + Act + Qtz

Subsequent field work has revealed a possible correlation between M_2 retrogression and predominantly subvertical high-grade ductile shear zones of local and regional extent. Structurally, these high-grade zones are dominated by transcurrent movement and a significant flattening deformation that induces a very strong augen gneiss fabric on certain Malene units. Biotite-rich blastomylonites that may be the result of K-metasomatism appear to have developed at this time, particularly along some gneiss-supracrustal contacts. The most prominent of these high-grade zones has become known as the "Godthåb straight belt" (cf (88)), where there has been local emplacement of felsic ("Qarassaq") dikes. These dikes were rotated into the shear zone but clearly crosscut the older Nûk gneiss fabric; a U-Pb zircon age of ~2660 Ma has been obtained for one of these dikes (7).

The nature of M_2 may have a significance beyond that of being a mere episode of retrograde metamorphism. For example, Fyfe (42), among others, has observed that during prograde high-temperature metamorphism that involves the progressive expulsion of volatiles, the deformation processes will be dominated by diffusion-controlled mechanisms. The presence of intergranular fluid allows for "ease of flowage" of large, deforming crystalline masses. Once metamorphism has progressed to the point where most of the fluids have been driven off, further deformation tends to be of the dislocation type, which will become concentrated in discrete zones.

This idea may explain certain aspects of the transition from the final stages of emplacement and deformation of the Nûk gneisses to the development of high-grade shear zones, which involves, on a larger scale, stabilization of newly formed continental crust (i.e., the large volume of Nûk gneiss). As noted above, the prograde metamorphism associated with emplacement of the Nûk gneiss precursors (M_1) is thought to have outlasted the final major deformation of these gneisses, which was characterized by transcurrent movement in steep linear belts. It may be that as the crust "dried out" during the final stages of M_1 (while the stress regime was still oriented so as to be causing transcurrent movement), subsequent deformation became dominated by dislocation mechanisms in discrete zones whose location was grossly controlled by the pattern of waning diffusion-dominated deformation. Thereafter, as the crust adjusted isostatically and cooled off due to lack of an external thermal input, the deformation was dominated by ductile shearing, mylonitization, and frictional sliding in discrete zones, such that renewed crustal heating was the effect. Any fluids that were present, whether they were circulating meteoric waters or derived from deeper crustal (or mantle?) sources, would have been involved in retrograde metamorphism controlled by physical access through shear zones, and their movement would have been driven by shear heating. These hypotheses need to be evaluated in the course of future field and laboratory study, although the nature of retrograde metamorphism in West Greenland has many features in common with the development of Invernian shear zones in the Lewisian Complex of Scotland (9).

M_3, which is characterized by the local development of andalusite + muscovite + chlorite, involves K-metasomatism and extensive hydrothermal alteration and was previously interpreted as a possible contact effect related to emplacement of the Qôrqut granite. It now seems more likely that M_3 also was controlled by the formation of shear zones, but in this case the associated structures are mid-Proterozoic (~1800 Ma) relatively shallow-level ductile-to-brittle transcurrent fault zones, of which the Kobbefjord Fault Zone is the best studied example (80).

Petrology

Mineral assemblages that characterize Malene main-stage metamorphism (M_1) are summarized schematically in Fig. 5.

Sillimanite is ubiquitous, but kyanite has been observed at only three localities, where it occurs as relic inclusions in garnet or cordierite, associated commonly with staurolite and rarely with corundum. In cordierite orthoamphibole gneiss, metamorphism proceeded by the progressive elimination of staurolite (1 → 2 → 3) and replacement of Fe-Mg amphibole by orthopyroxene. Lower pressure assemblages containing cordierite + cummingtonite are not observed to occur, but higher pressure assemblages Sill(Ky) + Oamph (3a) and Sill + Opx (4a) are preserved, albeit rarely, as relics in some samples. In rocks with semi-pelitic and pelitic compositions, muscovite-bearing assemblages are replaced by ones containing Sill + Kf. The very high temperature assemblage Garn + Crd + Kf + Bio (or Sill) has not been found. The distribution of these facies types is summarized in Fig. 4, where it can be seen that a "low-grade" staurolite zone is centered about Godthåb, with progressively higher-grade zones occurring outwards.

Mineralogical data from three localities spanning a range of metamorphic grade are illustrated in Fig. 6. Note that garnet compositions become more magnesian from (A) to (C), which is most probably caused by a

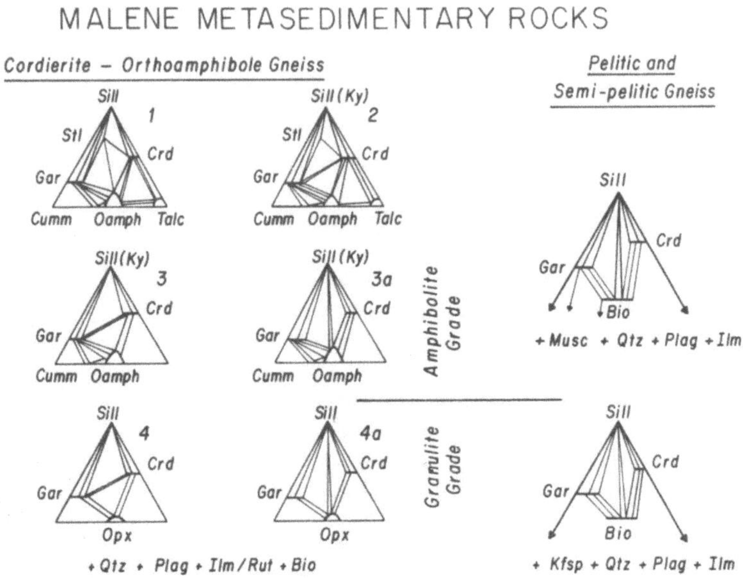

FIG. 5 - Metamorphic facies types observed in Malene metasedimentary rocks. Details of some assemblages are unclear, especially garn-cumm-oamph; the prograde orthoamphibole → orthopyroxene transition is not well understood.

pressure increase. In addition, note that in (C) the coexistence of garnet + cordierite is restricted to Kf-free assemblages, indicating that the reaction Bio + Sill = Garn + Crd + Kf has not taken place. The average pressures and temperatures calculated from these data, using a variety of geothermometers and geobarometers (39, 47, 52, 69, 71), are listed and shown graphically on a P-T grid in Fig. 7, where they can be compared to three experimentally determined reaction boundaries (2, 53, 76). These results complement those of previous studies (28, 30, 85) and indicate that prograde metamorphism of the Malene Supracrustals occurred over a temperature range of ∽550-850°C and a pressure range of ∽5-9 Kb.

SUMMARY

The results outlined in the preceding sections indicate that both the early Archean Isua and late Archean Malene main-stage metamorphic events are of the moderate-P (Barrovian) facies series type. This is the dominant type of regional metamorphism in orogenic belts of all

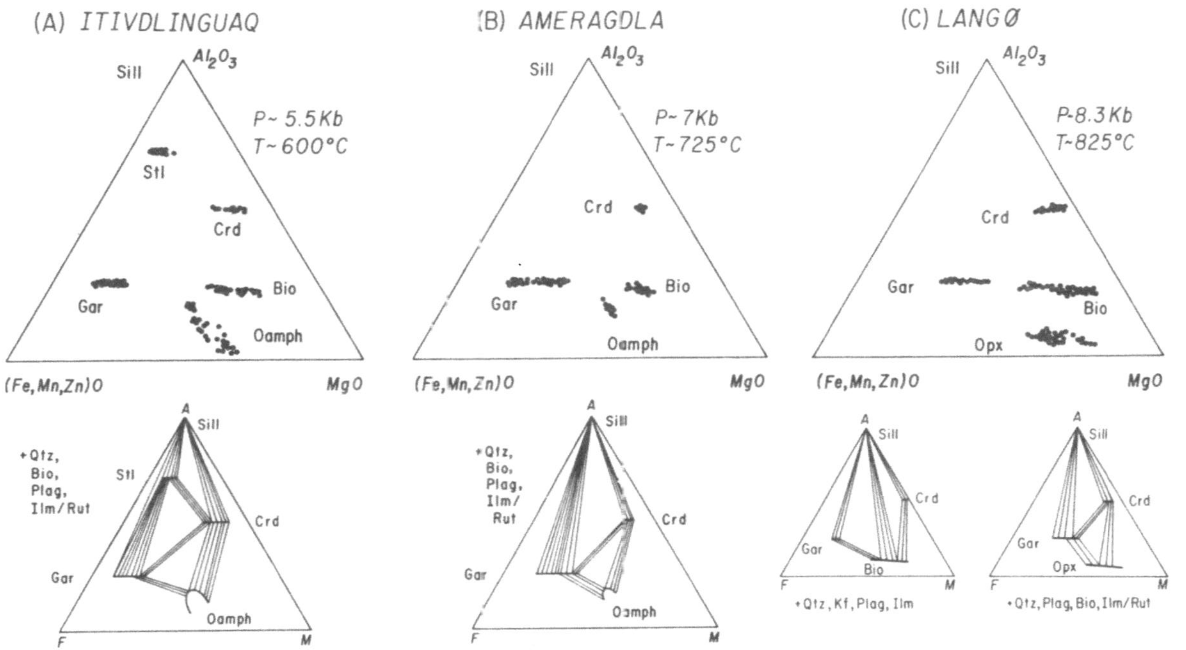

FIG. 6 - Mineral compositions based on selected microprobe analyses, and inferred phase relationships for three sets of Malene Supracrustal rocks corresponding to different grades of metamorphism; details of these results will be presented elsewhere.

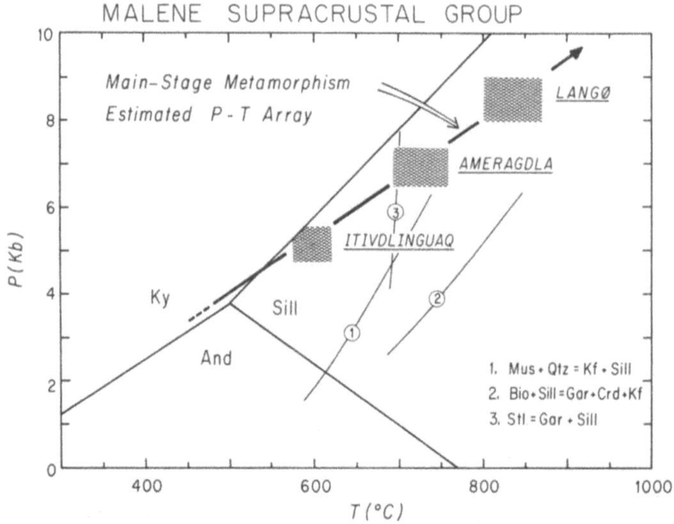

FIG. 7 - Estimated P-T conditions for Malene main-stage metamorphism. The aluminosilicate phase boundaries are those of Holdaway (54); reactions 1, 2, and 3 correspond to experimentally determined boundaries (2, 53, 76). Although the Malenes display evidence of prograde metamorphism, the arrow shown in the figure should not be construed as representing a single regional P-T trajectory. It is meant to imply a final P-T array established under prograde metamorphic conditions. In fact, many assemblages seem to have been established after a pressure decrease.

geological ages. Thus, there is nothing "unique" about Archean regional metamorphism as manifested in West Greenland, and, overall, it seems remarkably similar to that found in Phanerozoic systems such as the Appalachians and the Caledonides. The fact that only high temperature (>500°C) assemblages are found is a consequence of the present level of erosion rather than indicative of an abnormally high thermal gradient during metamorphism. Pressure estimates suggest that the depth of burial ranged from ~15 to ~30 km and indicate that the Archean crust of West Greenland was thick and probably of continental dimensions. Suggestions that the Archean crust was thick are not confined to West Greenland, and studies of metamorphic phase relations in supracrustal suites from other high-grade terrains lead to similar conclusions (e.g., (38, 51, 78)). In the opinion of the writer, a consensus will emerge as to the existence of thick Archean continental crust, and future studies will be carried out within such a conceptual framework.

DISCUSSION

The nature of Archean high-grade terrains, as exemplified by West Greenland, poses many intriguing problems for understanding continental evolution. Five examples are considered below; others, especially those concerned with the Archean sedimentary rock record, are best deferred for consideration elsewhere.

Crustal Thickening

If the rocks that are now exposed at the surface in high-grade terrains have been excavated from depths that were locally in excess of 25 km and which are presently underlain by normal crustal thicknesses of ca. 30 km, then it appears that a crustal thickness on the order of 60 km existed at some time in the past. Was such a thickened crust established in the Archean at the time of metamorphism, and by what mechanism were surficial rocks (i.e., the supracrustal suite) "buried" to such great depths?

The mechanism for crustal thickening inferred for the later Archean events in the Godthåbsfjord region of West Greenland (see Fig. 1) involved tectonic interleaving of Malene Supracrustal rocks with older granitoid Amîtsoq gneiss, accompanied by intrusion of Nûk gneiss (21, 61). In such a tectonic regime magmatic heat would represent an important thermal source for metamorphism, and it is relatively easy to envision how "stacking" of thrust slices could lead to deep burial of the supracrustal rocks, especially if the pile were simultaneously distended by Nûk gneiss material. A similar scenario is appealing for Isua times as well, in which magmatic emplacement of Amîtsoq gneiss provided a source of heat for metamorphism. However, preexisting sialic crust, which forms an integral part of the later Archean model, has yet to be recognized at Isua. Moreover, it is important to point out that a vast area of the West Greenland craton, extending well beyond the region where Amîtsoq gneiss is recognized (see Fig. 1), was affected by late Archean high-grade metamorphism involving inferred thickened crust. Hence, the presence of preexisting sialic material is not a requisite for crustal thickening or for the development of high-grade metamorphism.

The most probable sequence of events, then, appears to be initial magmatic crustal accretion accompanied by intense deformation involving the development of large-scale recumbent folds and nappes. High-grade terrains are viewed here as evolving in a regime of convergent tectonics. The belts of supracrustal rocks assume the position of highly deformed roof-pendants, septa, etc., that were engulfed by a composite plutonic suite.

The mechanism by which magmatic accretion occurred is not known; nor is it known whether it proceeded laterally or was synchronous throughout the region. It seems likely that the calk-alkaline orthogneisses were derived by partial melting of mafic rock; if so, then a large and unaccounted for quantity of this mafic material participated in the crustal accretion process. For want of a better explanation, one is almost compelled to appeal to island arc/subduction zone magmatism, although the kinematic and dynamic details of this process in the Archean need not have resembled those involved in the development of Mesozoic magmatic arcs.

Other possibilities should be discussed as well. For example, has Archean crust maintained a "constant" thickness over the course of geological time, in which loss of material by erosional uplift has been balanced by addition of material at the base of the crust? Although such concepts of magmatic underplating (e.g., (57)) may assist the process of crustal thickening per se, they cannot accommodate burial of the supracrustal rocks to the great depths inferred here. The same problem exists for a model in which deep crustal (high-grade) rocks are transported to the surface along listric thrust faults. The writer is unaware of any compelling geological evidence to indicate the presence of such faults, and the vast area of West Greenland affected by late Archean high-grade metamorphism would seem to argue against some type of cryptic tectonic control on the distribution of metamorphic facies types. Of course, future field work and/or geophysical surveys may force a reevaluation of this conclusion.

Another possibility involves a tectonic regime analogous to the situation in Tibet, where a doubling of crustal thickness has been achieved by continent-continent collision. Erosion of the upper plate exposes rocks of the lower one, which have been subjected to high-grade, deep-level metamorphism (70). Although this model provides an elegant explanation for the exposure of high-grade supracrustal rocks, it conveys a specific plate tectonic environment, which may be unrealistic when applied to the Archean, and also fares no better than a magmatic crustal thickening model in providing a solution for some alarming implications of the thermal regime indicated by metamorphic mineral parageneses, as discussed below.

Thermal Regimes

What is the relationship between metamorphic P-T arrays and ambient Archean thermal regimes? If the heat source for regional metamorphism was primarily magmatic intrusion, then we are looking at deep-level contact metamorphism developed on a vast scale. Theoretical considerations of some aspects of this process (87) indicate that thermal gradients provided by metamorphic mineral assemblages are higher than an "equilibrium geotherm." Values of ~30-40°C/km for the former are consistent with results summarized in this paper, suggesting that Archean continents were not very much hotter than in later times. However, are the present results for Greenland of general applicability to the Archean, and what are the consequences of extrapolating the estimated metamorphic P-T array downward into the inferred thick crust?

Examples of low-P "Abukuma-type" facies series regional metamorphism are widespread in the Archean, particularly in low-grade greenstone belts (e.g., (36)), so much so that several workers have concluded that low-P high-T metamorphism is symptomatic of the Archean (e.g., (43)). However, these generalizations have been made without any consideration for the pattern of regional metamorphism in high-grade terrains. Moreover, occurrences of moderate-P regional metamorphism - including the presence of kyanite - are known in Archean greenstone belts (40, 57, 75, 82, 83). Although some investigators (e.g., (4)) have tended to dismiss kyanite occurrences as due to local tectonic overpressures or unusual rock compositions, others (57, 82) have shown that kyanite-bearing assemblages comprise part of coherent regional metamorphic patterns. In the Pontiac Schist of the Superior Province, Canada, Jolly (57) showed that moderate-P facies series assemblages constituted part of an early metamorphic event, upon which low P facies series assemblages were superimposed. Laterally variable metamorphic facies series have been suggested by Percival (75) in the Slave Province, Canada, and by Saggerson and Turner (79) in the Rhodesian craton.

A comprehensive review of Archean metamorphism is beyond the scope of this paper, but it is clear that no metamorphic facies series is "diagnostic" of the Archean, and a diversity in patterns of metamorphism is recognized. What are lacking from the Archean metamorphic record are high-P, low-T lawsonite-glaucophane facies types; their absence may largely be a matter of poor preservation, as metamorphic rocks of the high-P facies series type are uncommon even in Paleozoic terrains.

An extrapolation of the Malene Supracrustal P-T array (Fig. 7) to deeper levels of the inferred thick crust suggests that temperatures would have exceeded 1000°C at depths greater than ~30 km. Most crustal rocks should be substantially molten under these conditions, and it is difficult to conceive of how such a thickened crust could have maintained its integrity for any length of time. Either the deeper portions of the thickened Archean crust were composed of extraordinarily refractory material (such as the residue of melting that produced the orthogneisses?), or the above extrapolation is seriously in error. In order to maintain a crustal thickness approaching 60 km, a thermal gradient substantially lower than that deduced from the metamorphic mineral parageneses is required. However, the same constraint applies to metamorphic P-T arrays in rocks of any age.

Heat Flux

If terrestrial heat production during the Archean was 3-4 times greater than the present value, why does there appear to be no evidence for extreme heat flux in the record of Archean metamorphism? Despite the problems discussed in the previous section, the results summarized here clearly are in conflict with values for Archean geotherms of 60-90°C/km that are favored by many investigators (26, 44, 50). A simple answer is that, like the present, processes other than conduction through continental crust represented the major mechanism for heat dissipation in the Archean (11). It seems essential that mantle convection was more vigorous and that the formation and destruction of oceanic (mafic) crust were more rapid than today; whether these processes were accompanied by plate tectonics in the modern sense remains unclear.

Ultramafic Rocks

Are some of the ultramafic rocks found in high-grade terrains remnants of Archean ophiolite suites, i.e., fragments of ancient upper mantle, or do they represent cumulates from magmas that formed the metabasaltic amphibolites with which they are associated in the field? The mineralogy of these bodies indicates a range in composition from dunite to harzburgite to spinel lherzolite; additional chemical study is needed to determine whether they are distinct from komatiites.

One strongly serpentinized dunite from Isua contains coexisting, intergrown T-chondrodite and Ti-clinohumite (32). The only other report of the coexistence of these phases is from a Tertiary kimberlite (3), whereas Ti-clinohumite is widespread in Alpine peridotites (84). The origin and implications of the Isua occurrence are unclear but may be profound.

Shear Zones

The formation of shear zones represents an important ingredient in the tectonic development of West Greenland. These were sites of fluid flux, metasomatism, retrograde metamorphism, and local melting. Such features may well exert an important control on the transfer of volatiles between various crustal (and possibly mantle) reservoirs. The isotopic composition of hydrous phases in the shear zones should be studied thoroughly so that the source(s) of fluids can be evaluated.

Acknowledgements. Portions of this research were supported by the National Science Foundation, the National Geographic Society, Harvard University, and the Greenland Geological Survey. Discussions with V.R. McGregor over a number of years have greatly enhanced my understanding of West Greenland geology. I would like to thank G.M. Smith for stimulating an interest in shear zones and their consequences, and especially J.L. Boak, who contributed in many ways to the results presented here. I would also like to thank R.D. Gee, H.D. Holland, S. Moorbath, R.C. Newton, and A.B. Thompson for reading and commenting on an earlier version of this manuscript.

REFERENCES

(1) Allaart, J. 1976. The pre-3760 m.y. old supracrustal rocks of the Isua area, central West Greenland, and the associated occurrence of quartz-banded ironstone. In The Early History of the Earth, ed. B.F. Windley, pp. 177-189. New York: Wiley.

(2) Althaus, E.; Karotke, E.; Nitsch, K.-H.; and Winkler, H.G.F. 1970. An experimental re-examination of the upper stability limit of muscovite + qtz. Neues. Jahrb. Min. Monatsch. 7: 325-336.

(3) Aoki, K.; Fujino, K.; and Akaogi, M. 1976. Titanochondrodite and titanoclinohumite derived from the upper mantle in the Buell Park kimberlite, Arizona, USA. Contr. Min. Petrol. 56: 243-253.

(4) Ayres, L.D. 1978. Metamorphism and the Superior Province of northwestern Ontario and its relationship to crustal development. In Metamorphism in the Canadian Shield, Paper 78-10, pp. 25-36. Ottawa: The Geological Survey of Canada.

(5) Baadsgaard, H. 1973. U-Th-Pb dates on zircons from the Early Pre-Precambrian Amîtsoq gneisses, Godthåb District, West Greenland. Earth Planet. Sci. Lett. 19: 22-28.

(6) Baadsgaard, H. 1976. Further U-Pb dates on zircons from the early Precambrian rocks of the Godthåbsfjord area, West Greenland. Earth Planet. Sci. Lett. 33: 261-267.

(7) Baadsgaard, H., and McGregor, V.R. 1981. The U-Th-Pb systematics of zircons from the type Nûk gneisses, Godthåbsfjord, West Greenland. Geochim. Cosmochim. Acta 45: 1099-1109.

(8) Baer, A.J. 1977. Speculations on the evolution of the lithosphere. Precambrian Res. 5: 549-560.

(9) Beach, A. 1976. The interrelationships of fluid transport, deformation, geochemistry and heat flow in early Proterozoic shear zones in the Lewisian complex. Phil. Trans. Roy. Soc. Lond. A280: 569-604.

(10) Beech, E.M., and Chadwick, B. 1980. The Malene supracrustal gneisses of northwest Buksefjorden: their origin and significance in the Archaean crustal evolution of southern West Greenland. Precambrian Res. 11: 329-356.

(11) Bickle, M.J. 1978. Heat loss from Earth: a constraint on Archaean tectonics from the relation between geothermal gradients and the rate of plate production. Earth Planet. Sci. Lett. 40: 301-315.

(12) Boak, J.L. 1982. Petrology and geochemistry of clastic metasedimentary rocks of the Isua Supracrustal Belt, West Greenland. Ph.D. Dissertation, Harvard University, Cambridge, MA.

(13) Boak, J.L., and Dymek, R.F. 1982. A Cr-Ba mineral bouillabaisse from the ca. 3800 Ma Isua and 3000 Ma Malene supracrustal rocks, West Greenland. Geol. Soc. Am. Abstr. Progr. 14: 446.

(14) Boak, J.L., and Dymek, R.F. 1982. Metamorphism of the ca. 3800 supracrustal rocks at Isua, West Greenland: implications for early Archaean crustal evolution. Earth Planet. Sci. Lett. 59: 155-176.

(15) Boak, J.L.; Dymek, R.F.; and Gromet, L.P. 1981. REE in early Archaean metasedimentary rocks from Isua, West Greenland: constraints on source terrains for the earth's oldest rocks. Geol. Soc. Am. Abstr. Progr. 11: 411-412.

(16) Boak, J.L.; Dymek, R.F.; and Gromet, L.P. 1982. Early crustal evolution: constraints from variable REE patterns in metasedimentary rocks from the 3800 Ma Isua supracrustal belt, West Greenland. In Lunar and Planetary Science XIII, pp. 51-52. Houston: The Lunar and Planetary Science Institute.

(17) Boak, J.L.; Dymek, R.F.; and Gromet, L.P. 1983. Petrology and rare earth element geochemistry of clastic metasedimentary rocks from the Isua supracrustal belt, West Greenland. Rapp. Grønlands geol. Unders. 112: 23-33.

(18) Bridgwater, D.; Collerson, K.D.; and Myers, J.S. 1978. The development of the Archaean gneiss complex of the North Atlantic region. In Evolution of the Earth's Crust, ed. D.H. Tarling, pp. 19-69. New York, London: Academic Press.

(19) Bridgwater, D.; Keto, L.; McGregor, V.R.; and Myers, J.S. 1976. Archaean gneiss complex in Greenland. In Geology of Greenland, eds. A. Escher and W.S. Watt, pp. 20-75. Copenhagen: The Geological Survey of Greenland.

(20) Bridgwater, D., and McGregor, V.R. 1974. Field work on the very early Precambrian rocks of the Isua area, southern West Greenland. Rapp. Grønlands geol. Under. 65: 49-54.

(21) Bridgwater, D.; McGregor, V.R.; and Myers, J.S. 1974. A horizontal tectonic regime in the Archaean of Greenland and its implications for early crustal thickening. Precambrian Res. 1: 179-197.

(22) Brown, M.; Burwell, A.D.; Friend, C.R.L.; and McGregor, V.R. 1981. The late Archaean Qorqut granite complex of southern West Greenland. J. Geophys. Res. 86: 10617-10632.

(23) Brothers, S., and Dymek, R.F. 1983. Ultramafic metamorphic rocks from the 3800 Ma Isua Supracrustal Belt, West Greenland. Geol. Soc. Am. Abstr. Progr. 15: 534.

(24) Chadwick, B., and Nutman, A. 1979. Archaean structural evolution in the northwest part of the Buksefjorden region, southern West Greenland. Precambrian Res. 9: 199-226.

(25) Coe, K. 1980. Nûk gneisses of the Buksefjorden region, southern West Greenland, and their enclaves. Precambrian Res. 11: 357-371.

(26) Collerson, K.D., and Fryer, B.J. 1978. The role of fluids in the formation and subsequent development of early continental crust. Cont. Min. Petrol. 67: 151-167.

(27) Compton, P. 1978. Rare earth evidence for the origin of the Nûk gneisses, Buksefjorden region, southern West Greenland. Contr. Min. Petrol. 66: 283-293.

(28) Dimroth, E. 1982. The oldest rocks on Earth: stratigraphy and sedimentology of the 3.8 billion year old Isua supracrustal sequence. In Sedimentary Geology of Highly Metamorphosed Precambrian Complexes, pp. 16-27. Moscow: Nauka Publishing House.

(29) Dymek, R.F. 1977. Mineralogic and petrologic studies of Archaean metamorphic rocks from West Greenland. Ph.D. Dissertation, California Institute of Technology, Pasadena, CA.

(30) Dymek, R.F. 1978. Metamorphism of Archaean Malene Supracrustals, Godthåb District, West Greenland. In Proceedings of the 1978 Archaean Geochemistry Field Conference, eds. I.E.M. Smith and J.G. Williams, pp. 339-342. Toronto: University of Toronto Press.

(31) Dymek, R.F. 1983. Fe-Ti oxides in the Malene Supracrustals and the occurrence of Nb-rich rutile. Rapp. Grønlands geol. Unders. 112: 83-94.

(32) Dymek, R.F.; Boak, J.L.; and Brothers, S. 1983. Tichondrodite (Ti-Ch) and Ti-clinohumite (Ti-Cl) in ca. 3800 Ma metadunite from Isua, West Greenland. Trans. Am. Geophys. Union (EOS) 64: 327.

(33) Dymek, R.F.; Boak, J.L.; Graubard, C.M.; and Smith, G.M. 1981. Field studies of metamorphosed Archaean supracrustal rocks, Godthåb-Isua area, West Greenland. In Reports of Research Sponsored by the National Geographic Society, Washington, D.C.: The National Geographic Society.

(34) Dymek, R.F.; Boak, J.L.; and Kerr, M.T. 1983. Green micas in the Archaean Isua and Malene supracrustal rocks, southern West Greenland, and the occurrence of barian-chromian muscovite. Rapp. Grønlands geol. Unders. 112: 71-82.

(35) Dymek, R.F.; Weed, R.; and Gromet, L.P. 1983. The Malene metasedimentary rocks on Rypeø, and their relationship to Amîtsoq gneisses. Rapp. Grønlands geol. Unders. 112: 53-69.

(36) Ermanovics, I.F., and Froese, E. 1978. Metamorphism of the Superior Province in Manitoba. In Metamorphism in the Canadian Shield, Paper 78-10, pp. 17-24. Ottawa: The Geological Survey of Canada.

(37) Ernst, W.G. 1981. The plate tectonic evolution of the Earth - A speculative account. Trans. Am. Geophys. Union (EOS) 62: 418.

(38) Erslev, E.A. 1981. Petrology and Structure of the Precambrian metamorphic rocks of the southern Madison Range, southwestern Montana. Ph.D. Dissertation, Harvard University, Cambridge, MA.

(39) Ferry, J.M., and Spear, F.S. 1978. Experimental calibration of the partitioning of Fe and Mg between biotite and garnet. Contr. Min. Petrol. 66: 113-117.

(40) Franklin, J.M.; Kasarda, J.; and Paulsen, K.H. 1978. Petrology and chemistry of the alteration zone of the Mattabi massive sulfide deposit. Econ. Geol. 70: 63-79.

(41) Friend, C.R.L.; Hall, R.P.; and Hughes, D.J. 1981. The geochemistry of the mid-Archaean (Malene) metavolcanics from Ivisârtoq and Ravns Storø, southern West Greenland. Spec. Publ. Geol. Soc. Aust. 7: 301-312.

(42) Fyfe, W.S. 1976. Chemical aspects of rock deformation. Phil. Trans. Roy. Soc. Lond. A283: 221-228.

(43) Fyfe, W.S. 1978. Crustal evolution and metamorphic petrology. In Metamorphism in the Canadian Shield, Paper 78-10, pp. 1-3. Ottawa: The Geological Survey of Canada.

(44) Fyfe, W.S. 1978. The evolution of the Earth's crust: modern plate tectonics to ancient hot spot tectonics. Chem. Geol. 23: 89-114.

(45) Gancarz, A.J. 1976. U-Th-Pb Studies of Archaean rocks from the Godthåb District, West Greenland. Ph.D. Dissertation, California Institute of Technology, Pasadena, CA.

(46) Gancarz, A.J., and Wasserburg, G.J. 1977. Initial Pb of the Amîtsoq gneiss, West Greenland, and implications for the age of the Earth. Geochim. Cosmochim. Acta 41: 1283-1302.

(47) Ghent, E.D.; Robbins, D.B.; and Stout, M.Z. 1979. Geothermometry, geobarometry and fluid compositions of metamorphosed calc-silicates and pelites, Mica Creek, British Columbia. Am. Min. 64: 874-885.

(48) Griffin, W.L.; McGregor, V.R.; Nutman, A.; Taylor, P.N.; and Bridgwater, D. 1980. Early Archaean granulite facies metamorphism south of Ameralik, West Greenland. Earth Planet. Sci. Lett. 50: 59-74.

(49) Hamilton, P.J.; O'Nions, R.K.; Bridgwater, D.; and Nutman, A. 1983. Sm-Nd studies of Archaean metasediments and metavolcanics from West Greenland and their implications for the Earth's early history. Earth Planet. Sci. Lett. 62: 263-272.

(50) Hargraves, R.B. 1976. Precambrian geologic history. Science 193: 363-371.

(51) Henry, D.J.; Mueller, P.A.; Wooden, J.L.; Warner, J.L.; and Lee-Berman, R. 1982. Granulite grade supracrustal rocks of the Quad Creek area, eastern Beartooth Mountains, Montana. In Precambrian Geology of the Beartooth Mountains, Montana and Wyoming, eds. P.A. Mueller and J.L. Wooden, Montana Bureau of Mines and Geology Spec. Publ. 84, pp. 147-156. Butte: Montana College of Mineral Science and Technology.

(52) Hensen, B.J., and Green, D.J. 1973. Experimental study of cordierite and garnet in pelitic compositions at high pressures and temperatures. III. Synthesis of experimental data and geological applications. Contr. Min. Petrol. 38: 151-166.

(53) Hoffer, E. 1976. The reaction sillimanite + biotite + quartz = cordierite + K-feldspar + H_2O and partial melting in the system K_2O-FeO-MgO-Al_2O_3-SiO_2-H_2O. Contr. Min. Petrol. 55: 127-130.

(54) Holdaway, M.J. 1971. Stability of andalusite and the aluminium silicate phase diagram. Am. J. Sci. 271: 97-131.

(55) Holland, J.G., and Lambert, R.St.J. 1975. The chemistry and origin of the Lewisian gneisses of the Scottish mainland: the Scourie and Inver assemblages and sub-crustal accretion. Precambrian Res. 2: 161-188.

(56) James, P.R. 1976. Deformation of the Isua block, West Greenland: a remnant of the earliest stable continental crust. Can. J. Earth Sci. 13: 816-823.

(57) Jolly, W.T. 1978. Metamorphic history of the Archaean Abitibi Belt. In Metamorphism in the Canadian Shield, Paper 78-10, pp. 63-78. Ottawa: The Geological Survey of Canada.

(58) Katz, M.B. 1976. Early Precambrian granulites-greenstones: transform mobile belts and ridge-rifts on early crust? In Early History of the Earth, ed. B.F. Windley, pp. 147-155. New York: Wiley.

(59) McGregor, V.R. 1968. Field evidence of very old Precambrian rocks in the Godthåb area, West Greenland. Rapp. Grønlands geol. Unders. 15: 31-35.

(60) McGregor, V.R. 1973. The early Precambrian gneisses of the Godthåb District, West Greenland. Phil. Trans. Roy. Soc. Lond A273: 343-358.

(61) McGregor, V.R. 1979. Archaean gray gneisses and the origin of continental crust: evidence from the Godthåb region, West Greenland. In Trondhjemites, Dacites and Related Rocks, ed. F. Barker, pp. 189-204. New York: Elsevier.

(62) McGregor, V.R., and Mason, B. 1977. Petrogenesis and geochemistry of metabasaltic and metasedimentary enclaves in the Amîtsoq gneisses, West Greenland. Am. Min. 62: 887-904.

(63) Michard-Vitrac, A.; Lancelot, J.; Allegre, C.J.; and Moorbath, S. 1977. U-Pb ages on single zircons from the early Precambrian rocks of West Greenland and the Minnesota River Valley. Earth. Planet. Sci. Lett. 35: 449-553.

(64) Moorbath, S.; O'Nions, R.K.; and Pankhurst, R.J. 1973. Early Archaean age for the Isua iron-formation, West Greenland. Nature 245: 138-139.

(65) Moorbath, S.; O'Nions, R.K.; and Pankhurst, R.J. 1975. The evolution of early Precambrian crustal rocks at Isua, West Greenland -geochemical and isotopic evidence. Earth Planet. Sci. Lett. 27: 229-239.

(66) Moorbath, S.; O'Nions, R.K.; Pankhurst, R.J.; Gale, N.H.; and McGregor, V.R. 1972. Further Rb-Sr age determinations on the very early Precambrian rocks of the Godthåb District, West Greenland. Nature Phy. Sci. 240: 78-92.

(67) Moorbath, S., and Pankhurst, R.J. 1976. Further Rb-Sr age and isotopic evidence for the nature of late Archaean plutonic events in West Greenland. Nature 262: 124-126.

(68) Moorbath, S.; Taylor, P.N.; and Goodwin, R. 1981. Origin of granitic magma by crustal mobilization: Rb-Sr and Pb/Pb geochronology and geochemistry of the late Archean Qôrqut Granite Complex of southern West Greenland. Geochim. Cosmoshim. Acta 45: 1051-1060.

(69) Newton, R.C. 1972. An experimental determination of the high pressure stability limits of magnesian cordierite under wet and dry conditions. J. Geol. 80: 398-420.

(70) Newton, R.C., and Perkins, D.T. 1981. Ancient granulite terrains - "eight Kbar metamorphism." Trans. Am. Geophys. Union (EOS) 62: 420.

(71) Newton, R.N., and Perkins, D.T. 1982. Thermodynamic calibration of geobarometers based on the assemblage garnet-plagioclase-orthopyroxene (clinopyroxene)-quartz. Am. Min. 67: 203-222.

(72) Nutman, A. 1982. Further work on the early Archaean rocks of the Isukasia area, southern West Greenland. Rapp. Grønlands geol. Unders. 110: 49-54.

(73) Nutman, A., and Bridgwater, D. 1983. Deposition of Malene supracrustal rocks on the Amîtsoq gneiss basement in outer Ameralik, southern West Greenland. Rapp. Grønlands geol. Unders. 112: 43-51.

(74) Nutman, A.; Bridgwater, D.; Dimroth, E.; Gill, R.C.O.; and Rosing, M. 1983. Early (3800 Ma) Archaean rocks of the Isua supracrustal belt and adjacent gneisses. Rapp. Grønlands geol. Unders. 112: 5-22.

(75) Percival, J.A. 1979. Kyanite-bearing rocks from the Hackett River Area, N.W.T.: implications for Archaean geothermal gradients. Contr. Min. Petrol. 69: 177-184.

(76) Richardson, S.W. 1968. Staurolite stability in a part of the system Fe-Al-Si-O-H. J. Petrol. 9: 468-488.

(77) Robinson, P.R.; Spear, F.S.; Schumacher, J.S.; Laird, J.; Klein, C.; Evans, B.W.; and Doolan, B.L. 1982. Phase relations of metamorphic amphiboles: natural occurrence and theory. In Reviews in Mineralogy. Amphiboles: Petrology and Experimental Phase Relations, eds. D.R. Veblen and P.H. Ribbe, vol. 9B, pp. 1-228. Washington, D.C.: Mineralogical Society of America.

(78) Rollinson, H.R.; Windley, B.F.; and Ramakrishnan, M. 1981. Contrasting high and intermediate pressure metamorphism in the Archaean Sargur Schists of Southern India. Contr. Min. Petrol. 76: 420-429.

(79) Saggerson, E.P., and Turner, L.M. 1976. A review of the distribution of metamorphism in the ancient Rhodesian Craton. Precambrian Res. 3: 1-54.

(80) Smith, G.M., and Dymek, R.F. 1983. A description and interpretation of the Proterozoic Kobbefjord Fault Zone, Godthåb District, West Greenland. Rapp. Grønlands geol. Unders. 112: 113-127.

(81) Thompson, A.B. 1976. Mineral reactions in pelitic schists. II. Calculation of some P-T-X (Fe-Mg) phase relations. Am. J. Sci. 276: 425-454.

(82) Thompson, P.N. 1978. Archean regional metamorphism in the Slave Structural Province - a new perspective on some old rocks. In Metamorphism in the Canadian Shield, Paper 78-10, pp. 85-102. Ottawa: The Geological Survey of Canada.

(83) Thurston, P.C., and Breaks, F.W. 1978. Metamorphic and tectonic evolution of the Uchi-English River subprovince. In Metamorphism in the Canadian Shield, Paper 78-10, pp. 49-62. Ottawa: The Geological Survey of Canada.

(84) Trommsdorff, V., and Evans, B.W. 1980. Titanian hydroxyl-clinohumite: formation and breakdown in antigorite rocks (Malenco, Italy). Contr. Min. Petrol. 72: 229-242.

(85) Wells, P.R.A. 1976. Late Archaean metamorphism in the Buksefjorden region, southwest Greenland. Contr. Min. Petrol. 56: 229-242.

(86) Wells, P.R.A. 1979. Chemical and thermal evolution of Archaean sialic crust, southern West Greenland. J. Petrol. 20: 187-226.

(87) Wells, P.R.A. 1980. Thermal models for the magmatic accretion and subsequent metamorphism of continental crust. Earth Planet. Sci. Lett. 46: 253-265.

(88) Windley, B.F. 1969. Evolution of the early Precambrian basement complex of southern West Greenland. Geol. Assn. Can. Spec. Paper 5: 155-161.

Patterns of Change in Earth Evolution, eds. H.D. Holland and A.F. Trendall, pp. 345-355.
Dahlem Konferenzen 1984. Berlin, Heidelberg, New York, Tokyo: Springer-Verlag.

Geothermal Gradients Through Time

A.B. Thompson
E.T.H. Zurich
8092 Zurich, Switzerland

Abstract. Temperature-depth curves (geotherms) on a global scale are controlled by the overall cooling of the Earth as limited by its thermal boundary layers. On global length scales and 10^9 a. time scales, the geothermal gradient in the continental lithosphere is likely to have decreased substantially, reflecting the progressive decrease in heat production from the radiogenic decay of K, Th, and U. The suboceanic lithospheric dT/dP may not have changed much since the Archean. On regional length scales, a thickening or thinning of the continental crust will cause perturbations of the steady state geotherms which, if exhumed after 10^7 a., will be reflected in the grade of metamorphic rocks and the types of magmatic rocks at the surface. If these rocks are not transposed upwards to the surface within 10^8 a., the thermal perturbations will have relaxed completely.

INTRODUCTION

The Earth is losing internal heat by three principal mechanisms: a) conduction through the continents, b) conduction through the ocean floor, and c) the creation and ageing of oceanic lithosphere. Because none of the present ocean floor is much older than $2x10^8$ a., we must rely upon observations on the continents to define temporal variations in the Earth's heat loss. Remnants of older oceans are found on some continents, but the bulk of the ocean floor has been subducted back into the mantle. Not all portions of the present continental crust are products of the same cumulative tectonothermal processes. Even with deep drill cores, structural mapping in deformed orogens, and crustal seismological investigations, our three-dimensional understanding of the continental crust and lithosphere is very poor. However, our knowledge of present-day geothermal gradients, above active and passive regions in the lithosphere and underlying mantle, is considerably aided by geophysical measurements of the heat flux, seismological structure, and electrical conductivity. Most of our deductions concerning changes in geothermal gradients through time come from observation of the rocks now at the continental erosion surface and from theoretical models of the planet's evolution. Notwithstanding these limitations, many new concepts concerning the tectonic and thermal evolution of the outer part of the Earth have appeared in recent years.

PRESENT-DAY GEOTHERMAL GRADIENTS

Beneath much of present continental area, away from obvious regions of thermal perturbation (subduction zones, magmatic activity), the thermal structure of the lithosphere is likely to be near equilibrium (e.g., (1), p. 302); the conductive heat loss may therefore be calculated in terms of a "steady state" geothermal gradient using the one-dimensional heat conduction equation for one uniform layer:

$$T_z - T_u = qz/K - Az^2/2K. \quad (1)$$

This can be generalized for any number of successive layers ((10), p. 9), for different values of temperature at the upper surface of a layer (T_u), of thickness z, conductivity K, and internal heat generation A; where q is the heat flux through the upper surface, and the temperature at depth z is T_z.

Comparison of the measured global surface heat flux with the results of parameterized convection calculations raises the question: "How much of the present surface heat flux is escaping primordial heat (i.e., non-radiogenic ((5), (6), p. 40, and (12))?"

Present-day temperature-depth curves have been calculated by Clark and Ringwood ((3), p. 53) appropriate to "steady state" geotherms beneath the Precambrian shields, continental regions outside the shields and beneath older ocean floor. These geotherms have been modified by other workers to take into account more recent data on the structure of the crust and lithosphere (1, 4, 17, 19, 24). Present-day geotherms are extrapolated back through time on the basis of certain mineralogical characteristics of exposed, or drilled, metamorphic and magmatic rocks of known radiometric or stratigraphic age. Through laboratory calibration of mineral compositions in specific assemblages, the pressure (P)-temperature (T) conditions of equilibration are deduced from observations on natural metamorphic and magmatic rocks.

SOME PETROLOGICAL CONSTRAINTS ON PAST GEOTHERMS

Metamorphism of the Earth's crust implies the presence of abnormal thermal conditions and consequently reflects perturbations of steady state geotherms. Several groups of metamorphic assemblages are considered, by analogy with current tectonothermal processes, to be diagnostic of similar events in the past. It has been suggested (14) that exposed glaucophane schist belts (Gla-Jad; high P-low T metamorphism) may be the remnants of fossil subduction zones and that andalusite-sillimanite facies series (And-Sil; low P-high T metamorphism) may have formed at depth in volcanic arcs. The kyanite-sillimanite facies series (Kya-Sil; medium P-medium to high T metamorphism) is considered to be indicative of past continental collisions (14). However, Gla-Jad facies P-T conditions are also generated following collision of >30 km continents; glaucophane schist belts by themselves are therefore not necessarily diagnostic of fossil subduction zones. Furthermore, And-Sil facies P-T conditions are generated following intracontinental thinning (22); they are therefore not necessarily an indication of ancient volcanic arcs.

Many Archean and Proterozoic terrains exhibit metamorphism in the (Kya-Sil) facies series and not just in the (And-Sil) facies series; some workers (1, 7) have therefore emphasized that metamorphic conditions in the Archean and Proterozoic continents were comparable to those in the Phanerozoic. Detailed analyses of P-T conditions as deduced from mineral geobarometry and thermometry in exposed continental rocks with Proterozoic or Archean radiometric ages have shown that conditions during metamorphism were not significantly different from those in stable continental terrains during the Phanerozoic or those that are considered diagnostic of currently active tectonothermal regimes (1, 7, 25, 26). This similarity cannot be directly interpreted in terms of specific geothermal gradients because the abnormal thermal conditions during metamorphism do not apply to "steady state" geotherms.

It is still a matter of controversy whether the apparent confinement of blueschist-eclogite (Gla-Jad) terrains to the Phanerozoic, and perhaps to the youngest Proterozoic, is really indicative of great crustal thickening (>60 km) in the last 10^9 years. Older high P-low T rocks could simply have been removed by erosion, or are just not exposed in Archean terrains (1, 4, 7, 20), or became overheated and turned into high P amphibolites and granulites in the deep crust. Likewise, the great abundance of granulite facies rocks in Archean and Proterozoic rocks compared to Phanerozoic rocks, may be related more to the fact that ancient rocks could have undergone more than one episode of thickening, which would have increased their chance of being exposed at the present erosion surface (23).

Certain features of magmatic rocks are also used to deduce temperature-depth relations. The peridotitic-komatiites preserved in some Archean terrains are of particular importance, because they indicate that temperatures in excess of 1650°C were locally achieved at shallow depths in the Archean upper mantle (9); a temperature of around 1200°C is required for the generation of present-day basaltic magmas. Pyroxenitic and basaltic komatiites have been found in Phanerozoic suites, and these seem to indicate temperatures around 1400°C at shallow depths in the oceanic mantle. The volume of high temperature lavas is such that they probably are products of thermal pulses of shorter duration than the conditions favorable to the production of the much more voluminous basalts. The "granitic" rocks of the Archean and Proterozoic are tonalitic and granodioritic and as such could well be products of the remelting of amphibolites (second-hand mantle). Such melting could occur near 850°C at 10kbar by dehydration-melting of amphibole assemblages without the presence of excess-H_2O fluid.

With the few exceptions noted above plus the Massif anorthosite suite in the Proterozoic, there are no other rock types particularly diagnostic of any period in Earth history by which constraints may be placed on the evolution of geotherms. Most rock types lack distinctive petrographic characteristics; their age can therefore be recognized only by radiometric dating. Detailed studies of radiometric ages has led some authors to conclude (e.g., (15), see also Moorbath, this volume) that the continental crust and lithosphere evolved during periods of intense crustal accretion from the mantle and that these periods were separated by long periods of quiescence. This periodicity is sometimes thought to be related to the "roll-over" time scale of the upper mantle ((6), p. 82).

GLOBAL GEOTHERMAL GRADIENTS ON 10^9-YEAR TIME SCALES

Processes which change the thickness of oceanic or continental crust or lithosphere will cause perturbations of "steady state" geotherms. It is thus necessary to attempt to distinguish long-term, global, geothermal gradients that characterize the overall cooling of the planetary interior from geothermal gradients that reflect shorter-term regional thermal events consequent upon deformation of the crust and lithosphere. The time scale of such deformations is usually much shorter than the time scale of local thermal re-equilibration by conduction. If we can deduce the tectonic style during any particular period of Earth history from geological observations on the continents, then we may be able to calculate the underlying geothermal gradients, both globally and regionally.

Changing Tectonic Styles in Earth History

Some recent studies (2, 26) have suggested that processes akin to modern plate tectonics were also operative in the Archean. The dominant mode of Archean continental thickening could therefore also have been through plate collision. In this view tectonics unique to the Archean, such as downsagging basins, basalt underplating, subcrustal accretion, and skimming orogeny ((26), p. 108) were only regionally important. The thermal features of these local events will be discussed below; they are only "models" that are frequently presented without physical basis.

Archean (4.0 to 2.7 b.y. ago), Proterozoic (2.7 to 0.6 b.y. ago), and Phanerozoic (0.6 b.y. to present) periods of Earth history appear to be characterized by three distinct series of geotectonic conditions preserved in the continental crust ((21), p. 383). High-grade gneisses and greenstone-granite terrains extend in some regions from the Archean into the Proterozoic. These rock types appear to be horizontally and not vertically stratified and may reflect different cratonic regimes, even in the Archean (25). Some tectonic models (2) consider that by the latter part of the Archean tectonic regime (3.5 to 2.7-2.5 b.y. ago) most of the present continental mass had differentiated. At this time the protocontinental material covered about one third of the Earth's surface and was mainly accreted by ocean-continent collision and arc-amalgamation. Continent-continent collision was minor; the granitic and granulitic fractions were therefore barely differentiated.

The Proterozoic tectonic regime is represented by large areas of stable continental cratons and the apparent end of the gneiss and greenstone-granite terrains characteristic of the Archean tectonic regime. Some workers (2) distinguish the Archean tectonic regime by the absence of geological evidence for rigid plate behavior. This is considered characteristic of the collision and rupture plate tectonics of the

Proterozoic. Isotopic studies (13, 15) show clear evidence for major magmatic activity at about 2.7 to 2.5 b.y. ago. These events are not synchronous worldwide and represent the first of three Proterozoic worldwide thermal pulses (also at 1.9 to 1.6 b.y. and 1.2 to 0.9 b.y.). It remains a matter of some controversy whether these pulses represent major periods of vertical crustal accretion from the mantle (13, 15, 18). Thickening of the continental crust would have resulted in major vertical motion in response to isostatic adjustment following the density redistribution in thickened crust and materially depleted upper mantle. Further cooling of thickened continental lithosphere during the Proterozoic could have induced mineral phase changes (such as gabbro to eclogite) in the lower continental crust. This would have depressed the surface height of the overlying crustal column, changed erosional and sedimentation processes at the surface, and possibly instigated subduction.

Thickening of the continents during the Proterozoic would have concentrated convection in the suboceanic areas to achieve an efficient cooling mode. This would have maintained a thin oceanic lithosphere and made the oceanic plate creation-destruction mechanism less efficient ((1), p. 302). Thus Archean oceanic ridges either spread faster or the total length of oceanic ridges was longer than today (2). The concentration of convection beneath the oceans would have tended to cause continental fragments to accumulate ((21), p. 395). The resulting Proterozoic continent-continent collisions would have resulted in differentiation of the granite and granulitic fractions of the continental crust. As a consequence of additional tectonothermal evolution during the Proterozoic and into the Phanerozoic, the strength but not the thickness of the continental and oceanic lithospheres would have become more comparable (4, 20, 21). Seafloor spreading and the breaking of the macrocontinental blocks would have become favorable and would have produced the precursors to the present, 200 m.y.-old configuration of continents and oceans.

The behavior of the Earth during the first 0.6 b.y. (Hadean, 4.6 to 4.0 b.y. ago) is not yet decipherable from observations on the continents. Some protocrustal material must have segregated at the same time as the separation of core from mantle. However, vigorous mantle convection in the first 600 m.y. would have recycled most of this mantle differentiate. Substantial cooling during the first 600 m.y. would have permitted thickening and stabilization of the oceanic lithosphere and allowed further differentiation to form protocontinents.

The Evolution of Steady State Geothermal Gradients

Some workers (e.g., (21)) have proposed that there are distinct average subcontinental and suboceanic "steady state" conduction geotherms for the five major time periods in Earth history. The subcontinental geotherms in particular are considered by Tarling (21) to have become progressively less steep from the Hadean through to the Phanerozoic. However, other workers (e.g., (7)) consider that because Archean metamorphic mineral assemblages are indicative of geobarometric pressures of 8 to 10 kbar., Archean mountains had elevations comparable to those of Phanerozoic and recent mountain belts. Elevations are limited by the strongly temperature-dependent creep-strength of the lithosphere; hence similar elevations imply that Archean continental thermal regimes were similar to those of the present day (7). If the Archean subcontinental thermal regimes were indeed similar to today's, then the creation and subduction of Archean oceanic plates could have been much greater than predicted by most current thermal models. Alternatively, if the Archean heat flux was about the same as that at present, then the time required for mantle convection to respond to changes in heat supply was comparable to the time scale of the decay of the principal radiogenic elements in the system (7). The decay times for K, U, and Th are well established, but their distribution and quantity are poorly known. Although the estimates according to Lambert (11) are as well constrained as possible, they could be in error by a factor of two (7). As changes in the average geothermal gradient may be a rather abrupt phenomenon, it is probably appropriate to consider these spasmodic and dramatic changes in the global pattern together with the shorter-term and regional perturbations to steady state geothermal gradients.

REGIONAL GEOTHERMAL GRADIENTS ON 10^8-YEAR TIME SCALES

Points along geotherms may in principle be constrained by comparing the mineral assemblages in metamorphic and magmatic rocks with the pressure (P)-temperature (T) stability ranges determined experimentally. However, a rock in complete equilibrium will only reflect a single point on a time-dependent P-T curve. To understand further the relationship of such P-T points to prevailing "steady state" geotherms, we need to understand, first, how these values can be attained, and second, how they may be modified en route to the surface.

Crustal Thickening and Exhumation of Metamorphic Rocks

To return metamorphic rocks to the erosion surface requires the operation of tectonic mechanisms that are able to transport large continental blocks

(>100 km^3), or to produce repeated continental thickening such that erosion-isostasy can exhume them.

Tectonic processes resulting in the thickening or thinning of continental crust and the resulting perturbations to the steady state geotherms have been considered by several workers. As shown elsewhere (22, 23), large-scale thickening by thrusting or by vertical stretching would only rarely have resulted in high enough crustal temperatures at a depth of 20–30km to cause metamorphism in the Kya-Sil P-T range. Crustal thinning by intracontinental extension could lead to Kya-Sil P-T conditions, and subsequent magmatic accretion could result in And-Sil P-T conditions. Calculated P-T time paths for crustal thickening and thinning demonstrate that these processes can be accompanied by thermal changes of several hundred degrees in time periods on the order of 10^7 a. when hot and cold rock units are transposed or when isotherms are relatively extended or compressed. If the thermal relaxation lasts on the order of 10^8 a., thermal perturbations are eliminated. The length scale of regions affected by such thermal perturbations are obviously constrained by the mechanical properties of the crust and lithosphere, as well as by the relative rates of the moving plates. The thin viscous shell model for continental deformation (8) suggests that collisions may affect the continental lithosphere for distances up to 1000 km from the plate boundaries; this proposal is well supported by geological observations.

The time dependence of geotherms deduced from metamorphic and some intrusive magmatic rocks is of considerable importance. Cation, anion, and isotopic closures are strongly temperature-dependent. It is therefore quite likely that mineral assemblages in metamorphic rocks record conditions during uplift rather than at the maximum depth following a tectonic event (16, 23). Thus, deduced geotherms are strongly time-dependent on 10^6 to 10^7 a. time scales, even though the overall perturbation to the geotherm and the exhumation may exceed 10^8 a.

SOME OUTSTANDING PROBLEMS AND SUGGESTIONS

1. What are reasonable geological criteria by which we can decide if tectonothermal phenomena are regional or global?

2. What are the relative length scales and time scales of continental deformation compared to thermal readjustment?

3. Can we quantify the dimensions of exhumed high-grade metamorphic terrains? Can we determine if the underlying metamorphic material is usually higher- or lower-grade?

4. If global tectonothermal events are not synchronous, how can we quantify spasmodic and drastic steps in the evolution of the mantle?

5. If mountain belts are reactivated, how do we quantify the intensity and magnitude of the successive tectonothermal events?

6. Can long-term steady state geotherms be constrained by P-T equilibration estimates on metamorphic granulites, mantle xenoliths, and high temperature magmas? Such rocks may only reflect abnormally large thermal perturbations from the desired geotherm.

REFERENCES

(1) Bickle, M.J. 1978. Heat loss from the Earth: A constraint on Archaean tectonics from the relation between geothermal gradients and the rate of plate production. Earth Planet. Sci. Lett. 40: 301-315.

(2) Burke, K.C.A.; Dewey, J.F.; and Kidd, W.S.F. 1976. Dominance of horizontal movements, arc and microcontinental collision during the later permobile regime. In The Early History of the Earth, ed. B.F. Windley, pp. 113-129. New York: Wiley & Sons.

(3) Clark, S.P., and Ringwood, A.E. 1964. Density distribution and constitution of the mantle. Rev. Geophys. 2: 35-88.

(4) Davies, G.F. 1979. Thickness and thermal history of continental crust and root zones. Earth Planet. Sci. Lett. 44: 231-238.

(5) Davies, G.F. 1980. Thermal histories of convective Earth models and constraints on radiogenic heat production in the Earth. J. Geophs. Res. 85: 2517-2530.

(6) Elder, J. 1976. The Bowels of the Earth. Oxford: Oxford University Press.

(7) England, P.C., and Bickle, M. 1982. Constraints on Archaean geothermal regimes and their implication for Earth thermal history models. Proceedings of Planetary Volatiles Conference, October 1982, Colorado, Lunar Planetary Institute.

(8) England, P.C., and McKenzie, D.M. 1982. A thin viscous shell model for continental deformation. Geophys. J. Roy. Astro. Soc. 70: 295-321.

(9) Green, D.H. 1981. Petrogenesis of Archaean ultramafic magmas and implications for Archaean tectonics. In Precambrian Plate Tectonics, ed. A. Kroener, pp. 469-489. Amsterdam: Elsevier.

(10) Jaeger, J.C. 1965. Application of the theory of heat conduction to geothermal measurements. In Terrestrial Heat Flow, ed. W.H.K. Lee, pp. 7-23. Am. Geophys. U. Geophys. Monog. 8.

(11) Lambert, R.St.J. 1976. Archean thermal regimes, crustal and upper mantle temperatures, and a progressive evolutionary model for the Earth. In The Early History of the Earth, ed. B.F. Windley, pp. 363-373. New York: Wiley & Sons.

(12) McKenzie, D.P., and Weiss, N.O. 1975. Speculations on the thermal and tectonic history of the Earth. Geophys. J. Roy. Astro. Soc. 42: 131-174.

(13) McLennan, S.M., and Taylor, S.R. 1982. Geochemical constraints on the growth of the continental crust. J. Geol. 90: 347-362.

(14) Miyashiro, A. 1980. Metamorphism and plate convergence. In The Continental Crust and Its Mineral deposits, ed. D.W. Strangway, pp. 591-605. Geol. Asso. Can. Spec. Paper 20.

(15) Moorbath, S. 1978. Age and isotope evidence for the evolution of continental crust. Phil. Trans. Roy. Soc. Lond. A. 288: 401-413.

(16) Oxburgh, E.R., and England, P.C. 1980. Heat flow and the metamorphic evolution of the Eastern Alps. Eclogae geol. Helv. 73: 379-398.

(17) Ringwood, A.E. 1975. Composition and Petrology of the Earth's Mantle. New York: McGraw-Hill.

(18) Ringwood, A.E. 1982. Phase transformations and differentiation in subducted lithosphere: Implications for mantle dynamics, basalt petrogenesis and crustal evolution. J. Geol. 90: 611-643.

(19) Sclater, J.G.; Jaupart, C.; and Galson, D. 1980. The heat flow through oceanic and continental crust and the heat loss of the earth. Rev. Geophys. Space Phys. 18: 269-311.

(20) Sleep, N.H., and Windley, B.F. 1982. Archean plate tectonics: Constraints and inferences. J. Geol. 90: 363-379.

(21) Tarling, D.H. 1978. Plate tectonics: Present and past. In Evolution of the Earth's Crust, ed. D.H. Tarling, pp. 361-408. London: Academic Press.

(22) Thompson, A.B. 1981. The pressure-temperature (P,T) plane viewed by geophysicists and petrologists. Terra Cognita. Spec. Issue: 11-20.

(23) Thompson, A.B., and England, P.C. 1984. Pressure-temperature-time paths of regional metamorphism of the continental crust: II. Some petrological constraints from mineral assemblages in metamorphic rocks. J. Petrol., in press.

(24) Tozer, D.C. 1973. The concept of a lithosphere. Geofisica Internac. 13: 363-388.

(25) Watson, J.V. 1978. Precambrian thermal regimes. Phil. Trans. Roy. Soc. Lond. A. 288: 431-440.

(26) Windley, B.F. 1976. New tectonic models for the evolution of Archaean continents and oceans. In The Early History of the Earth, ed. B.F. Windley, pp. 105-111. New York: Wiley & Sons.

Patterns of Change in Earth Evolution, eds. H.D. Holland and A.F. Trendall, pp. 357-370.
Dahlem Konferenzen 1984. Berlin, Heidelberg, New York, Tokyo: Springer-Verlag.

Variation in Tectonic Style with Time: Alpine and Archean Systems

M.J. Bickle
Dept. of Earth Sciences, University of Cambridge
Cambridge CB2 3EQ, England

Abstract. Geophysical modelling of Archean tectonics during times of higher heat production and heat loss than today predicts that plate-tectonic motions should have operated more rapidly in the Archean. The Archean geological record is still too poorly understood to accept or reject this prediction with confidence. Large volumes of calc-alkaline rocks and evidence of horizontal tectonics are the features of Archean terrains most suggestive of a plate-tectonic regime. However, it is puzzling that there are no obvious major sedimentary, volcanic, metamorphic, and structural facies belts in granite-greenstone terrains related to collision orogeny. Features that distinguish Archean from modern terrains include the basalt–dominated greenstone sequences dispersed around gregarious composite granitoid batholiths and the prevalence of granulite facies gneiss belts. These features are difficult to interpret since there is still considerable controversy and uncertainty concerning their basic geological relationships.

INTRODUCTION

Archean crust (> 2500 Ma) preserved in both the distinctive granite-greenstone and the high-grade gneiss terrains exhibits a tectonic style that is not shared by younger crust. The change in tectonic style is usually related to secular changes that reflect the decrease in radiogenic heat production within the Earth and the growth of the continental crust.

The multiplicity of Archean tectonic models in the literature (29) demonstrates that it is not yet possible to define Archean tectonic processes confidently on the basis of the fragmentary and poorly understood Archean record. The approach adopted here is to review extrapolations of plate-tectonic processes to earlier periods of Earth history during which the higher heat production and presumably heat loss were more rapid, and to compare the features of Alpine systems that are related to plate-tectonics with possible analogues in Archean terrains.

HEAT LOSS - IMPLICATIONS FOR ARCHEAN TECTONICS

At present the Earth loses about half its heat during plate-creation and subduction. It is theoretically possible to calculate the proportion of heat loss by plate-formation (or analogous convective processes) by monitoring continental thermal gradients and calculating the heat lost by conduction through the lithosphere. In the Archean heat production was probably between two and three times greater than at present; if plate-tectonics or an analogous convective heat loss mechanism were unavailable, because - for example - thin, hot plates were too buoyant to be subducted, all the Earth's heat must have been lost by conduction through the lithosphere. This implies that thermal gradients in the upper part of the continental crust were ~60°C/km 2500 Ma ago and ~100°C/km some 3500 Ma ago. Gradients of this magnitude are well in excess of those in many metamorphic Archean terrains and preclude the stable existence of the 30-40 km thick Archean cratons that we see today (4,8). It seems inescapable, therefore, that there were some heat loss mechanisms in the Archean analogous to plate-tectonics. The only possible alternatives seem to be an Earth in which the continental lithosphere was insulated in some way from the convecting mantle or an Earth from which the loss of heat was lower in the Archean than at present.

The presence of magnesium-rich komatiitic lavas in Archean terrains indicates that temperatures in the mantle were 200°C to 250°C higher during the Archean than at present. Komatiitic lavas with liquid compositions more magnesian than ~22% MgO are restricted to the Archean; the most magnesian liquids contained 33% MgO and were probably extruded at temperatures up to about 1650°C (1). An increase in mantle temperature of ~200°C would produce a doubling of terrestrial heat loss (22). However, a general increase of this magnitude in mantle temperature with the implied increase in heat convected to the base of the Archean continental lithosphere is not easily reconciled with current estimates of continental thermal gradients in Archean high-grade gneiss terrains or with the thickness of crust developed within Archean orogenic belts (England and Bickle, in preparation).

Models of heat loss during the Archean therefore predict that a significant fraction of heat was lost by a mechanism analogous to plate-tectonics. The operation of any such mechanisms implies the recycling of large volumes of volcanic material, and hence the formation and subduction of some sort of oceanic crust. The contradiction between the mantle temperatures implied by komatiitic lavas and the theoretical modelling of mantle convection on the one hand (which implies a doubling of convected heat in the mantle) and the evidence from metamorphic terrains regarding the thermal state and thickness of the continents on the other hand (which suggests near equality of Archean and modern thermal gradients) is important. Continents may have been insulated from the convecting mantle; tectonic environments of Archean high-grade gneiss belts may be special, perhaps analogous to modern trench complexes cooled by underlying subducted lithosphere, or the metamorphic and tectonic models may be in error. The truth of any of these propositions would have significant implications for our understanding of mantle and crustal tectonics.

ALPINE TECTONICS

One aspect of ancient crustal environments that must not be overlooked in comparisons with modern geology is that continental crust is only preserved after being cycled through a number of geological environments (the Wilson cycle (10)). The features distinctive of the modern plate-tectonic (Wilson) cycle include formation of oceanic crust by seafloor spreading, destruction of oceanic crust and formation of major calc-alkaline volcanic-plutonic provinces along island-arc or continental margins above subduction zones, and finally the modification of continental margins by continental collision associated with overthrusting and metamorphism. The various elements that participate in this cycle are not equally well preserved. Oceanic crust preserved as disrupted ophiolite bodies forms only a small proportion of modern orogenic belts and is progressively more difficult to find in older orogenic belts, even where good local preservation indicates its former presence. The major calc-alkaline volcanic-plutonic Andean type belts have a high chance of being preserved. During the final phase of the cycle, i.e., continental collision, compressional structures (thrusts and folds) should be superimposed on the rocks formed during the earlier parts of the cycle. Distinctive and asymmetric linear belts of sedimentary, metamorphic, and structural facies are characteristic of each phase of the Wilson cycle; their linearity and asymmetry is an important consequence of plate-tectonic mechanisms.

ARCHEAN TECTONIC ENVIRONMENTS

Archean crust can be subdivided into granite-greenstone terrains and high-grade gneiss terrains (29). However, cratonic sedimentary and volcanic basins with ages of ~2700 to 3000 Ma do exist in South Africa (Witwatersrand and Pongola) and Australia (Fortescue Group), and these must have formed contemporaneously with some granite-greenstone and gneiss terrains.

The granite-greenstone terrains contain distinctive basalt-dominated sedimentary-volcanic greenstone sequences disposed in synformal zones between domal granite-gneiss batholiths (for maps of many granite-greenstone terrains see (9)). In some greenstone terrains (e.g., Barberton, South Africa; Rhodesian Craton, Zimbabwe; and the Pilbara, Western Australia) batholiths are ovoid in plan and conform to the classic "gregarious batholith" style noted by McGregor (21). In other greenstone terrains (notably the Superior Province, Canada, and the Yilgarn Block, Western Australia) the domal batholiths have a distinctly linear trend; greenstone belts up to 700 km in length and up to 200 km wide occur in belts within a heterogeneous granitoid crust.

The dominance of basaltic volcanics and the presence of the more magnesium-rich komatiitic volcanics, lend a rather uniform aspect to these supracrustal sequences in comparison with modern supracrustal associations. The dominance of volcanics might signal a fundamental difference between Archean and present-day tectonic processes; a hotter Archean mantle might, for instance, have erupted rather more copiously in any given setting. On the other hand, it might indicate that greenstone belts are samples of a particular volcanic-dominated environment; back-arc marginal basins, continental margin rift sequences, and oceanic crust are the most popular modern analogues.

Sedimentary sequences consisting mainly of turbidites or braided alluvial deposits together with minor volcanics occupy most of the upper parts of the stratigraphic sections of greenstone sequences. The Superior Province, Canada, is an important exception. There three major (1000 km long by 50 to 150 km wide) belts consisting of metasedimentary sequences and a range of gneissic and plutonic granitoid rocks separate more typical granite-greenstone terrains. These metasedimentary sequences are thought to be lateral equivalents of volcanic-dominated greenstone sequences. Major turbidite belts occur in some other greenstone terrains (e.g., Pilbara), but their stratigraphic position and their age are often poorly known. Deposition of sediments in both volcanic arc (18) and in passive continental margin environments (e.g., Barberton and the Pilbara (12)) have been proposed for greenstone terrains.

The analysis of the tectonic setting of granite-greenstone terrains is invariably hindered by uncertainties regarding a number of basic tectonic relationships; some of the more important problems are outlined below.

The high-grade gneiss terrains, the other main component of Archean crust, consist of monotonous sequences of quartzo-feldspathic gneiss almost certainly derived from tonalitic calc-alkaline plutonic rocks. The quartzo-feldspathic gneisses are intercalated with foliated metasedimentary and metavolcanic supracrustal rocks and with disrupted layered igneous complexes which include distinctive anorthosite bodies. The metamorphic grade ranges from amphibolite to medium and high pressure granulite facies. The intercalation of supracrustals and gneissic plutonic rocks is argued to be a result of major horizontal overthrust style tectonics (e.g., (23)).

GRANITE-GREENSTONE TERRAINS - PROBLEMS IN GEOLOGICAL INTERPRETATION

Basement

Perhaps the most intriguing problem of granite-greenstone tectonics is the nature of their original basement. Basal contacts of the greenstone sequences are generally intrusive or tectonic; in many greenstone terrains these contacts are poorly exposed. The radiometric age of the dominantly tonalitic or granodioritic components of the batholiths is very often equal - within the analytical errors - to the radiometric age of the volcanic rocks within the greenstone belts.

The basement problem is particularly significant for an understanding of Archean tectonics, because of the suggestion that the basic volcanic units in the greenstone stratigraphy represent disrupted oceanic crust (7). However, the lack of exposed basement in granite-greenstone terrains is to be expected by analogy with modern plutonic terrains (13). Arguments regarding the oceanic nature of greenstone sequences must therefore be based on their stratigraphy and on similarities to sequences whose basal contacts are preserved.

Unconformities with older granitoid rocks beneath greenstone successions are preserved in a number of localities (e.g., (2, 5)). The ~2700 Ma

Belingwe greenstone belt in Zimbabwe has a typical greenstone stratigraphy, and identifiable clastic debris from the older basement are limited to the meter above the basal unconformity. The absence of debris of older basement in sedimentary units interbedded with the volcanics is not compelling evidence against the presence of an underlying granitoid basement, as has, for example, been argued by Lowe (20). Angular unconformities within greenstone terrains are quite common, and many workers do not consider the stratigraphic successions in the older sequences to be significantly different (see (14) for an alternative viewpoint).

The stratigraphy of greenstone sequences provides the most compelling evidence against their origin as oceanic crust. Their stratigraphy almost invariably includes sedimentary horizons and felsic volcanic horizons interbedded within the monotonous basaltic volcanics. Although sequences may be repeated by thrusting, enough primary contacts between sediments and volcanics are preserved to rule out a simple, purely igneous, oceanic crust stratigraphy. The rocks do not exhibit retrogressive metamorphic assemblages characteristic of the seafloor with its early alteration of deeper crustal layers to amphibolite facies. In a number of greenstone terrains, volcanic rocks are interbedded with shallow water sedimentary rocks (3); on isostatic grounds this is also inconsistent with an oceanic origin for greenstone terrains.

One further line of evidence suggests the presence of a widespread older continental basement underlying greenstone terrains. Middle or lower crustal granulite facies gneissic rocks exposed by late uplift along the Kapuskasing Zone in the Superior Province have U-Pb zircon ages 10 to 100 Ma older than adjacent high level greenstones (24). The older basement of greenstone terrains may only be preserved in the lower crust; intrusive plutons may well be separating greenstones from their basement.

Despite these arguments, a simatic or oceanic crustal origin for greenstone belts is popular. The vast areas of structurally complex, poorly exposed, and little studied areas of greenstone may indeed contain relict Archean oceanic crust, but unequivocal examples of such crust have yet to be described.

Batholiths

The second fundamental problem of granite-greenstone terrains is the cause, significance, and structural consequences of the batholiths. The granitoid areas are often distinctly heterogeneous both in rock type and structure. They often include high-strain gneisses and migmatites with intercalated supracrustal rocks, probably of greenstone belt origin, and a range of foliated to undeformed younger plutonic phases ranging from tonalite to granite. In some granite-greenstone terrains the ages of the various components of the batholiths span hundreds of millions of years; this composite nature makes them quite unlike modern calc-alkaline batholiths.

Greenstone belts almost invariably exhibit an increase in both metamorphic grade and strain towards the batholiths; very often contacts contain intercalated gneissic granitoid and amphibolite that is frequently extensively migmatized (4). Lower strain contacts with intrusive relations against early components of the batholith are preserved locally. It is generally accepted that the deformation and subsequent metamorphism took place while the rocks were solid, i.e., after magmatic intrusion of the granitoid. However, considerable controversy surrounds the timing and the structural and metamorphic consequences of batholith formation. Some workers (e.g., (25)) have concluded that the diapiric rise of granitoid batholiths is responsible for the major part of the strain and metamorphism observed both within the batholiths and in the surrounding greenstone sequence. Others (6) have concluded that the high-strain deformations and the early regional metamorphism are related to an earlier tectonic phase that was superposed on the granite-greenstone terrain, and that crustal thickening during this phase triggered the diapiric instability that was responsible for the formation of the batholiths. The complexity of the limited structural data available for Archean granite-gneiss batholiths makes it difficult to choose between these competing hypotheses.

Stratigraphy and Spatial Variation in Greenstone Terrains

The spatial variation of sedimentary, volcanic, metamorphic, and structural facies within tectonic provinces should provide critical evidence regarding the nature of tectonic processes. Numerous publications on Archean granite-greenstone tectonics suggest that there are such variations, but it has proved difficult to substantiate any of them, given the paucity of detailed facies analysis and particularly the lack of a clear-cut stratigraphic correlation across hundreds of kilometers of greenstone terrains (see, for example, the classic paper by Goodwin and Ridler (15) and the subsequent discussion by Walker (26)). In some granite-greenstone terrains individual greenstone stratigraphies or even stratigraphic groups separated by regional unconformities have been correlated on a regional scale (e.g., (17, 28)), and considerable stratigraphic continuity is apparent in the less deformed areas. However, the lack of sufficiently precise chronological control makes it difficult to establish the significance of regional facies variations.

Another factor that complicates the mapping of the structural, metamorphic, volcanic, and sedimentary facies belts is the limited area of many granite-greenstone terrains. Well studied areas in Barberton, South Africa (250 km x 150 km), the Pilbara, Western Australia (450 km x 100 km), the Rhodesian Craton, Zimbabwe (600 km x 250 km), the Karnataka Subprovince, India (500 km x 300 km), and the Slave Province, Northern Canada (500 km x 400 km) are barely comparable in size with the relatively small European Alps (900 km x 300 km). Only the Yilgarn Craton in Western Australia (1000 km x 900 km) and the Superior Province in Canada (1500 km x 1000 km) are large enough to exhibit complete sections across orogenic belts. Recent geochronological research has shown that these larger cratons consist of a number of linear age provinces ~1000 km + long and ~100 km or more wide. These age provinces together with the linear paragneiss and metasedimentary belts in the Superior Province and the major turbidite belts in the Barberton and the Pilbara areas (7) are perhaps the best evidence for the presence of linear Archean orogenic zones on a scale much larger than individual greenstone belts or batholiths.

Archean Calc-alkaline Provinces

Large volumes of calc-alkaline rocks are preserved in granite-greenstone terrains and in high-grade gneiss belts. In many granite-greenstone terrains calc-alkaline felsic volcanic rocks extruded at distinct felsic volcanic centers are similar in chemistry and in age to the dominant tonalite and granodiorite components of the batholiths (e.g., (16, 19)).

Both volcanic and plutonic Archean calc-alkaline rocks are petrographically and chemically similar to the volcanic and plutonic calc-alkaline igneous rocks formed in Andean-type continental margins (if allowance is made for the movement of the more mobile elements during high-grade metamorphism). The only significant chemical differences between Archean and modern calc-alkaline rocks is the relative depletion of heavy rare earth elements in some Archean plutonic and felsic volcanic rocks and the relative enrichment of these rocks in the transition metals, particularly Ni and Cr (9). The significance of these subtle chemical differences between modern and Archean calc-alkaline rocks are difficult to evaluate since there is still considerable uncertainty regarding the genesis of modern calc-alkaline rocks.

A more important difference between Archean granite-greenstone and modern calc-alkaline orogenic belts is that the greenstone supracrustal sequences do not contain sedimentary facies that are characteristic of the highly unstable Andean orogenic belts. If the greenstone sequences developed during an earlier, unrelated tectonic cycle, then what is the significance and tectonic setting of the calc-alkaline felsic volcanics? The supracrustal assemblages preserved in high-grade gneiss belts differ from those in greenstone belt sequences; they are characteristic of passive continental margin environments, although amphibolite derived from mafic lavas is also common. Felsic volcanic material is notably rare. Windley (29) has summarized a number of models for greenstone and gneiss belt genesis and has suggested that greenstone and high-grade gneiss belts form in Archean calc-alkaline belts similar to modern Andean belts but that they occupy different parts of the arc; gneiss belts are thought to have been derived from continental margins and granite-greenstone terrains from back-arc basins. These models imply a rather critical timing of the plutonism, mafic and felsic volcanism, and deformation and metamorphism in Archean terrains; the reality of this timing remains to be tested.

Collision Tectonics

The relative importance of horizontal and vertical tectonics in granite-greenstone terrains is a matter of controversy in theories of Archean tectonics. The presence of recumbent folding or major overturning of supracrustal successions on the scale of 10 km or more has been described in a number of widely spaced areas in most granite-greenstone terrains. The recent mapping of classic recumbent structures in the well exposed and well studied Barberton greenstone belt (11) illustrates the main contention of Burke et al. (7), i.e., that stratigraphic repetition and overthrusting may have been overlooked in many greenstone belts. The interpretation of these areas of recumbent folding is hampered by the scale of the overturning and by the geological complexity that invariably limits the mapping of these recumbent structures. To date overthrust belts have not been mapped on a regional scale in Archean granite-greenstone terrains. It has been suggested that recumbent folding is limited in extent and that it is a consequence of mature diapirism (25).

The genesis of high-strain gneisses within granitoid batholiths is a problem related to the tectonics within the greenstone sequences. Bickle et al. (6) presented evidence that gneisses within the Shaw Batholith in the Pilbara Block were derived from granodiorite that was originally intrusive into the greenstone sequence and were deformed during horizontal deformation events which intercalated greenstone and gneiss on a scale of 10^2 km. This has two important implications. First, high-strain banded gneisses are relatively widespread within the granitoid crust of granite-greenstone areas and might provide evidence for the widespread horizontal tectonics sought by Burke et al. (7). Second, it suggests a tectonic transition between granite-greenstone terrains and high-grade gneiss terrains where horizontal structures are developed on a scale comparable to that of modern overthrust belts. The contrast between the assemblage of supracrustal rocks in granite-greenstone terrains and in high-grade gneiss belts has been noted above, but transitional assemblages do occur. The Isua belt in West Greenland is an example of such an assemblage.

The nature of the tectonic setting in which the high-grade gneiss belts were generated is a matter of some importance. Some investigators have argued that the horizontal tectonism was synchronous with the emplacement of tonalitic magmas (e.g., (29)) and that it may be related to shallow-dipping subduction rather than to collision tectonics. Wells (27) has argued that the evidence from the nature of the metamorphism is consistent with a plutonic environment rather than collision-related tectonic burial.

The final and perhaps most significant evidence regarding the role of horizontal tectonics in the Archean comes from the one COCORP seismic reflection profile across the Archean granite-greenstone and gneiss terrain in central Minnesota (Gibbs et al., in preparation). This shows that there are prominent reflectors with a dip of 30° that separate the ~2700 Ma Superior Province from the ~3500 Ma Minnesota River valley gneiss terrain. The reflectors have been interpreted as major thrust faults which originated during a late Archean collision between the Superior Province in the north and the older crust to the south.

The COCORP reflection profile confirms the relatively shallow nature suggested by gravity surveys for both the granitoid batholiths and the typical greenstone belts within the granite-greenstone terrains. Below a zone that ranges in depth between 4 and 7 km, the reflections correlated with a granitoid batholith and a greenstone belt are replaced by complex, stronger, but more coherent and often shallowly dipping reflections; I interpret these as reflections from high-grade granitoid gneissic rocks that are intercalated with foliated supracrustal rocks.

The seismic reflection study suggests that the structure of the Archean crust is rather more complex than indicated by the surface geology in granite-greenstone terrains.

CONCLUSION

Our present knowledge is clearly insufficient to preclude plate-tectonics as the main tectonic mechanism in the Archean. The predominance of calc-alkaline rocks and the evidence for major horizontal collision-type structures are the two aspects of Archean terrains most suggestive of a plate-tectonic regime. The main distinctive features of the Archean terrains are the basic volcanic-dominated greenstone sequences disposed around granitoid batholiths and the widespread granulite facies gneissic belts in the high-grade regions. Uncertainties still surround the basic geological relationships between these terrains and preclude a firmly based interpretation of these features. We do not know whether the batholiths are due to thermally driven instability in the crust, to a tectonically overthickened crust, or to the greater density of the volcanic-dominated supracrustal layers. We do not know whether the komatiite and basalt-dominated supracrustal sequences reflect a more copious volcanicity due to a hotter Archean mantle, nor whether greenstone belts sample a volcanic-dominated tectonic environment analogous to some modern environments. We do not know whether it was greater crustal thickening in the Archean that has resulted in the exposure of deeper Archean granulite facies layers in the crust than in Phanerozoic orogenic belts.

REFERENCES

(1) Arndt, N.T., and Nisbet, E.G. 1982. Komatiites. London: George Allen & Unwin.

(2) Baragar, W.R.A., and McGlynn, J.C. 1976. Early Archaean basement in the Canadian Shield: a review of the evidence. Geol. Surv. Can. Paper 76-14.

(3) Barley, M.E.; Dunlop, J.S.R.; Glover, J.E.; and Groves, D.I. 1979. Sedimentary evidence for an Archaean shallow-water volcanic-sedimentary facies, Eastern Pilbara Block, Western Australia. Earth Planet. Sci. Lett. 32: 74-84.

(4) Bickle, M.J. 1978. Heat loss from the Earth: a constraint on Archaean tectonics from the relation between geothermal gradients and the rate of plate production. Earth Planet. Sci. Lett. 40: 301-315.

(5) Bickle, M.J.; Martin, A.; and Nisbet, E.G. 1975. Basaltic and perioditic komatiites and stromatolites above a basal unconformity in the Belingwe greenstone belt, Rhodesia. Earth Planet. Sci. Lett. 27: 155-162.

(6) Bickle, M.J.; Bettenay, L.F.; Boulter, C.A.; Groves, D.I.; and Morant, P. 1980. Horizontal tectonic interaction of an Archaean gneiss belt and greenstones, Pilbara Block, Western Australia. Geology 8: 525-529.

(7) Burke, K.; Dewey, J.F.; and Kidd, W.S.F. 1976. Dominance of horizontal movements, arc and microcontinental collisions during the later permobile regime. In The Early History of the Earth, ed. B.F. Windley, pp. 113-129. London: Wiley Interscience.

(8) Burke, K., and Kidd, W.S.F. 1978. Were Archaean continental geothermal gradients much steeper than those of today? Nature 272: 240-241.

(9) Condie, K.C. 1981. Archaean Greenstones. Amsterdam: Elsevier.

(10) Dewey, J.F., and Bird, J.M. 1970. Mountain belts and the new global tectonics. J. Geophys. Res. 75: 2625-2647.

(11) de Witt, M.J. 1982. Gliding and overthrust nappe tectonics in the Barberton greenstone belt. J. Struct. Geol. 4: 117-136.

(12) Eriksson, K.A. 1982. Sedimentation patterns in the Barberton Mountain Land, South Africa, and the Pilbara Block, Australia: evidence for Archaean rifted margins. Tectonophysics 81: 179-193.

(13) Gastil, R.G.; Phillips, R.P.; and Allison, E.C. 1975. Reconnaisance geology of the state of Baja California. Geol. Soc. Am. Mem. 140.

(14) Glikson, A.Y. 1979. Early Precambrian tonalite-trondhjemite sialic nuclei. Earth-Sci. Rev. 15: 1-73.

(15) Goodwin, A.M., and Ridler, R.H. 1970. The Abitibi orogenic belt. Geol. Surv. Can. Paper 70-40: 1-30.

(16) Hawkesworth, C.J., and O'Nions, R.K. 1977. The petrogenesis of some Archaean volcanic rocks from southern Africa. J. Petrol. 18: 487-520.

(17) Hickman, A.H. 1981. Crustal evolution of the Pilbara Block, Western Australia. Spec. Publ. Geol. Soc. Aust. 7: 57-69.

(18) Hyde, R.S. 1980. Sedimentary facies in the Archaean Timiskaming Group and their tectonic implications, Abitibi greenstone belt, north eastern Ontario, Canada. Precambrian Res. 12: 161-195.

(19) Krogh, T.E.; Davis, D.W.; Nunes, P.D.; and Corfu, F. 1982. Archaean evolution from precise U-Pb isotopic dating. Abstract. Geological Association of Canada/Mineralogical Association of Canada, Joint Meeting, Winnipeg 1982.

(20) Lowe, R.D.R. 1982. Comparative sedimentology of the principal volcanic sequences of Archaean greenstone belts in South Africa, Western Australia and Canada: implications for crustal evolution. Precambrian Res. 17: 1-29.

(21) MacGregor, A.M. 1951. Some milestones in the Precambrian of southern Africa. Proc. Geol. Soc. S. Afr. 54: 27-71.

(22) McKenzie, O.P., and Weiss, N.O. 1980. The thermal history of the Earth. In The Continental Crust and Its Mineral Deposits, ed. D.W. Strangway. Geol. Assoc. Can. Spec. Paper 20: 575-590.

(23) Myers, J.S. 1976. Granitoid sheets, thrusting and Archaean crustal thickening in West Greenland. Geology 4: 265-268.

(24) Percival, J.A., and Card, K.D. 1982. Evolution of contrasting high- and low-grade Archaean terranes in southern Superior province. Abstract. Geological Association of Canada/Mineralogical Association of Canada, Joint Meeting, Winnipeg 1982.

(25) Schwerdtner, W.M., and Lumbers, S.B. 1980. Major diapiric structures in the Superior and Grenville provinces of the Canadian Shield. In The Continental Crust and Its Mineral Deposits, ed. D.W. Strangway. Geol. Assoc. Can. Spec. Paper 20: 149-180.

(26) Walker, R.G. 1978. A critical appraisal of Archaean basin-craton complexes. Can. J. Earth Sci. 15: 1213-1218.

(27) Wells, P.R.A. 1976. Late Archaean metamorphism in the Buksefiorden region, southwest Greenland. Contrib. Min. Petrol. 56: 229-242.

(28) Wilson, J.F.; Bickle, M.J.; Hawkesworth, C.J.; Martin, A.; Nisbet, E.G.; and Orpen, J.L. 1978. Granite greenstone terrains of the Rhodesian Archaean craton. Nature 271: 23-27.

(29) Windley, B.F. 1977. The Evolving Continents. London: J. Wiley.

Patterns of Change in Earth Evolution, eds. H.D. Holland and A.F. Trendall, pp. 371-386.
Dahlem Konferenzen 1984. Berlin, Heidelberg, New York, Tokyo: Springer-Verlag.

Variation in Tectonic Style with Time (Variscan and Proterozoic Systems)

K. Weber
Geologisch-Paläontologisches Institut und Museum
der Universität Göttingen
3400 Göttingen, F.R. Germany

Abstract. The Precambrian and the early Paleozoic records contain evidence for numerous intracontinental orogenies. These evolve out of ensialic rift zones whose character and scale strongly influence the subsequent orogenic developments. There is a continuous gradation from weakly deformed, nonmetamorphic auloacogenes to highly metamorphic mobile belts. A number of examples are used to illustrate the characteristic features of sedimentary, magmatic, metamorphic, and structural evolution during the rifting stage. A matter of special importance in this context is the granulite facies metamorphism which took place during early Paleozoic time in what later became the Variscan orogene in Europe. It was concurrent with the continuous accumulation of sediment at the surface and with widespread preorogenic igneous activity. A model is offered for discussion that treats the granulite facies metamorphism and the contemporaneous sedimentation and emplacement of preorogenic granites as the result of widespread rifting within continental lithosphere.

INTRODUCTION

In a number of Precambrian and early Paleozoic orogenies, sedimentary, magmatic, metamorphic, and structural developments combine to suggest an intracontinental (ensialic) setting. In other cases subduction of oceanic lithosphere took place; in these there is a good deal of evidence for a Wilson cycle of tectonic history. This is suggested, for example, by ophiolites in the Caledonides, at Bou Aggar in Morocco, in the Arabian Shield, and in NE Africa. The late Precambrian to early Paleozoic tectonic development of the Hoggar-Iforas region in the Central Sahara has been interpreted by Caby et al. (11) in terms of a complete Wilson cycle. However, the fact that blueschist metamorphic rocks are not found in the Precambrian mobile belts suggests that Precambrian orogenes may have had thermal histories and a tectonic evolution different from those of Phanerozoic orogenes.

Precambrian-Paleozoic tectonic evolution appears to be characterized by more frequent intracontinental orogeny than the Mesozoic-Cenozoic history of the Earth. The intracontinental orogenes develop from continental rifts on which varying degrees of crustal shortening have been imposed at a later time. A variety of intermediate cases between aulacogenes and orogenes are therefore encountered. The 2200-1700 Ma old Athapuscow Aulacogene in the Great Slave Lake region of Canada, for example, contains a thick pile of sediments which are unmetamorphosed and only weakly deformed. Hoffman et al. (16) interpret this area as an abandoned rift which began at a triple junction and extended into the interior of a continent. On the other hand, the early Proterozoic Labrador Geosyncline evolved (according to Dimroth (13)) beyond the aulacogene stage. Here, an alpine-type LP/HT metamorphic belt apparently evolved out of a fault-bounded basin. The Damara orogene and the European Variscides can be regarded as representing further stages of this development. In this discussion below, sedimentary, magmatic, metamorphic, and structural characteristics are examined, all of which can be interpreted as evidence for intracontinental rifting processes. Such rifting processes can be regarded as including the forerunners, or the initial stages of development, of subsequent orogenic processes which lead to crustal shortening; these processes are not treated in this discussion.

INTRACONTINENTAL RIFTING PROCESSES

Martin and Porada (24) and Porada (29) have been able to demonstrate that in the Damara Orogen of Namibia the characteristically geosynclinal stage was preceded by a graben stage of development. Martin and Porada distinguish three grabens that are parallel to the general strike of the orogen. These contain thick sequences of clastic sediments and ignimbrites. The northern graben is filled with more than 6000 m of ignimbrites with subordinate andesites, bostonites, volcanic breccias, and felsitic tuffs; the southern graben was filled by approximately 5000 m of the pelitic-dolomitic Duruchaus Formation, which includes thick, evaporitic, playa-lake deposits. These are now albitolite-chert horizons, brown dolomites rich in albite and occasionally in coarse tabular microcline, cauliflower quartz, dark biotite-rich horizons with scapolite porphyroblasts, and tourmaline-rich beds that contain up to 50% tourmaline (5, 7). Primary evaporite minerals have not been preserved. However, their former presence can be inferred from the morphology of their pseudomorphs and from their decay minerals. Gypsum has been replaced by quartz and microcline; all of the other pseudomorphs consist of crypto- to microcrystalline albite with varying amounts of dolomite, calcite, tourmaline, paragonite, or chlorite. The primary minerals were shortite, borax, northupite, leucosphenite, gaylussite, trona, thenardite, gypsum, and - in the tourmaline-rich beds - probably colemanite and ulexite. The metamorphosed playa-lake deposits are rich in fluid inclusions whose salinity ranges from 30 to 55% NaCl (5).

The formation of diagenetic evaporite minerals, syneresis structures, and evaporite breccia was followed by discordant brecciation of the Duruchaus Formation and intrusion into higher levels. The brecciation and intrusion are related to a partial mobilization of the evaporite carbonate horizons under conditions of high pore fluid pressure during folding and nappe formation. Similar meta-evaporite horizons have been found in the northern part of the Damara Orogen (39). Behr et al. (7) and Weber and Ahrendt (38) have shown that these evaporites played an important role during the structural development of the Damara Orogen. This is particularly true in the Naukluft Nappe Complex; there a siliceous dolomite whose composition is quite similar to that of the Duruchaus Formation has been interpreted by Behr et al. (6, 7) and Weber and Ahrendt (38) to be a discordant intrusion, introduced under high pore fluid pressure into the base of a nappe sequence; this intrusion probably acted as a lubricant during an early stage of transposition of the Naukluft Nappe Complex.

Sabkha and playa environments seem to have been quite widespread in Proterozoic time. Sabkha and playa sediments from the Upper Proterozoic grabens of the Adelaide palaeorift in South Australia have been described by Rowlands et al. (32). The Katangan Supergroup of Zambia and Zaire which is approximately as old as the Damara Sequence and the Adelaide system has been interpreted by Bartholomé et al. (2) and by Renfro (31) as former sabkha sediments. The upper Proterozoic Kada-Udokan Copperbelt in Siberia has been compared by Rayner and Rowlands (30) to the evaporite environments of the Adelaide Belt.

Derrick et al. (12) have proposed that the Proterozoic Leichhardt River Fault Trough in the Mt. Isa region of Queensland (Australia) was formed by the extension and sagging of continental crust. The fault trough is set in an 1865-1800 Ma old basement of granites and acid volcanics; the trough was filled by 10 km of epicontinental clastics, minor dolomites, red beds, and marginal fanglomerates. Up to 6 km of subaerial to shallow subaqueous basalt are also present. The stromatolite-bearing shallow water sediments of the Mt. Isa group are approximately 4500 m thick; they are the youngest material in the Leichhardt River Fault Trough. Quartz-albitolites (A-Marker Horizon), cauliflower quartz, halite casts (hopper cube moulds), and pseudomorphs after anhydrite laths are common. These evaporitic horizons are interpreted by McClay and Carlile (25) and Neudert and Russell (28) as representing a hypersaline environment of the sabkha type. At Mt. Isa as well as in the Adelaide Belt, the Damara Orogen, the Katanga-, and the Kadar-Udokan Copperbelt, these evaporite sequences are of great importance to an understanding of the copper mineralization.

Sedimentary, magmatic, metamorphic, and structural characteristics indicate that there was intracontinental tectonic development in many other regions. In what follows, certain important features suggestive of such a setting for the Variscan Orogen in Europe are described. In conclusion, some comparisons with the modes of evolution of Precambrian intracontinental orogenes are attempted.

Jäger (18) has presented a comprehensive account of radiometric dates in Central and Western Europe and has come to the conclusion that oceanic sedimentation in late Precambrian and early Paleozoic time was followed by three phases of orogeny: the Cadomian, the Caledonian, and the Hercynian events. The increase in the $^{87}Sr/^{86}Sr$ ratios of granitic magmas from Cadomian to Hercynian time suggests to Vidal et al. (35) that the Variscan crustal segment evolved as a closed system, and that Variscan magmatism arose principally by remelting of sialic crust. The $^{87}Sr/^{86}Sr$ pattern of the Variscan crust is very different from that observed in magmatic rocks associated with most of the Mesozoic and Cenozoic subduction zones.

There are certain peculiarities which are important for the geodynamic interpretation of the Variscides and which make it difficult to regard the "Caledonian" event within the Variscan realm as representing an orogenic event in which significant crustal shortening was achieved:

Between 500 and 400 Ma ago, i.e., during the Ordovician and Silurian time, enormous quantities of granitic melts of calk-alkaline to peralkaline composition were produced. It is evident that the deformation and metamorphism took place with these rocks in the solid state. Therefore, it seems unlikely that intrusion and metamorphism can be related to the same structural event. These granites were emplaced in the pre-Variscan crust and at a later date - during Devonian time - deformed into orthogneisses. These orthogneisses are widespread in the Variscan basement. The alkaline to calk-alkaline granites in the Alto Alentejo in Portugal, in the Hesperian Massif of northern Spain, in the southern part of the Armorican Massif and the French Massif Central, in the Black Forest and the Vosges, in the Variscan Basement of the Central and Eastern Alps, in the Münchberg Massif, and the Mid-German Crystalline Rise are all examples of this type of rock unit.

The Ordovician-Silurian magmatism is contemporaneous with a granulite facies metamorphism which was in progress at depth in many regions where sedimentation continued at the surface during the same range of time. Two regions provide classic examples of granulite facies metamorphism concurrent with stratigraphic continuity during the Lower Paleozoic. These are the Granulite Gebirge in Saxony and the Hesperian Massif of northwestern Spain. In the areas around the Granulite Gebirge the stratigraphic succession ranges from Upper Proterozoic to Devonian. Evidence for a pregranulite migmatization (Anatexis I) is preserved in metatectic structures; Anatexis I was followed around 450 Ma b.p. by granulite facies metamorphism at 8 Kb and 700-800°C (4, 18, 19, 37, 40).

In the region of the Hesperian Massif the sedimentary record is practically continuous from late Precambrian to mid-Devonian time. Local gaps can be attributed to block faulting. Ordovician granulite facies metamorphism which has taken place at 10-11 Kb and about 850°C (21) corresponds with a widening of the sedimentary basin and a period of block faulting. The Moldanubian granulites in Lower Austria were also formed during Ordovician time (446 ± 35 Ma; (1)), at approx. 11 Kb and 760°C. Additional examples of Ordovician granulite facies metamorphism are known from the Ivrea Zone (17) and from the Massif Central where an Ordovician age is likely for the metamorphism of the leptyno-amphibolitic group.

It is not a simple matter to explain the coincidence of granulite facies metamorphism deep in the crust, the intensive pretectonic igneous activity, and the development of continuous sedimentary sequences at the surface. It is made more difficult if one assumes that the process of producing granulite facies metamorphism is connected with orogenic crustal shortening. The fact that the early Paleozoic granitoids predate crustal shortening and also the widespread evidence of more or less uninterrupted Lower Paleozoic stratigraphic successions in many parts of the Variscan Europe would tend to discredit any such assumptions.

The following proposals for continental rift metamorphism during Ordovician time are based mainly on the model developed by Den Tex and his co-workers (10, 33, 34) for the Hesperian Massif. They proposed that a continental rift developed on an anomalous mantle, and that the heat transferred into the lower crust produced granulite facies metamorphism.

Work on recent passive continental margins has shown that rifting in continental crust promotes ductile stretching in the lower crust; this contrasts with the brittle reaction of the upper 10 to 15 km of the crust where grabens form. This view was introduced by McKenzie (26, 27); it has been widely accepted and might provide a basis for understanding the contemporaneity of "Caledonian" granulite facies metamorphism, sedimentation, and preorogenic igneous activity. The continental crust is underlain by lithospheric mantle. Rifting of the continental crust must have some association with rifting in the lithospheric mantle which leads the way to the rifting of the overlying crust and promotes the ascent of hot asthenospheric material. Such convective supply of heat may heighten temperatures at the crust-mantle boundary. Partial melts of tholeiitic composition may be transformed, after solidification, into eclogites in the higher parts of the lithospheric mantle or at the crust-mantle boundary, or else they may invade the lower crust and become metagabbros or amphibolites.

The P-T conditions of granulite facies metamorphism, 7-11 Kb and 700-850°C, suggest that granulites can be produced at the base of 25-40 km thick continental crust. However, the granulite facies metamorphism in almost all of these Ordovician granulites follows on a pregranulite migmatization which corresponds to Anatexis I. This is an understandable relationship, for the granulite facies dewatering of deep crustal rocks is a gradual process. The expulsion of water coincides with the introduction of CO_2 derived from the mantle or the decay of carbonates. Lead isotope ratios in K-feldspars from several metamorphic and granitic rocks in the southern Black Forest suggest a very early formation of the basement rocks and have been interpreted by Kober and Lippolt (20) as the result of crust-mantle interaction during Anatexis I: mantle lead was injected upwards out of a degassing mantle region during the genesis of "Caledonian" magma.

Increasing temperatures, and expulsion of water from the granulites, led to the formation of calk-alkaline granite magmas above the granulite horizon. These melts intruded the higher crust and were later transformed into orthogneisses during crustal shortening.

A further indication of crustal stretching can be derived from the nature of primary granulite fabrics. Fold structures which predate the amphibolite facies overprinting are extremely rare in the granulites (40). Behr's (2) analyses of the quartz fabrics of the Saxonian Granulite Gebirge have shown that the discoidal quartzes of the core granulites display small circle configurations with c-axes normal to the metamorphic layering. Lister and Dornsiepen (23) interpret these fabrics to mean that the small circle pattern in the core granulites are typical of strain histories intermediate between axially symmetrical shortening and plane strain. Assuming that the granulite facies metamorphic layering was horizontal at the outset, the highly symmetrical quartz fabric might be indicative of crustal stretching.

HIGH-GRADE METAMORPHISM IN PRECAMBRIAN METAMORPHIC ROCKS

Grambling (14) has compared approximately one hundred metamorphic mineral parageneses older than 1000 Ma and has examined them for any indication of secular trends. He was able to show that the average geothermal gradients were lessened from Archean to late Precambrian time and that average metamorphic pressures increased. In the Archean, average metamorphic geotherms and pressures stood at 54°C/km and 4.1 Kb, in the early Proterozoic 47°C/km and 4.3 Kb, and in the late Precambrian 35°C/km and 6 Kb. The major difference between Precambrian and Phanerozoic metamorphic conditions is the absence of blueschist metamorphism in pre-1000 Ma old rocks. The secular trends in the Precambrian rocks are interpreted by Grambling to reflect gradual cooling and thickening of the Earth's crust, at least in orogenic belts.

The minimum metamorphic geotherms of about 20°C/km remained constant during the whole time, whereas the maximum preserved geotherms declined from 95°C/km in the Archean, to 77°C/km in the early Proterozoic, and to 50°C/km in the late Proterozoic. High metamorphic pressures are not restricted to younger rocks. However, their frequency increases significantly from older to younger metamorphic terrains.

High-grade metamorphic rocks are widespread in Proterozoic and also in Archean terrains. They indicate the existence of an approximately 30 km thick continental crust. Structural analyses of Precambrian regions show that high-grade metamorphic rocks were brought tectonically into a near-surface position as in Phanerozoic time.

Crustal thicknesses in the range of 30 km give an explanation for high-grade metamorphic pressures but not for the corresponding temperatures. High-grade metamorphic temperatures indicate a perturbation of continental geotherms at least in Phanerozoic, and possibly also during Proterozoic times. It can be argued that, during the early stages of continental crust development, high-grade metamorphic conditions were general at lower crustal levels because of the missing or very thin shell of underlying lithospheric mantle. The Archean geothermal regime and the characteristics of Archean tectonics must be seen in this context. The bimodal volcanics of the Greenstone Belts and the genetically related tonalitic-trondhjemitic plutonism may not be related to island arcs or active continental margins, but to intracontinental rift environments. The upper boundary of the Archean asthenosphere was positioned near the crust-mantle boundary, and crustal growth took place by magmatic underplating. The Archean and most of the Proterozoic mobile belts show no typical signatures of Wilson cycles since their initiation is connected by a sufficiently thick and stiff lithosphere. The development of such a lithosphere is a gradual process which began during the late Archean and approached the present stage during Proterozoic and Phanerozoic times.

P-T PATH OF THE GRANULITE FACIES ROCKS

Further evidence of a "preorogenic" granulite facies metamorphism in the lower crust emerges from the P-T path of the granulites. In the course of the orogenic crustal shortening which followed granulite facies metamorphism, all of the Variscan granulites, and possibly the Proterozoic granulites as well, received an amphibolite facies overprint. They suffered a retrograde metamorphism which took them through the P-T field of Barrovian-type metamorphism and into the P-T field of Abukuma-type metamorphism (42). This P-T path is illustrated in Fig. 1. A general tendency, which Grambling (14) has also shown to occur in Precambrian metamorphites, is for high-grade metamorphic rocks to be produced along lower geotherms than medium and low-grade metamorphism. According to Zwart and Dornsiepen (42), the Variscan granulites went through the kyanite-sillimanite field along a geothermal gradient of 30°C/km and through the andalusite-sillimanite field along a geothermal gradient of 60°C/km. This variation in the metamorphic gradients took place during more or less isothermal decompression and is secular. The high-P-T metamorphism is Ordovician, the Barrovian-type metamorphism is Lower Devonian (Acadian) - and is preserved in many regions of the Variscan basement as relics within prograde metamorphic sequences - and the Abukuma-type metamorphism is of Carboniferous age. This secular change in the conditions of metamorphism is difficult to understand if the radiometric dates are unrealistic, and if one ascribes the granulites in the Variscan basement to Cadomian orogenesis (42) or attributes them to a Caledonian orogenesis affecting the pre-Variscan basement (41). In these cases it is implied that a Cadomian or a Caledonian orogen was eroded to the level of granulitic basement, and that the rocks which were first metamorphosed during Variscan time had post-Cadomian or even post-Caledonian educts. These suggestions are inconsistent with the stratigraphic evidence in the weakly metamorphic successions; in many Variscan regions these suggest a more or less continuous accumulation from late Precambrian through the Devonian or Carboniferous.

An alternative view of the case is more plausible. All prograde metamorphic reactions are dewatering reactions, i.e., endothermic processes which consume heat. The continental rift zones, which apparently had an association with granulite facies metamorphism at depth, contained a thick fill of sediments. As these sediments went through prograde metamorphism, they retarded the rise of isotherms. This buffering effect was enhanced by the widespread development of granitic melts above the level of the granulite facies. The rate of metamorphism is determined by the enthalpy of reaction, the heat capacity, and the net input of heat into metamorphic piles. The net input of heat can be understood as the sum of heat which enters the system and which is generated inside the metamorphic pile minus the heat which leaves by advection and conduction. At the beginning of orogenic crustal-shortening the front of metamorphism had reached a middle level of the crust where the kyanite-sillimanite parageneses

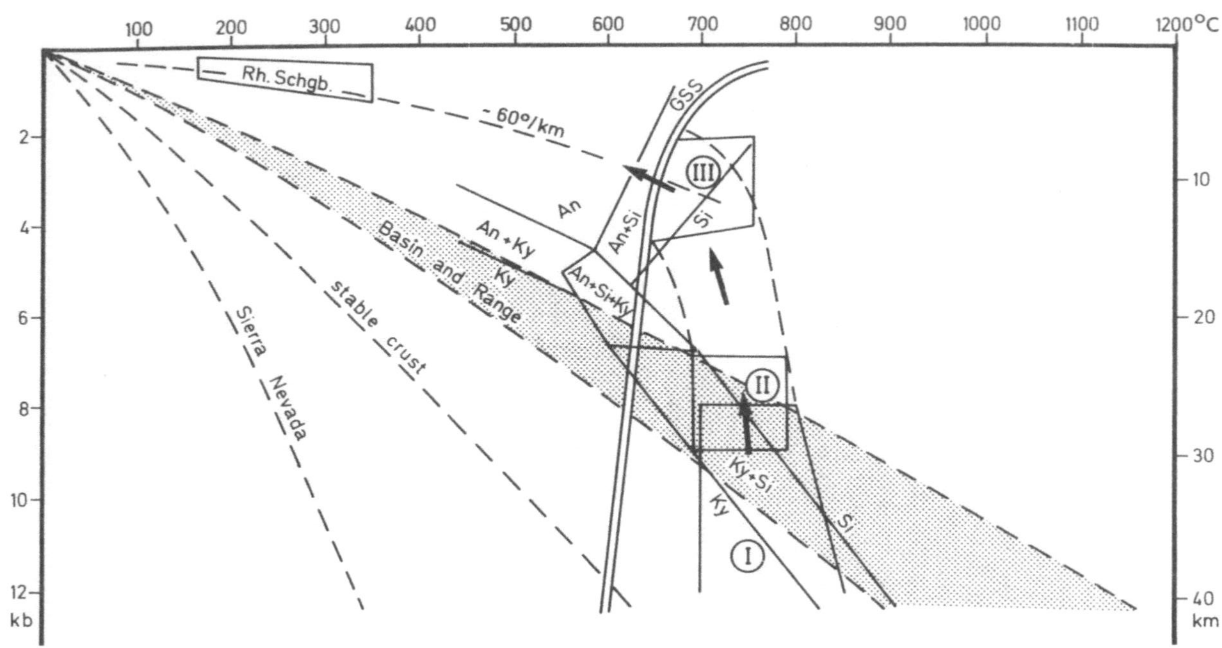

FIG. 1 - I PT field of early Variscan granulites (modified after Zwart and Dornsiepen (42)).
II P-T field of Barrovian-type metamorphism overprinting early Variscan granulites (after Zwart and Dornsiepen (42)).
III P-T field of Abukuma-type metamorphism overprinting early Variscan granulites (after Zwart and Dornsiepen (42)).
Geotherms in the Sierra Nevada, the stable crust, and the Basin and Range after Lachenbruch and Sass (22).
GSS: Granite-saturated solidus.

were produced under conditions of a low geothermal gradient. The crustal imbrication involved in orogenic shortening brought these higher-grade metamorphic rocks to high crustal levels where, under a more or less isothermal relaxation of pressure, they took on an Abukuma-type overprint. In the case of the granulites, this overprinting is retrograde. In the surrounding country rocks, it is prograde and often associated with anatexis (Anatexis II).

This, of course, does nothing to explain why the very weakly metamorphosed rocks, as, for example, the rocks of the Rheinische Schiefergebirge, which were never deeply buried and which first encountered deformation and metamorphism at a late stage of orogeny, became exposed to Abukuma-type conditions. It is necessary to take the view that in addition to "socle" effects and possibly enhanced radioactive production of heat, there were further sources of heating, because it cannot be presumed that the synorogenic heat flow was derived entirely from the processes involved in the rifting of continental lithosphere.

The model of Variscan orogenic evolution developed by Weber (36, 37) and Weber and Behr (40), like the model for the Damara Orogen developed by Marin and Porada (23), invoked delamination processes of the kind proposed by Bird (8, 9) which, according to Weber (36, 37) and Weber and Behr (40), would lead to subduction of lithospheric mantle below continental crust. The mass of subducted lithospheric mantle is balanced by upwelling hot asthenospheric material under continental crust; this material behaves like an Andrews-Sleep cell. This produces a synorogenic delivery of heat to the continental crust above. Processes of this kind were effective from the Devonian onward, had a strong bearing on the metamorphic history set out here, and may also have been responsible for production of the huge volumes of syn- to posttectonic granites whose geochemical characteristics in the Damara Orogen and in the European Variscides show them to be remelted metamorphic crust.

CONCLUDING REMARKS

The problem of a secular variation in tectonic style during Precambrian to Variscan time is treated here on the basis of the evidence of the ensialic orogenes which are frequent in this age range. There is a wide range in the degree to which regional structural development has proceeded and in the critical influence exerted on this development by magmatic and metamorphic processes. The range of products ranges from aulacogenes, with their very slight deformation, negligible metamorphism, and igneous activity as represented by the Athapuscow Aulacogene to extremely highly deformed, highly metamorphosed orogenes with abundant granite as represented by the Damara Orogen. Granulite facies metamorphism which is thought to be associated with continental rifting in the development of the European Variscides, is regarded as a common feature of intracontinental orogenesis. It is not necessarily present in all ensialic orogenesis. A case in point could be the Damara Orogen, in which Haack et al. (15) were able to show on the basis of $\delta^{18}O$ values that the granites are remelted crust, and that granulites can be excluded from the list of likely source rocks.

The thermal development of an intracontinental rift is heavily influenced by the primary thickness of the crust as well as by the scale and the rate of lithospheric rifting. Granulite facies metamorphism at pressures between 7 and 12 Kb requires a primary crustal thickness in the range of 25-40 km; it also probably requires low spreading rates and relatively slight crustal thinning. A thinner initial crust and/or the rapid spreading rates implied by more strongly developed preorogenic crustal thinning do not provide the conditions that are prerequisites of granulite facies continental rift metamorphism.

Acknowledgements. I owe thanks to H. Ahrendt, H.J. Behr, W. Franke, H. Martin, and H. Porada for many helpful discussions and to S.C. Matthews for translating the manuscript.

REFERENCES

(1) Arnold, A., and Scharbert, H.G. 1973. Rb-Sr Altersbestimmungen an Granuliten der südlichen Böhmischen Masse in Österreich. Schweiz. Min. Petr. Mitt. 53: 61-78.

(2) Bartholomé, P.; Evrard, P.; Kateksha, J.; Lopez-Ruiz, J.; and Ngongo, M. 1973. Diagenetic ore-forming processes at Kamoto, Katanga, Republic of Congo. In Ores in Sediments, eds. G.C. Amstutz and A.J. Bernard, vol. 3, pp. 21-41. International Union Geological Sciences, Ser. A. Berlin: Springer-Verlag.

(3) Behr, H.J. 1961. Beiträge zur petrographischen und tektonischen Analyse des Sächsischen Granulitgebirges. Freiberger Forsch.-H. C. 119: 1-146.

(4) Behr, H.J. 1980. Polyphase shear zones in the granulite belts along the margins of the Bohemian Massif. J. Struct. Geol. 2(1-2): 249-254.

(5) Behr, H.J.; Ahrendt, H.; Martin, H.; Porada, H.; Röhrs, J.; and Weber, K. 1983. Sedimentology and mineralogy of Upper Proterozoic Playa-Lake Deposits. In Intracontinental Fold Belts - Case Studies in the Variscan Belt of Europe and the Damara Orogen of Namibia, eds. H. Martin and W. Eder, pp. 577-610. Berlin: Springer-Verlag.

(6) Behr, H.J.; Ahrendt, H.; Porada, H.; and Weber, K. 1983. The Sole Dolomite at the base of the Naukluft Nappe Complex. In Evolution of the Damara Orogen, ed. R.M. Miller. Spec. Publ. Geol. Soc. S. Afr., in press.

(7) Behr, H.J.; Ahrendt, H.; Schmidt, A.; and Weber, K. 1980. Saline horizons acting as thrust planes along the southern margin of the Damara Orogen (Namibia/SW-Africa). Geol. Soc. Lond. Spec. Publ. 9: 167-172.

(8) Bird, P. 1978. Initiation of intracontinental subduction in the Himalaya. J. Geophys. Res. 83(No. B10): 4975-4987.

(9) Bird, P. 1979. Continental Delamination and the Colorado Plateau. J. Geophys. Res. 84(No. B13): 7561-7571.

(10) Calsteren, van P.W., and Tex, E. Den. 1978. An early Palaeozoic continental rift in Galicia (NW Spain). In Tectonics and Geophysics of Continental Rifts, eds. J.B. Ramberg and E.R. Neumann, pp. 125-132. Dordrecht: D. Reidel Publishing Co.

(11) Caby, R.; Bertrand, J.M.L.; and Black, R. 1981. Pan-African closure and continental collision. In Precambrian Plate Tectonics, Development in Precambrian Geology, ed. A. Kröner, vol. 4, pp. 407-434. Amsterdam: Elsevier.

(12) Derrick, G.M. 1982. A Proterozoic rift zone at Mount Isa, Queensland, and implications for mineralisation. BMR J. Austral. Geol. Geophy. 7: 81-92.

(13) Dimroth, E. 1981. Labrador Geosyncline: Type example of early Proterozoic cratonic reactivation. In Precambrian Plate Tectonics. Development in Precambrian Geology, ed. A. Kröner, vol. 4, pp. 351-354. Amsterdam: Elsevier.

(14) Grambling, J.A. 1981. Pressures and temperatures in Precambrian metamorphic rocks. Earth Planet. Sci. Lett. 53: 63-68.

(15) Haack, U.; Hoefs, J.; and Gohn, E. 1982. Constraints on the origin of Damara Granites by Rb/Sr and $\delta^{18}O$ data. Contrib. Mineral. Petrol. 79: 279-289.

(16) Hoffman, P.; Dewey, J.F.; and Burke, K. 1976. Aulacogenes and their genetic relation to geosynclines, with a proterozoic example from Great Slave Lake, Canada. In Modern and Ancient Geosynclinal Sedimentation, eds. R.H. Dott and R.H. Shaver. Soc. Econ. Pal. Min. Spec. Publ. 19: 38-55.

(17) Hunziker, J.C., and Zingg, A. 1980. Lower Palaeozoic Amphibolite to Granulite Facies Metamorphism in the Ivrea Zone (Southern Alps, Northern Italy). Schweiz. Min. Petr. Mitt. 60: 181-213.

(18) Jäger, E. 1977. The evolution of the central and west European continent. In La chaîne varisque d'Europe moyenne et occidentale. Coll. Intern. C.N.R.S. 243: 227-239.

(19) Jäger, E., and Watznauer, A. 1969. Einige Rb/Sr-Datierungen an Granuliten des Sächsischen Granulitgebirges. Monatsber. Deut. Akad. Wiss. Berlin 11: 420-426.

(20) Kober, B., and Lippolt, H.J. 1983. Lead isotopes in K-feldspars and rocks from the southern Schwarzwald, SW-Germany. Abstract. Terra cognita 3(No. 2-3): 199.

(21) Kuijper, R.P. 1979. U-Pb systematics and the petrogenetic evolution of infracrustal rocks in the Paleozoic basement of western Galicia (NW Spain), Verhandlung nr. 5, pp. 1-110. Amsterdam: ZWO Laboratorium voor Isotopen-Geology.

(22) Lachenbruch, A.H., and Sass, J.H. 1978. Models of an extending lithosphere and heat flow in the Basin and Range province. Geol. Soc. Am., Mem. 152: 209-250.

(23) Lister, G.S., and Dornsiepen, U.F. 1982. Fabric transitions in the Saxony granulite terrain. J. Struct. Geol. 4: 81-92.

(24) Martin, H., and Porada, H. 1977. The intracratonic branch of the Damara orogen in South West Africa. I. Discussion of geodynamic models. Precambrian Res. 5: 311-338.

(25) McClay, K.R., and Carlile, D. 1978. Mid Proterozoic sulphate evaporites at Mt. Isa Mine, Queensland, Australia. Nature 274: 240-241.

(26) McKenzie, D. 1978. Active tectonics of the Alpine-Himalayan belt: the Aegean Sea and surrounding regions. Geophys. J. R. Astron. Soc. 55: 217-254.

(27) McKenzie, D. 1978. Some remarks on the development of sedimentary basins. Earth Planet. Sci. Lett. 40: 25-32.

(28) Neudert, M.K., and Russell, R.E. 1981. Shallow water and hypersaline features from Middle Proterozoic Mt. Isa sequences. Nature 293: 284-286.

(29) Porada, H. 1983. Geodynamic model for the geosynclinal development of the Damara Orogen, Namibia/South West Africa. In Intracontinental Fold Belts - Case Studies in the Variscan Belt of Europe and the Damara Belt of Namibia, eds. H. Martin and W. Eder, pp. 503-5042. Berlin: Springer-Verlag.

(30) Rayner, R.A., and Rowlands, N.J. 1980. Stratiform copper in the Late Proterozoic Boorloo Delta, South Australia. Min. Depos. (Berlin) 15: 139-149.

(31) Renfro, A.R. 1974. Genesis of evaporite-associated stratiform metalliferous deposits - a sabkha process. Econ. Geol. 69: 33-45.

(32) Rolands, M.J.; Blight, P.G.; Jarvis, D.M.; and von der Borch, C.C. 1980. Sabkhas and playa environments in late Proterozoic grabens, Willouran Ranges, South Australia. J. Geol. Soc. Austral. 27: 55-68.

(33) Tex, E. Den. 1981. A geological section across the Hesperian Massif in western and central Galicia. Geol.en Mijnbouw 60: 33-40.

(34) Tex, E. Den. 1982. Dynamothermal metamorphism across the continental crust/mantle interface. Fortschr. Min 60(1): 57-80.

(35) Vidal, P.; Auvray, B.; Charlot, R.; and Cogné, J. 1981. Precambrian relics in the Armoricain Massif: their age and role in the evolution of the western and central European Cadomian-Hercynian Belt. Precambrian Res. 14(1): 1-20.

(36) Weber, K. 1981. The structural development of the Rheinische Schiefergebirge. Geol. en Mijnbouw 60(1): 149-159.

(37) Weber, K. 1983. The Variscan events: from early Palaeozoic continental rift metamorphism to Lower Devonian through Upper Carboniferous crustal shortening. Geol. Soc. London Spec. Publ., in press.

(38) Weber, K., and Ahrendt, H. 1983. Mechanisms of nappe emplacement at the southern margin of the Damara Orogen (Namibia). Tectonophys. 92: 253-274.

(39) Weber, K., and Ahrendt, H. 1983. Structural development of the Ugab Structural Domain of the northern Zone of the Damara Orogen. In Intracontinental Fold Belts - Case Studies in the Variscan Belt of Europe and the Damara Orogen of Namibia, eds. H. Martin and W. Eder, pp. 699-720. Berlin: Springer-Verlag.

(40) Weber, K., and Behr, H.J. 1983. Geodynamic interpretation of the mid-European Variscides. In Intracontinental Fold Belts - Case Studies in the Variscan Belt of Europe and the Damara Orogen of Namibia, eds. H. Martin and W. Eder, pp. 427-472. Berlin: Springer-Verlag.

(41) Ziegler, P.A. 1982. Geological Atlas of Western and Central Europe. Shell International Petroleum Maatschappij B.V.

(42) Zwart, H.J., and Dornsiepen, U.F. 1978. The tectonic framework of Central and Western Europe. Geol. en Mijnbouw 57: 627-654.

Standing, left to right:
Alfred Kröner, Dennis Gee, Keith O'Nions, Hans Ahrendt, Wolfgang Frisch, Bob Dymek, Kevin Burke.

Seated, left to right:
Frank Richter, Ron Oxburgh, Alan Thompson, Mike Bickle, Klaus Weber.

Patterns of Change in Earth Evolution, eds. H.D. Holland and A.F. Trendall, pp. 389-406.
Dahlem Konferenzen 1984. Berlin, Heidelberg, New York, Tokyo: Springer-Verlag.

The Long-term Evolution of the Crust and Mantle
Group Report

A.B. Thompson and F.M. Richter, Rapporteurs
H. Ahrendt, R.D. Gee
M.J. Bickle, A. Kröner
K.C. Burke, R.K. O'Nions
R.F. Dymek, E.R. Oxburgh
W. Frisch, K. Weber

INTRODUCTION

Our group had the task of considering the evidence relating to very long-term changes in the evolution of the Earth. At present the terrestrial geological record begins at about 3.8 Ga, by which time the surface of the planet had already become the site of deposition of sedimentary rocks. The preceding 15% of Earth history must have contained a period of dramatic and perhaps unidirectional change during which the planet evolved from its primordial condition, dominated by condensation or accretion, probably through a phase of widespread melting, to one in which surface processes akin to those of today were able to operate.

The group chose not to speculate at length about events during the first 0.7 Ga of Earth history but to concentrate on the period covered by the terrestrial record. The surface features and the processes that shaped the world 3.8 Ga ago probably differed from those of the present day. However, this conclusion rests largely on a somewhat subjective evaluation of phenomena that are understood poorly and on a history that is poorly preserved. In the limited time available, the group thought that it would be most useful to identify those phenomena that do not appear to be characteristic of the entire history of the Earth since 3.8 Ga ago. It is likely that long-term changes are reflected in variations of the relative abundance of various features of the Earth. We have attempted to identify these changes, but the problem of sampling and of poor preservation are serious, and firm conclusions must await a much more rigorous and quantitative review of the evidence.

CHANGE OF TECTONIC STYLE WITH TIME

General Question: Does tectonic style change gradually over long time periods, or are these changes concentrated in periods of short duration and great intensity?

We begin with geological features that may be considered characteristic of particular geological time periods.

Archean Granite-greenstone Belts

Supracrustal sequences dominated by tholeiitic basalt with locally abundant komatiites with up to 30% MgO and also, in some cases, calc-alkaline volcanics are preserved between heterogeneous granitoid crust often exposed in ovoid or elongate "batholiths". The latter form linear or two-dimensional systems of domes and batholiths. Large granite-greenstone provinces (e.g., the Superior Province in Canada), contain major linear metasediment-gneiss belts up to 100 km wide and up to 1000 km long. Associated sediments are dominantly graywackes most likely derived from volcanic sources.

Archean High-grade Gneiss Terrains

These areas contain extensive bodies of complexly deformed granitoid gneiss. These are derived from calc-alkaline igneous precursors that were typically metamorphosed at the upper amphibolite or granulite facies. Supracrustal successions of abundant metabasaltic amphibolite and lesser metasediments are tectonically intercalated with these gneisses. The metasediments may include quartzite, metapelite, and carbonate which may represent a shelf facies association. Other important rock types include lenses of ultramafic material and calcic anorthosite representing plagioclase cumulates from differentiated mafic igneous complexes. The gneiss terrains are characterized by convergent tectonics as revealed by the presence of thrusts, recumbent folds, and nappes which are broadly synchronous with plutonism (e.g., W. Greenland).

Proterozoic Belts

The earliest types of Proterozoic tectonic activity from 2.2 to 1.7 Ga are characterized by belts of the following characteristics: initial development by rifting of Archean blocks; deposition of early shelf sediments, extensive (>100 km) marine and freshwater sediments, banded-iron formations, evaporites, followed by thick graywacke sequences with volcanics (some Mg-rich) - in troughs. The rifts are located on extant crustal segment boundaries (an ensialic floor is now present in these intracratonic fold belts). Compressional tectonics are correlatable with lateral changes in metamorphic grade and granitoid plutonism.

It has been proposed that some Proterozoic orogenic belts, which made their appearance at about 2.2-1.8. Ga, are intracratonic. This does not mean that clear rift tectonics only began in the Proterozoic, as the Pongola of South Africa and Swaziland (~3.1 Ga old) represents a rift-fill as may the Great Dike (~2.6 Ga old). Belts such as the Wopmay orogen have been interpreted as associated clearly with plate tectonic processes (13).

Phanerozoic Collision Belts (Caledonides, Alps, and Himalayas)

Evidence for the operation of various parts of the Wilson cycle (1) is well established in some Phanerozoic orogenic belts. The European Alps show both the ocean-opening stage related to the splitting of continental Pangea and the formation of the Penninic ocean, and the ocean-closing stage in subduction-reconvergence followed by the collision of the African plate, perhaps in response to the opening of the Atlantic Ocean. The stratigraphic-sedimentary sequence of the Alps has preserved characteristics of the whole process. The presence of oceanic crust is recorded by ophiolitic slices, and subduction is suggested by their metamorphism to blueschist-eclogite facies. Continental collision is reflected by calc-alkaline plutonism and Barrovian-type facies in the Lepontine Alps. Large-scale (10^2 km) nappe thrusts appear to be diagnostic of a collision belt.

Phanerozoic Variscan Belts (Hercynides)

The absence of distinct ophiolitic material from several areas of the European Variscides has been suggested to be evidence for Phanerozoic ensialic orogeny. The grade of metamorphism, locally reaching granulite facies, several periods of deformation, and calc-alkaline to alkaline intrusion and ophiolite slices in some areas appear to be otherwise similar to the Phanerozoic collision belts. If continental stretching was followed rapidly by collision before the appearance of intervening ocean floor, it is conceivable that adiabatic decompression of the upward movement of lithosphere could have provided the heat necessary for metamorphism, orogeny, and melting of the thinned crust to produce the observed plutonism.

Phanerozoic Continental Marginal Belts (Circumpacific)

Such belts are typified by extensive island-arc or Andean volcanism and calc-alkaline plutonism at the boundary between an extensive ocean area and a continent. Many microcontinental masses with variable rock types assembled in an allochthonous collection could represent transported exotic blocks (19). Long lasting subduction produced by continuous sinking of old, dense oceanic lithosphere could transport microcontinental masses to the allochthon. Thickening of the accreted material can apparently induce metamorphism and melting in the accreted wedge.

Relations of High- to Low-grade Terrains in the Archean

Although there are cases where granite-greenstone belts and the associated high-grade gneiss terrains differ in age and in the nature of the associated sedimentary facies, evidence has appeared for unbroken prograde transitions from one to the other (16). This could suggest that the high-grade gneisses form the basement of the granite-greenstone belts. Although submarine pillow lavas are among the extrusive materials in the greenstone belts, the structure of the present oceanic crust has not been matched successfully with that of greenstone belts. The metamorphic grade of greenstone belts, even in sequences as much as 10 km thick, is much less than in recent seafloor sections (up to amphibolite facies).

There is little evidence for deep-water deposition in the sedimentary sequences on stable Archean cratons. This observation eliminates neither the possibility that there were large and deep oceans in the Archean, nor the possibility of plate tectonics involving oceanic crust. High- and low-grade Archean terrains clearly have a remarkable facility for survival, and there remains the distinct possibility that other Archean crustal components were returned to the mantle or were otherwise destroyed. Volcanic rocks in greenstone belts were initially only metamorphosed to low grades; they were therefore well insulated from the Archean mantle.

Tectonic styles in Archean provinces are difficult to investigate. They sometimes give evidence of thrusting in tectonic packets (Bickle, this volume). The apparent absence of linear metamorphic belts in the Archean may be real.

Presently available paleomagnetic data for the Kaapvaal, Yilgarn, and Superior cratons indicate minimum latitudinal velocities of between 1.5. and 4.5. cm a^{-1} during the Archean (3.4 to 2.5 Ga ago) and are thus not significantly different from those of today. These paleomagnetic data also indicate that some crustal blocks move independently of each other.

Ensialic Belts in the Proterozoic

The extensive Proterozoic platform sedimentary deposits (quartzites, evaporites, shelf limestones) suggest that there were much larger areas of continent (10^7 km^2) during this period than during the Archean. Paleomagnetic results from different Proterozoic blocks seem to provide evidence for the drift of continental masses at that time. Unequivocal oceanic crust as accepted for the Phanerozoic is not yet recognized in Proterozoic crust older than about 0.96 Ga (Arabian Shield and Eastern Desert, Egypt, where coherent ophiolite complexes with sheeted dykes have been described (8)).

Many Proterozoic belts may reflect the operation of non-Wilson cycle tectonics since there are no exposed ocean floor remnants. However, evidence that would permit matching preexisting structures and stratigraphy across fold belts is lacking. Demonstration of the continuity of lowermost stratigraphic unit across belts and the occurrence of shallow-water sediments flooring basaltic complexes would reinforce the failed-rift or lithospheric stretching models of ensialic orogeny. The incompleteness of the geological record and the inadequacy of exposures may account for the inability to detect collisional sutures. Paleomagnetic methods are probably not sensitive enough to detect small lateral movements of Proterozoic crustal blocks.

The massif anorthosite suite (with syenites, mangerites, and alkali and Rapakivi granites) characterizes a Proterozoic (ensialic?) belt running from the Urals through Fennoscandia, Canada, to S. California. The resemblance of some of these magmatic types to igneous material in recent rift systems may imply that special melting conditions, rather than special source material, is necessary for their formation.

The preservation of Proterozoic orogenic belts interpreted as ensialic (Capricorn Orogen in Australia (12), Labrador Trough (6)), and others interpreted to be ensimatic (Wopmay Orogen, Canada (13)) suggests that either aborted or complete Wilson cycles were operating in the Proterozoic.

Which Geological Phenomena Are Restricted in Time?

Komatiites with a high MgO content (22 to 30 wt%) are apparently restricted to the Archean. Massif anorthosite suites are largely restricted to the Proterozoic. Serpentinite-ophiolite-sheeted dyke complexes older than 0.96 Ga have not been found. Blueschists are apparently confined to the Phanerozoic, although rare examples as old as 1 Ga are claimed to have been found.

Which Geological Data Are Evidence for the Beginning of Plate Tectonics?

The oldest well documented ophiolite is about 1 Ga old; the proposition that ocean floor basalts were characteristic of much older ocean floor rests only on geochemical indicators. The difficulty of answering this "non-question" is increased by multiple definitions of plate tectonics and the need to reach agreement regarding the diagnostic value of features such as major linear calc-alkaline intrusive belts for the operation of the Wilson cycle.

Some Suggestions for New Observations

a) Determination of the nature of the floor of Proterozoic fold belts; was it sialic or ophiolitic? The answer would help us to distinguish between the several orogenic types.

b) Petrological and geochemical investigation of Archean and Proterozoic (meta-)sediments, as they could provide an average of rock types not otherwise available in the exposed sections. Sampling should be on a continent-wide basis, tied to present regional geology and to the regional geology at their time of formation (to delineate provinces).

c) Systematic investigations of structure and stratigraphy at the boundaries of Archean granite-greenstone belts and high-grade gneiss terrains.

d) Determination of the nature of the granites in Archean batholiths; are they diapiric? These studies should complement petrological-geochemical investigations regarding their depth or origin and the mechanical state of the Archean crust and lithosphere.

e) Determination of the parent age of serpentinites in Archean greenstone belts; was the parent a dunite or a harzburgite?

f) Investigation of the basement of Archean terrains to determine the stratigraphic significance of apparent unconformities.

GEOLOGICAL OBSERVATIONS TO CONSTRAIN THEORIES OF THE EVOLUTION OF THERMAL REGIMES

General Question: How can we use the characteristics of exposed rocks to deduce thermal conditions at any particular time in Earth history?

Metamorphic Rocks

The process of metamorphism itself requires perturbation of steady state thermal conditions. The presence of metamorphic rocks at the Earth's surface further requires a mechanism to elevate them; the simplest of these mechanisms is erosional uplift of overthickened crust. Although some high-grade metamorphic material is unquestionably uplifted along listric thrust faults or by diapiric rise within the crust, erosional uplift is a more likely mechanism for exhuming many high-grade terrains. Supporting evidence for this proposition includes the lack of obvious thrusts and their associated metamorphic zones, the very wide and uniform extent and grade of high-grade terrains, and the abundant evidence for crustal thickening (folding, thrusting, and igneous intrusion) prior to the crystallization of the final metamorphic fabric.

Erosion-controlled exhumation of tectonically thickened crust is likely to result in some heating in the early stages of uplift, because temperature increases due to radiogenic heat production are relatively fast compared to erosion (10). Pressure-temperature (P-T) conditions recorded in metamorphic rock are therefore likely to be more recent than those attained in thickened crust during metamorphism prior to erosion. If the P-T conditions recorded are indeed those of the highest temperature achieved during metamorphism, then deduced geobarometric pressures would be less by up to 1/2 in extreme cases than the pressure at the depths achieved by tectonic thickening (11). Because many metamorphic rocks contain "disequilibrium features," such as chemical zonation of minerals, they may be capable of yielding segments of P-T time paths; these are potentially much more helpful in reconstructing the response of the crust to tectonic processes than individual P-T points obtained from "perfectly equilibrated" samples.

High pressure metamorphism and the blueschist-eclogite association

The blueschist-eclogite association, which is largely confined to the Phanerozoic, is often considered to be diagnostic of subduction-zone metamorphism. The thrust-thickening models of Oxburgh and Turcotte (17) have shown that the appropriate P–T conditions can certainly be achieved in continental collision zone metamorphism. These P–T conditions are also achieved in homogeneous thickening models if the initial geotherm has a low slope (11), and blueschist-eclogite facies could be preserved to the surface if uplift is rapid and immediately follows collision. Thus the presence of the entire ophiolite suite metamorphosed to blueschist-eclogite facies is required to identify subduction-zone metamorphism with certainty.

Granulites and abnormal P–T conditions

P–T conditions inferred from mafic to pelitic granulites from widely dispersed areas with a range of geological ages reveal only a limited range of P–T conditions of equilibration during granulite facies metamorphism (8-10kb; 700-900°C (18)). Although the significance of this result is unclear, it may have profound implications for the nature and mechanisms of crustal thickening. The pressure estimate yields a minimum value of Archean crustal thickness and implies an instantaneous thickness >30 km for the crust in high-grade Archean gneiss terrains (see Dymek, this volume). As outlined by England and Thompson (11), only crust thickened to about 70 km could cause rocks to pass through the granulite P–T region in a single exhumation episode. Other mechanisms of attaining granulite P–T conditions include rapid successive thickening episodes or rapid increase in heat input into the base of the crust, either by magma injection or by extension of the lithosphere beneath already thickened crust. Similar increased thermal inputs are required for the regional generation of high-T and low-P (andalusite-sillimanite) conditions.

The apparent rarity of Phanerozoic granulites may be related to the difficulty of exhumation. The lack of preservation of pre-Phanerozoic blueschist-eclogites could be related to erosional destruction at the surface following rapid exhumation.

Other common metamorphic associations and P–T conditions

While steady state temperatures at the base of a 30-35 km crust may be appropriate to Barrovian (kyanite-sillimanite) metamorphism, crustal thickening is normally required to induce transport of these metamorphic rocks to the surface. Clearly the notion of a "hot and thin" Archean crust cannot apply to the Archean high-grade gneiss terrains (Dymek, this volume) which retain a memory of metamorphic P–T conditions very similar to those found in Phanerozoic orogenic belts. With the possible exception of the blueschist-eclogite association, the mineral geobarometric P-T conditions deduced for the Archean and Proterozoic match the range of conditions during the Phanerozoic.

Mineral compositions in metamorphic rocks are frequently altered by diffusional processes during uplift. The inferred P-T conditions tend to be influenced by continuous and non-simultaneous cation closure along P-T time paths and therefore cannot be used to deduce geotherms without an independent measure of time. Multiple P-T conditions are likely to be recorded by metamorphic rocks and if, like isotopic closure, they occur at different temperatures following the maximum temperature achieved along the path, then deduced pressure, temperature, and the apparent age are all functions of time.

Magmatic Rocks

Basalts and calc-alkaline intrusives are found in all parts of the geological record since 3.8 Ga ago. Even alkali basalts and trachytes are found in the Archean Superior Province (Timiskaming (5) and in the Archean granities of Finland (14)).

As recent alkali basalts are often found in elevated areas (seamounts or elevated continental areas), their apparent scarcity in the geologic record could be related to removal by erosion. Archean and Proterozoic sediments and volcanics could be preferentially preserved as the elevated regions become eroded away.

Komatiites and Archean mantle temperatures

High-HgO komatiites appear to be restricted to the Archean and may form from 1% to 30% of an ultramafic komatiite-basaltic komatiite-tholeiite association. This association suggests that in the Archean upper mantle the tholeiite through peridotitic-komatiite association was generated at temperatures between 1300° and about 1700°C. The suggestion that there was simply a komatiite-producing layer in the Archean mantle beneath a "hot and thin" Archean crust is therefore hardly warranted, since such trapped layers would cause melting in the overlying material.

Archean greenstones and granites

There is little field evidence to suggest that Archean granitoids are produced by in situ melting of greenstones.

The isotopic composition of several elements in Archean granites (sensu stricto) (Moorbath, unpublished) shows that they were not derived from contemporaneous basaltic parents by partial melting. Isotopic data for Archean tonalites are not inconsistent with an origin of these rocks by partial melting of a basaltic parent, but greenstones were not necessarily the source rocks.

The relationship between Archean granulites and the partial melting of (hydrated?) basaltic material is not known. Melting at 700°- 800°C in the Archean lower crust could buffer temperatures so that overlying granulites would not become hotter. Alternatively, the granulites could represent the more refractory residues after the removal of more fusible portions.

GEOCHEMICAL EVOLUTION OF THE EARTH

A large number of topics fall under the heading of geochemical evolution or geochemical evidence for change. The sedimentary record was discussed in part because certain features (REE, LIL, V, Cr, Ni, etc., concentrations) appear to some workers to be different in Archean rocks, and also because it may indicate whether the rocks that have survived are reasonably representative. At the other extreme, data for particular isotopic systems provide evidence for the operation of particular evolutionary processes.

The Sedimentary Rock Record

The "distinctive" concentration of REE, LIL, V, Cr, Ni, and others in Archean sedimentary rocks as compared to more recent sedimentary composites may be more a function of sampling bias (i.e., more volcanoclastics in greenstone belts) than an indication of temporal change (7)). Early sediments are particularly important; as yet no claim for significant changes in composition or for the presence of evolutionary patterns can be made. The need for further study of these sediments is apparent once one begins to wonder just how unrepresentative the preserved record of nonsedimentary rocks may be. Isotopic studies of Archean sediments may contain evidence for an even greater age of the source terrain (see **CONCLUSION**).

Noble Gases

O'Nions (this volume) gives an introduction to the way in which noble gases can be used to argue for early degassing of the mantle.

The presence of a ^{129}Xe isotope anomaly in mid-ocean ridge basalt (MORB) is most suggestive of a rapid initial (<100 Ma) mantle degassing event, and high $^{40}Ar/^{36}Ar$ ratios in the same rocks are consistent with the early formation of the atmosphere. Observations of the isotopic composition of helium confirm that terrestrial degassing is still going on and allow a quantitative estimate of some fluxes (Holland, this volume). The bimodal distribution of "mantle helium," in addition to the isotopic signatures of Ar and Xe, point to the long-term (perhaps permanent) existence of two isolated geochemical reservoirs. The first, characterized by high $^{4}He/^{3}He$, $^{40}Ar/^{36}Ar$, and $^{129}Xe/^{130}Xe$ ratios, can be thought of as a depleted MORB source. The second, having low $^{4}He/^{3}He$, $^{40}Ar/^{36}Ar$, and $^{129}Xe/^{130}Xe$ ratios, indicates a relatively undepleted "primitive" reservoir which may be the source of ocean-island basalts such as those found in Iceland and Hawaii. Furthermore, $^{4}He/^{3}He$ - $^{87}Sr/^{86}Sr$ systematics point to another source, viz, recycled crustal or sedimentary materials, for apparently "enriched" ocean-island basalts such as Tristan da Chuna and the Azores. The long-term existence of such reservoirs, particularly as suggested by the noble gas isotopic signatures, do not appear consistent with the proposed late accretion (by Cl chondrites or Jovian-system type materials) of volatiles, unless some special mechanism of separating individual components (i.e., the noble gases from water and CO_2) prior to accretion was involved.

The fact that we observe primitive helium isotope ratios in such places as Hawaii and Iceland means that the layers must be able to "leak" in places while maintaining their large-scale isolation. The need for "leaks" is based on data from presently active areas, and thus there is currently no information regarding the existence of similar leaks in the past.

Bulk Earth

Our ability to characterize past thermal regimes depends critically on estimates of the bulk composition of the Earth, especially the total abundance of the long-lived heat producing elements U, Th, and K. As with many things geochemical, ratios are not strongly disputed (Th/U = 3.8, K/U = 10^4); the key question is the amount of uranium or of some element whose ratio to uranium is reasonably well-known. If Ca is used as a guide, the preferred value for the concentration of U in the silicate portion of the Earth is 20 ppb. The uncertainty in this estimate is difficult to assess, but various statements during this workshop suggest that the U content is "no less than 15 ppb" and "certainly no more than 30 ppb."

If the estimate of 20 ppb for U and the above ratios are correct, the total heat flux out of the Earth exceeds present-day heat production by a factor of two; this makes an important demand on models for the thermal regime of the Earth. One fourth to one third of the Earth's uranium is present in the continental crust. The balance is distributed in the mantle; the concentration of uranium may only be 5 ppb in the upper mantle which has been depleted by the extraction of U into the continents; the "primitive," undifferentiated lower mantle may contain 20 ppb U.

Isotopic Evolution of the Mantle/crust System

Given a parent-daughter ratio (i.e., Sm/Nd, assumed chondritic) and an initial isotopic ratio (taken from meteorites), one can define the course of bulk Earth evolution curves, i.e., those which we would observe if no differentiation had taken place. Departures from this curve are evidence for and measures of differentiation and can be used to date the process of differentiation. The present mantle as sampled by mid-ocean ridge basalts clearly reflects a history of depletion; the average age of depletion is about 2 Ga. One might choose also to call this number the mean age of the continental crust, as it is generally regarded that the reservoir is enriched when the mantle is depleted. As pointed out by O'Nions (this volume), the meaning of this number is far from clear. It would be the "real" mean age of the continents only if there were no recycling of continental or continent-derived material back into the mantle. In effect the mean age reflects the relative rate of differentiation and recycling; on the basis of this type of data it is impossible to rule out a constant continental volume and a decreasing recycling rate. The present volume-age relation for continental rocks may speak more of survival than of production.

Isotopic measurements of Archean igneous rocks, especially their initial isotopic ratios (their isotopic composition when emplaced), add an extra dimension if one assumes that the initial ratios are representative of the mantle at the time of emplacement. For example, the Isua rocks (W. Greenland) suggest that the mantle had already been depleted for a significant length of time before their emplacement. The arguments for this are at present quite fragile. The data base is limited, and very small departures from the bulk Earth evolution curve take on an exaggerated importance in early Earth history. If the existing data are taken at face value, continental material in the sense of a reservoir enriched in incompatible elements existed well before the time of the Isua rocks. The Isua rocks and others of similar age have a special importance, not just for their own sake, but because they contain evidence regarding the evolution of the Earth more than 3.8 Ga ago.

If the oldest rocks are survivors of a much more extensive crust, the question of their representativeness becomes all the more important. At present the likelihood of a portion of the crust surviving is related to its composition. If the same was true during the Archean, then the surviving fragments are probably a highly biased sample of the crust of that time.

MODELS FOR THE EARLY HISTORY OF THE EARTH

Several important aspects of the thermal and chemical evolution of the Earth involve model calculations. Some models are used to translate one set of observations (for example, on the strength of the lithosphere) into a constraint on some other property of the system (for example, the prevailing thermal regime). In other cases, analogues such as numerical or laboratory experiments are used to characterize convective heat transfer; these characterizations are then applied to describe the thermal evolution of a convective planet with decaying heat sources.

Accretion and Meteorite Bombardment

There is increasing agreement among scientists modelling the accretion of the Earth that it was rapid, in the sense that about 98% of the mass was in place within about 100 Ma, and that the initial state of the Earth is what can be called "hot"; convection was probably vigorous during and/or after the accretion phase. A time scale of about 100 Ma may be an upper limit to be consistent with the isotopic composition of Xenon in the atmosphere.

A late heavy bombardment event lasting until perhaps 4.0 Ga is generally accepted and could have added a significant portion of the volatile inventory of the Earth.

Noble metals in mantle-derived ultramafic rocks preserve approximately chondritic relative proportions; this is consistent with the addition of a meteoritic component early in Earth history, but after core formation (e.g., (3, 4, 15)).

The very early history of the Earth is relevant to our deliberations, because it sets the initial thermal conditions and determines the mechanisms for disrupting (but not recycling) very early crustal material. Sufficient water was probably present at the Earth's surface, so that wet-melting played a role in the geochemical recycling that is apparently required by the data for the isotopic evolution of the mantle.

Evolution of the Lithosphere

Several types of observations can be related to the thermal structure of the (continental) lithosphere during the Archean. The properties of sedimentary basins are one example. Where basin formation results from stretching and thinning of the crust, the rate of subsidence and the ultimate thickness of sediments is related to the properties of the crust and lithosphere. It is still difficult to apply such concepts to the Archean, because the key parameters for the relevant sedimentary sequences are not yet well enough known. Clearly, there is a need for additional studies of Archean sediments.

A second line of reasoning begins with the height of mountains (9). The maximum height of present-day mountains and plateaus, such as in Tibet, is probably limited by the strength of the lithosphere, which is very sensitive to its temperature structure. If mountains exceed a certain height, the lithosphere beneath them deforms rapidly, and their elevation is reduced. The hotter the lithosphere, the lower its strength, and the smaller the maximum supportable height of mountains. If we could estimate the height of Archean mountains, we could develop a crude but useful thermometer for the lithosphere.

The height of Archean mountains is not observable, but the burial pressure of rocks today at the erosional surface can be estimated using geobarometers or metamorphic grade. The burial pressure can, by simple isostatic arguments be related back to the height of the original topography. A very common observation is that granulites presently exposed show pressures of 8 to 10kbars which corresponds to a depth of 25 to 30 km. This is not very different from what one might expect to find in Tibet once the plateau has been reduced by erosion. The implication is that the height of Archean mountains was similar to that of mountains today, and that the thermal regime was also similar to that of the present day. The nature of the underlying assumptions make this a very qualitative estimate, but a very interesting one, nevertheless. A similar conclusion can be reached in a somewhat different way. Several Archean terrains (e.g., the Superior and Yilgarn Provinces (>2.5 Ga)) have a "perfectly normal" crustal thickness today and would have melted had the thermal regime since its emplacement been very much higher than today's (2). This is also a very qualitative statement, but one which is simple and compelling.

Thermal Evolution Models

Several of the thermal and geochemical estimates discussed above make special demands on models for the thermal evolution of the Earth. The principal ones and their implications are listed below.

a) Bulk composition: 4.5 Ga ago the rate of the release of radiogenic heat was probably more than four times greater than at present. Secular cooling of the Earth may be contributing as much as half the present-day heat flow (as the ratio of measured heat flow to the calculated contemporaneous rate of heat production is about two). Such a large contribution due to secular cooling is most easily achieved if mantle convection is layered; even then the temperature of the upper mantle must have changed by about 200° (~1550 to ~1350) during the last $3x10^9$ years. The inability of whole mantle convection models to account for the apparent present-day secular cooling is demonstrated in Richter's paper (this volume).

b) Komatiites: The eruption of komatiites containing more than 30 wt% MgO requires mantle temperatures in excess of 1700°C. This is strong evidence for a hotter Archean mantle. If komatiitic magmas developed in regions that were hotter than average, perhaps in regions of rising convection currents, there is no serious conflict between temperature constraints imposed by komatiites and the proposal that an "average" mantle temperature of 1550°C was needed for secular cooling. The mean mantle temperature and the variations about the mean must have decreased with time; hence it is not surprising that komatiites are restricted to early periods of Earth history.

c) Lithospheric thermal structure: If the conclusion based on the "height of Archean mountains" is accepted, together with the implication that the thermal structure of the lithosphere in the Archean was similar to that of today, then thermal models run into serious trouble. They predict that the interior temperature has changed by a few hundred degrees and that the surface temperature gradient was steeper in proportion to the higher heat flux. This suggests that the lithosphere was thinner, perhaps by as much as a factor of three, $3.0x10^9$ years ago. This proposition is valid even in regionalized models that try to "protect" continental areas by postulating the existence of oceanic areas that remove heat very efficiently from the Earth during the creation of new oceanic lithosphere at ridges.

The difficulty seems to be more than a simple inadequacy of present thermal modelling. The komatiite constraint and the demands of an unchanging lithospheric thermal structure are in conflict; one demands change, the other seems to preclude it.

There is one other feature of the thermal evolution models worth noting. Surface temperature gradients must have been steeper over most, if not all, of the surface of the planet. There must therefore have been widespread melting of the Earth at depths of about 50 km. Melting is common in the oceanic regions today; its pervasiveness presumably increases as we go back into the past. Continental areas are spared the full share of heat flux by the more efficient heat loss in the oceans; however, widespread melting under the continental lithosphere is predicted during the Archean. This is so even when an oceanic style of heat transfer is assumed to have prevailed (with the specific assumption that the characteristic horizontal size of the oceanic plates is the same as at present). How the Earth coped with such widespread melting, even if melting was restricted to oceanic areas, is an important question, one that transcends the details of thermal evolution models.

Since the end of the workshop, detrital zircons with an age of around 4.15 Ga have been discovered with the ion-microprobe facility at the Australian National University in Canberra. Such zircons must have been derived from granitoid rocks, and further measurements will hopefully verify their apparent age. These results could compress still more tightly the earliest period of Earth history by the end of which the processes that generate basalts and granitoids had become established in its outer parts.

CONCLUSION

Unidirectional changes in Earth processes have certainly taken place during the past 3.8 Ga, the most profound of which involved the growth and stabilization of the continents, the evolution of an oxidizing atmosphere, and the diversification of life forms. It was the admittedly subjective view of our group, however, that these changes should not blind us to the fact that many processes have apparently not changed at all. The really surprising thing is just how little change can be documented to have taken place during the last 80% of Earth history.

REFERENCES

(1) Burke, K., and Dewey, J. 1974. Hot spots and continental breakup: implications for collisional orogeny. Geology 2: 57-60.

(2) Burke, K., and Kidd, W.S.F. 1978. Were Archean continental geothermal gradients much steeper than those of today? Nature 272: 240-241.

(3) Chou, C.-L. 1978. Fractionation of siderophile elements in the earth's upper mantle. Proceedings of the Lunar and Planetary Science Conference, vol. 13, pp. 219-230, Houston, Texas.

(4) Chou, C.-L.; Shaw, D.M.; and Crockett, J.H. 1983. Siderophile trace elements in the earth's oceanic crust and upper mantle. J. Geophys. Res. 88: A507-A518.

(5) Cooke, D.L., and Moorhouse, W.W. 1969. Timiskaming volcanism in the Kirkland Lake area, Ontario, Canada. Can. J. Earth Sci. 6: 117-132.

(6) Dimroth, E. 1972. The Labrador geosyncline revisited. Am. J. Sci. 272: 487-506.

(7) Dymek, R.F.; Boak, J.L.; and Gromet, L.P. 1983. Average sedimentary rock rare-earth element patterns and crustal evolution: Some observations and implications from the 3800 Ma Isua supracrustal belt, West Greenland. In Archean Geochemistry - Early Crustal Genesis Field Workshop. Houston: Lunar Planetary Institute.

(8) El Bayoumi, R.M. 1982. Ophiolites and melange complex of Wadi Ghadir area, Eastern Desert, Egypt. Precambrian Res. 16: A17-A18.

(9) England, P.C., and Bickle, M.J. 1982. Constraints on Archaean geothermal regimes and their implication for Earth thermal history models. In Lunar Planetary Institution Proceedings of the Planetary Volatiles Conference, October 1982, Colorado.

(10) England, P.C., and Richardson, S.W. 1977. The influence of erosion upon the mineral facies of rocks from different metamorphic environments. J. Geol. Soc. Lond. 134: 201-213.

(11) England, P.C., and Thompson, A.B. 1984. Pressure-temperature-time paths of regional metamorphism: Part I, Heat transfer during the evolution of regions of thickened continental crust. J. Petrol., in press.

(12) Gee, R.D. 1979. Structure and tectonic style of the Western Australian shield. Tectonophysics 58: 327-369.

(13) Hoffman, P.F. 1980. Wopmay orogen: a Wilson cycle of early Proterozoic age in the northwest of the Canadian shield. Geol. Assoc. Can. Spec. Paper 20: 523-549.

(14) Jahn, B.-M.; Auvray, B.; Blais, S.; Capdevila, R.; Cornichet, J.; Vidal, F.; and Hameurt, J. 1980. Trace element geochemistry and petrogenesis of Finnish greenstone belts. J. Petrol. 21: 201-244.

(15) Morgan, J.W.; Wandless, G.A.; Petrie, R.K.; and Irving, A.J. 1981. Composition of the earth's upper mantle. I. Siderophile trace elements in ultramafic nodules. Tectonophysics 75: 47-67.

(16) Newton, R.C., and Hanson, E.C. 1983. The origin of Precambrian granulite terrains evidence from the transition zones. In Proterozoic Geology, ed. G.L. Medaris, Jr. Geological Society of America Memoir 214, in press.

(17) Oxburgh, E.R., and Turcotte, D.L. 1974. Thermal gradients and regional metamorphism in overthrust terrains with special reference to the eastern Alps. Schweiz. Min. Petr. Mitt. 54: 641-662.

(18) Perkins, D.P. III, and Newton, R.C. 1981. Charnockite geobarometers based on coexisting garnet-pyroxene-plagioclase-quartz. Nature 292: 144-146.

(19) Wilson, J.T. 1968. Static or mobile Earth: The current scientific revolution. In Gondwanaland Revisited, American Philosophical Society Proceedings, vol. 112, pp. 309-320, Philadelphia, Pennsylvania.

List of Participants

AHRENDT, H.
Geologisch-Paläontologisches Institut
und Museum der Universität Göttingen
3400 Göttingen, F.R. Germany

Field of research: Structural and isotope geology

ALVAREZ, W.
Dept. of Geology and Geophysics
University of California
Berkeley, CA 94720, USA

Field of research: Asteroid impact as cause of Cretaceous-Tertiary extinctions: Magnetic reversal stratigraphy, driving mechanism for plate tectonics, Mediterranean (esp. Apennine) structure and tectonics

BERGER, W.H.
Scripps Institution of Oceanography
La Jolla, CA 92093, USA

Field of research: Ocean history

BICKLE, M.J.
Dept. of Earth Sciences
University of Cambridge
Cambridge CB2 3EQ, England

Field of research: Archean tectonics and Earth evolution

BIRKELUND, T.
Institute of Historical Geology
and Paleontology
1350 Copenhagen K, Denmark

Field of research: Stratigraphy and evolution of boreal Jurassic and Cretaceous faunas, mainly cephalopods; sudden faunal changes

BREY, G.P.
Abt. Kosmochemie
Max-Planck-Institut für Chemie
6500 Mainz, F.R. Germany

Field of research: Geochemistry of neogene volcanics of Germany, metal-silicate-oxide partition coefficients at high pressures and temperatures, high pressure phase relationships of pyroxenes

BURKE, K.C.
Dept. of Geological Sciences
State University of New York
Albany, NY 12222, USA

Field of research: The tectonics of the Earth interpreted in terms of cycles and ocean opening and closing (Definition of tectonics: "The large-scale evolution of planetary lithospheres")

DEFFEYES, K.S.
Geology Dept., Princeton University
Princeton, NJ 08540, USA

Field of research: Resources - origin of oil, uranium, and silver deposits; diagenesis of sediments; low-temperature equilibria

DYMEK, R.F.
Dept. of Geological Sciences
Harvard University
Cambridge, MA 02138, USA

Field of research: Petrology, mineralogy, and geochemistry, with a special interest in Precambrian geology and Earth history

FRISCH, W.
Geologisches Institut
der Universität Tübingen
7400 Tübingen, F.R. Germany

Field of research: Geodynamic evolution of the Alpine orogen

FÜCHTBAUER, H.
Geologisches Institut
der Ruhr-Universität Bochum
4630 Bochum 1, F.R. Germany

Field of research: Sedimentary petrology, especially diagenesis

FÜTTERER, D.K.
Alfred-Wegener-Institut
für Polarforschung
2850 Bremerhaven, F.R. Germany

Field of research: Antarctic marine geology/sedimentology

GEE, R.D.
Geological Survey of
Western Australia
Perth, W.A. 6000, Australia

Field of research: Geological evolution of Precambrian crust; tectonic, magmatic, and sedimentary cycles in the Australian Precambrian

HOLLAND, H.D.
Dept. of Geological Sciences
Harvard University
Cambridge, MA 02138, USA

Field of research: Geochemistry

HOLSER, W.T.
Dept. of Geology, University of Oregon
Eugene, OR 97403, and
University of New Mexico
Albuquerque, NM 87131, USA

Field of research: Geochemical history of the ocean, evaporite rocks

HSÜ, K.J.
Geological Institute
Swiss Federal Institute of Technology
8092 Zurich, Switzerland

Field of research: Paleoceanography, sedimentology, isotope geochemistry

JENKINS, W.J.
Woods Hole Oceanographic Institution
Woods Hole, MA 02543, USA

Field of research: Noble gas isotope geochemistry

KNOLL, A.H.
Biological Laboratories
Harvard University
Cambridge, MA 02138, USA

Field of research: Precambrian paleontology and paleobotany

KRÖNER, A.
Institut für Geowissenschaften
Johannes-Gutenberg-Universität
6500 Mainz, F.R. Germany

Field of research: Evolution of the Precambrian lithosphere, geotectonics, Archean paleomagnetism, geochemistry on non-orogenic volcanic rocks, isotope geochemistry and geochronology

KULKE, H.G.
c/o DEMINEX
Postfach 100 944
4300 Essen 1, F.R. Germany

Field of research: Petroleum geology, sedimentology

LASAGA, A.C.
Dept. of Geosciences
Pennsylvania State University
University Park, PA 16802, USA

Field of research: Geochemical cycles, water-rock interactions, bonding in minerals, diffusion and crystal growth

LIPPS, J.H.
Dept. of Geology
University of California
Davis, CA 95616, USA

Field of research: Paleobiology of mass extinctions in the marine realm, paleobiology of foraminifera

McLAREN, D.J.
Dept. of Geology
University of Ottawa
Ottawa, Ontario K1N 6N5, Canada

Field of research: The relations between punctuated Earth history, evolution, and biostratigraphy

MOORBATH, S.
Dept. of Geology and Mineralogy
Oxford University
Oxford OX1 3PR, England

Field of research: Geochronology; isotope geochemistry; evolution of continental crust in space and time, with special emphasis on the Precambrian

O'NIONS, R.K.
Dept. of Earth Sciences
University of Cambridge
Cambridge CB2 3EQ, England

Field of research: Isotope geochemistry

OXBURGH, E.R.
Dept. of Earth Sciences
University of Cambridge
Cambridge CB2 3EQ, England

Field of research: Tectonics

PADIAN, K.
Dept. of Paleontology
University of California
Berkeley, CA 94720, USA

Field of research: Evolutionary biology, with particular emphasis on major features of vertebrate evolution and the evolutionary process

RAUP, D.M.
Dept. of Geophysical Sciences
University of Chicago
Chicago, IL 60637, USA

Field of research: Evolutionary paleobiology

RICHTER, F.M.
Dept. of Geophysical Sciences
University of Chicago
Chicago, IL 60637, USA

Field of research: Geophysical sciences

SARNTHEIN, M.
Geologisch-Paläontologisches Institut
und Museum der Universität Kiel
2300 Kiel, F.R. Germany

Field of research: Marine geology - paleoclimatology (atmospheric circulation), deep-water paleoceanography

SCHMINCKE, H.-U.
Institut für Mineralogie
der Ruhr-Universität Bochum
4630 Bochum 1, F.R. Germany

Field of research: Petrology, volcanology

SEILACHER, A.
Geologisches Institut
der Universität Tübingen
7400 Tübingen, F.R. Germany

Field of research: Paleobiology

SHOEMAKER, E.M.
U.S Geological Survey
Flagstaff, AZ 86001, USA

Field of research: Small bodies in the solar system, accretion of the planets and emplacements of the Oort cloud of comets, and bombardment history of the Earth

SMIT, J.
Geological Institute
University of Amsterdam
1018 VZ Amsterdam, The Netherlands

Field of research: Stratigraphy and sedimentology, with emphasis on minor and major disconformities in Earth history

STECKLER, M.
Lamont-Doherty Geological Observatory
of Columbia University
Palisades, NY 10964, USA

Field of research: Lithospheric mechanics, evolution of sedimentary basins

THOMPSON, A.B.
E.T.H. Zurich
8092 Zurich, Switzerland

Field of research: Application of aspects of petrology to evaluation of dynamic processes in the Earth

TOON, O.B.
Ames Research Center, NASA
Moffett Field, CA 94035, USA

Field of research: Atmospheric sciences

TRENDALL, A.F.
Geological Survey of
Western Australia
Perth, W.A. 6000, Australia

Field of research: Development of time scales for early Earth history, the significance of iron formation for the geochemical evolution of the Earth

VALETON, I.
Geologisch-Paläontologisches Institut
der Universität Hamburg, Geomatikum
2000 Hamburg 13, F.R. Germany

Field of research: Environmental conditions of weathering and ore formation (bauxites, laterites of Fe, Ni,Cr, etc.); periods of intense weathering during Earth history, sedimentology of redbeds, phosphates, glauconite, clay minerals

WALLISER, O.H.
Geologisch-Paläontologisches Institut
und Museum der Universität Göttingen
3400 Göttingen, F.R. Germany

Field of research: The influence of abiotic global events on the evolution of the biosphere, time-specific features

WEBER, K.
Geologisch-Paläontologisches Institut
und Museum der Universität Göttingen
3400 Göttingen, F.R. Germany

Field of research: Structural geology

WEFER, G.
Geologisch-Paläontologisches Institut
und Museum der Universität Kiel
2300 Kiel, F.R. Germany

Field of research: Carbonate sedimentology, isotope paleoecology

WETZEL, A.
Geologisches Institut
der Universität Tübingen
7400 Tübingen, F.R. Germany

Field of research: Sedimentology and physical properties of mudstones

WÖRNER, G.
Institut für Mineralogie
der Ruhr-Universität Bochum
4630 Bochum 1, F.R. Germany

Field of research: High level magma chambers, chemical evolution of alkaline magmas, intra-plate continental volcanism

WILSON, J.F.
Dept. of Geology
University of Zimbabwe
Harare, Zimbabwe

Field of research: Granite-greenstone terrains in Zimbabwe

Subject Index

Author Index

Dahlem Workshop Reports

Life Sciences Research Reports (LS)

LS 31 Microbial Adhesion and Aggregation.
Editor: K. G. Marshall (1984, in press)

LS 30 Leukemia.
Editor: I. L. Weissman (1984, in press)

LS 29 The Biology of Learning.
Editors: P. Marler, H. S. Terrace (1984, in press)

LS 28 Changing Metal Cycles and Human Health.
Editor: J. O. Nriagu (1984)

LS 27 Minorities: Community and Identity.
Editor: C. Fried (1983)

LS 26 The Origins of Depression: Current Concepts and Approaches.
Editor: J. Angst (1983)

LS 25 Population Biology of Infectious Diseases.
Editors: R. M. Anderson, R. M. May (1982)

LS 24 Repair and Regeneration of the Nervous System.
Editor: J. G. Nicholls (1982)

LS 23 Biological Mineralization and Demineralization.
Editor: G. H. Nancollas (1982)

LS 22 Evolution and Development.
Editor: J. T. Bonner (1982)

LS 21 Animal Mind – Human Mind.
Editor: D. R. Griffin (1982)

LS 20 Neuronal-glial Cell Interrelationships.
Editor: T. A. Sears (1982)

Physical, Chemical and Earth Sciences Research Reports (PC)

PC 5 Patterns of Change in Earth Evolution.
Editors: H. D. Holland, A. F. Trendall (1984)

PC 4 Atmospheric Chemistry.
Editor: E. D. Goldberg (1982)

PC 3 Mineral Deposits and the Evolution of the Biosphere.
Editors: H. D. Holland, M. Schidlowski (1982)

Springer-Verlag Berlin Heidelberg New York Tokyo

MIX
Papier aus verantwortungsvollen Quellen
Paper from responsible sources
FSC® C105338

If you have any concerns about our products,
you can contact us on
ProductSafety@springernature.com

In case Publisher is established outside the EU,
the EU authorized representative is:
Springer Nature Customer Service Center GmbH
Europaplatz 3, 69115 Heidelberg, Germany

Printed by Libri Plureos GmbH
in Hamburg, Germany